The Art of Sound Reproduction

The Art of Sound Reproduction

John Watkinson

Focal Press
An imprint of Butterworth-Heinemann Ltd
225 Wildwood Avenue, Woburn, MA 01801-2041
Linacre House, Jordan Hill, Oxford OX2 8DP

  A member of the Reed Elsevier plc group

OXFORD   JOHANNESBURG   BOSTON
MELBOURNE   NEW DELHI   SINGAPORE

First published 1998
Transferred to digital printing 2004
© John Watkinson 1998

British Library Cataloguing in Publication Data
A catalogue record for this book is available from the British Library

Library of Congress Cataloguing in Publication Data
A catalogue record for this book is available from the Library of Congress

ISBN 0 240 51512 9

Typeset by Laser Words, Madras, India

Contents

Chapter 3 Sound and psychoacoustics   74

Chapter 4 Sources of sound   111

Preface

Sound reproduction has always been a challenging area in which to work owing to the large number of engineering and artistic disciplines it embraces. First class results require a working knowledge of all of these disciplines and it can be very rewarding when something goes well. On the other hand lack of awareness in one or more areas can prove embarrassing. I spend a good deal of time teaching digital audio to a wide range of audiences. I have consistently found that there is no difficulty for those with a good understanding of analog techniques. Unfortunately such an understanding is hard to come by.

With certain exceptions very few audio engineers have any formal training. Most have graduated from the university of life and whilst this suffices for practical tasks it is unlikely that any theoretical understanding will be gained in this way. The prevalence in the industry of old wives' tales and the acceptance of exotic components of doubtful worth is an unfortunate consequence of that general lack of theoretical background. Knowing what to do but not why is no use in a changing world.

The rate of change in audio technology has accelerated recently. The onward march of digital techniques first embraced production and then moved on to consumer devices and program delivery. The rapid change has made it difficult for those working in this field to keep up with the technology and even harder for those just setting out on a career.

This book is designed to make life a little easier by providing all of the theoretical background necessary to understand sound reproduction. There is also a good deal of practice. All specialist terms are defined as they occur. As the subject is interdisciplinary I have attempted to use both the musical and the physical term where possible.

As in all of my books I have kept to plain English in preference to mathematics wherever possible. Facts are not stated, but are derived from explanations which makes learning the subject much easier. This is the only viable approach when the range of disciplines to be covered is so wide.

It is not so long ago that digital audio was considered a specialist subject to be treated with a certain amount of awe. This is no longer the case; it has moved into the mainstream of audio recording, production and distribution. Consequently the approach taken in this volume is to consider analog and digital audio as alternatives throughout and to stress the advantages of both.

Audio is only as good as the transducers employed, and consequently micro-phone and loudspeaker technology feature heavily here. It has been a great pleasure to explode a few myths about loudspeakers.

The newcomer will have no difficulty with the initial level of presentation whereas the more experienced reader will treat this as a refresher and gain more value from the explanations of new technology. There is an old saying that those who believe education and entertainment to have nothing in common probably know little about either. Consequently I have not let the advanced technology in this volume completely exclude light heartedness. The audio industry is heavily populated with people who are larger than life and it can be very entertaining. A book which failed to prepare its readers for all aspects of its subject would be poor value indeed.

Needless to say this book reflects a great deal of research. I am indebted to Richard Salter for lightening that burden by sharing his knowledge so freely. The section on multitrack mixing consoles could not have been written without his help.

John Watkinson
Burghfield Common, England, January 1997

Chapter 1

Introduction

In this chapter sound reproduction is outlined beginning with a short history. There are many different types of audio with differing degrees of compatibility between them and these are compared here. It is important to bear in mind that audio has many applications and the criteria for each one may be quite different. Many of the subjects introduced in this chapter will be amplified elsewhere in the book.

1.1 A short history

Accurate sound reproduction is a very young subject because it depends heavily on other enabling technologies which are themselves recent. Of these the most obvious is electronic amplification. Prior to that prospects were limited, but many of the fundamentals which we regard as modern were in place before amplification.

The earliest technology of interest to us is the pipe organ. Large organs with many ranks of pipes posed a formidable control problem. The organist would select pipe ranks by the use of stops which would connect or couple his key presses to different pipes depending on the stop combination. Figure 1.1 shows that this is literally a combinational logic problem which a digital designer would instantly recognize. The organ solved the control problem initially by a combination of mechanical and pneumatic parts. Controlling airflow with a small valve was an early way of obtaining power gain in a control system, predating electronic amplifiers. There was little chance that mechanical systems of this kind could be made touch-sensitive or variable and to this day organ control is purely binary. The key is either pressed or it isn't and the pipe gets a fixed airflow or is cut off. In practical organs it is impossible to prevent a certain amount of air leakage and there is always a background hiss making the organ one of the first devices to have a signal-to-noise ratio.

When electricity was discovered organ control became electrical with switch contacts on the keys and solenoids to control the airflow. The coupling logic became electrical using relays, later electronic logic coupling was employed. The Solid State Logic audio manufacturer began business in this way. Electrical control made it possible to play the organ from a remote

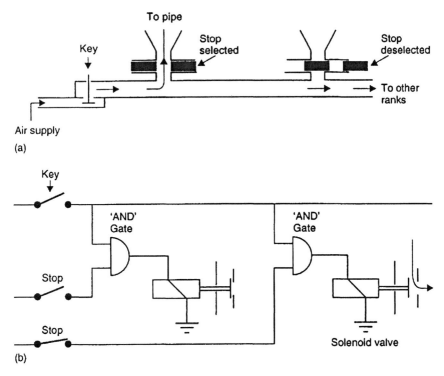

Figure 1.1 (a) In the pipe organ one key can operate pipes in many different ranks according to which stops are selected. This is a combinational logic application which was initially solved mechanically. (b) Later organs used electronic logic.

console. The wiring in between was a forerunner of MIDI (Musical Instrument Digital Interface).

The binary nature of organ playing made it possible to record the performance by punching holes in a moving card corresponding to the key pressed. The cards could be played by supplying air pressure to one side of the card and using the punched holes as valves. Figure 1.2 shows that with pneumatic bellows amplification a large organ could be controlled by a lightweight punched card. The cards were joined together with fabric hinges so that a long recording could be made and they were propelled through the reader by an air motor. City streets in the Netherlands are graced to this day by portable pipe organs working on this principle as they have for hundreds of years.

Punched cards designed for organ control were adopted for machine control during the industrial revolution and later for storing data in early digital computers. It is possible to see a direct lineage from the street organ to digital audio.

It is much more difficult to record arbitrary kinds of sound, rather than the commands required to control an instrument. The first major step was the invention of the telephone in 1876 by Alexander Graham Bell. The invention came about as part of his efforts to teach the deaf to speak. Figure 1.3 shows the main components of the telephone, the first sound reproduction system.

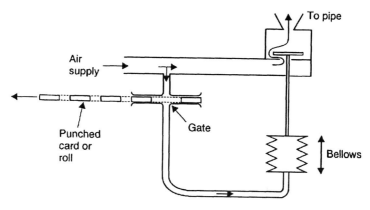

Figure 1.2 Programmable organs used punched cards or rolls. Perforations admitted air to a bellows which operated the pipe valve.

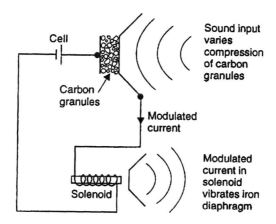

Figure 1.3 Main components of the telephone. The microphone relies on vibrations altering compaction hence resistance of carbon granules. Varying current changes attraction of iron earpiece diaphragm.

These components are still present today. The sound falls upon a microphone causing it to modulate the flow of current from a battery in an electric circuit. The current flow is said to be an *analog* of the sound. The use of an electric current allows simple transmission over considerable distance. The modulated current then operates an earpiece which recreates the sound more or less. The microphone and the earpiece are called transducers from the Latin 'to lead across' meaning that real sound in the acoustic domain crosses to the electrical domain or vice versa.

Audio transducer design improved dramatically after the development of the valve or vacuum-tube amplifier. This meant that the transducer could be optimized for quality rather than to transfer as much power as possible. Using amplification, microphones having outputs in the order of millivolts were no longer a problem. Microphones based on metallic ribbons or coils moving in magnetic fields were developed, followed by microphones in which the diaphragm was one plate of a variable capacitor; possibly the ultimate in low moving mass.

Amplification also allowed breakthroughs in loudspeaker design. The first quality loudspeaker having a moving coil was developed by Rice and Kellog[1] in 1925 shortly after the valve amplifier. Electrostatic loudspeakers were also developed using the force experienced by a charge in an electric field.

The subsequent development of the transistor and then the integrated circuit made it possible to reduce the size of the audio circuitry. In audio quality terms the transistor was not necessarily a blessing as the small circuitry was often accompanied by small and inferior loudspeakers.

Amplification also revolutionized radio communications and made sound broadcasting practical. The first sound broadcasts used amplitude modulation (AM) of the carrier wave in the MF and HF bands. This is shown in Figure 1.4(a). AM broadcasts are prone to interference and do not reproduce the whole frequency range. Shortly after the Second World War, frequency modulation (FM) was developed as shown in (b). This is less susceptible to interference and has a wider frequency range, but needs a wider spectrum requiring a move to the VHF band which is only operative within line-of-sight of the transmitter. Adding a further subcarrier made it possible to transmit stereo in VHF-FM.

Terrestrial television sound channels were mostly delivered using a mono FM subcarrier, but in some countries a digital stereo system known as NICAM was introduced in the late 1980s. Recently, trial broadcasts of digital radio, also called DAB (digital audio broadcasting), have begun.

Telephones and radio broadcasts only reproduce sound in another place. Recording allows sound to be reproduced at another time. The first sound recorders were purely mechano-acoustical as the technology of the day did not permit anything else. The cylinder recorder is shown in Figure 1.5(a). This has a diaphragm which moves according to the incident sound. The diaphragm is connected to a stylus which makes a groove in the surface of a rotating cylinder. The sound modulates the depth of the groove which is therefore an analog of the sound. The cylinder might be covered with a soft

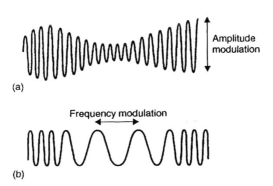

Figure 1.4 (a) In amplitude modulation (AM), the carrier signal is multiplied by the audio signal. (b) In frequency modulation (FM), the audio signal changes the carrier frequency leaving the amplitude unchanged.

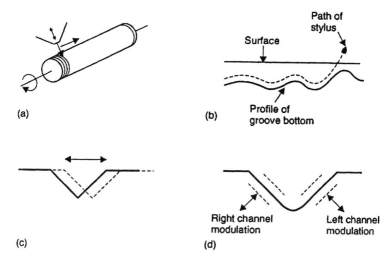

(a)

(b) Surface Path of stylus Profile of groove bottom

(c)

(d) Right channel modulation Left channel modulation

Figure 1.5 (a) The cylinder recorder stores the vibration waveform of the diaphragm in the varying depth of a helical groove formed as the cutter is driven along the cylinder axis. (b) Vertical or hill-and-dale modulation is prone to the stylus being thrown from the groove by transients. (c) Lateral modulation improves tracking. (d) Orthogonal modulation of groove walls allows stereo recording with reasonable channel separation.

metal foil or a layer of wax. The cutter is driven along the cylinder by a leadscrew. On replay the groove modulations are transferred to the stylus and actuate the diaphragm.

The cylinder recorder was a limited success because the cylinders could not be duplicated. Applying the modulated groove principle to the surface of a disk resulted in a recording which could be duplicated by pressing. Figure 1.5(b) shows that the first disks used vertical stylus motion, the so-called hill-and-dale recording. This soon changed to lateral recording (c) which tracked better. Stereo was possible with independent orthogonal (at 90°) modulation of the two groove walls shown in (d).

The first disks were also purely mechano-acoustic, and recording and replay required a large horn which would couple as much energy as possible between the air and the stylus. Figure 1.6 shows that once amplification became

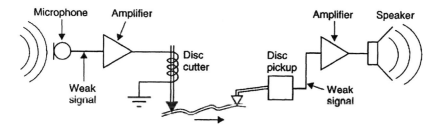

Figure 1.6 The invention of electronic amplification transformed sound recording because it was then only necessary to reproduce *information* rather than power. Amplifiers between microphone and cutter and between pickup and speaker allow both microphone and pickup to be optimized for signal accuracy not output.

available it was then possible to use a microphone and an amplifier to drive an electromechanical transducer in the disk cutter. In the player another transducer known as the pickup produced a low level electrical signal which was amplified to drive an electro-acoustic transducer.

The first disks were purely mechanical and significant power was obtained from the groove wall. In order to limit the pressure exerted by the stylus this had to be quite large. Figure 1.7 shows that the short wavelength response limit of a stylus is reached when the stylus is too big to follow fine detail in the groove. To reproduce high frequencies, the groove speed has to be high. Consequently early disks turned at 78 rpm and playing times were short. Once electrical pickups and amplifiers were developed, there was no need to transfer so much power from the groove and a much smaller stylus could be used. A slower speed could then be used and 33.33 rpm^2 became the standard for 12-inch microgroove records, with 45 rpm being used for 7-inch records. A further ingredient was the switch from shellac to vinyl which had a finer structure and produced less noise. The microgroove vinyl disk was very popular until it was eclipsed by the compact disc.

Early experiments were made by Poulsen in the 1890s with magnetic recording using soft iron wire driven between two spools. Again this development was waiting for the valve amplifier to make it practicable. Magnetism inherently suffers hysteresis which distorts the sound, and high-frequency bias was developed to overcome it.

Soft iron wire with the right magnetic characteristics was flimsy and prone to breakage. The solution was to make a composite recording medium where the weak magnetically optimal material is supported by a mechanically optimal substrate. This is the advantage of tape. In the first tapes paper was the substrate but this soon gave way to cellulose acetate and then polyester. The analog tape recorder as it is known today was essentially developed during the Second World War in Germany. The story of its development despite the difficulties caused by hostilities is quite remarkable.

Since the Second World War the analog recorder has been refined and its sound quality improved almost to the physical limits. The work of Dolby had a great impact on analog tape because suitable complementary signal processors before and after the tape recorder gave a significant reduction in tape hiss. The open reel tape recorder was not a great success with the consumer as it was too fiddly to operate. Open reel recorders remained popular in professional circles, but the consumer took to the tape cassette instantly. The Compact Cassette introduced by Philips in 1963 was originally designed for dictating machines, but continuous development has improved the sound quality considerably.

Low-frequency   High-frequency
tracking   mistracking

Figure 1.7 Disk reproduction reaches an upper frequency response limit when the modulation of the groove is too small for the stylus to follow.

In analog systems the sound pressure waveform is represented variously by electrical voltage in a conductor, the strength of magnetic flux on a tape, or by the velocity of a stylus traversing a groove in a disk. It is inherent in any analog system that the sound quality is only perfect if the analog is perfect. In real equipment the analog is never perfect. In the absence of an input signal all real transmission or recording systems still output a signal which is their inherent noise. In the presence of the signal the noise is added to the signal at the output. If the analog is not perfectly linear, the waveform will be distorted. In a complex analog system the defects of each successive stage are superimposed.

Digital systems are quite different because the information being carried is discrete; that is to say it consists of whole numbers only. In well-engineered systems whole numbers can be transmitted and recorded an indefinite number of times without any alteration to their value whatsoever. The technique of error correction ensures that if any numbers are altered from their original value they can be identified and put right. Consequently, in digital systems there is no superimposition of degradation and digital transmissions and recordings have no quality; they simply reproduce the original numbers.

In order to use this technique for audio, the sound waveform has to be converted to the digital domain in an analog-to-digital converter (ADC). The time axis of the audio waveform is broken into evenly spaced units called *samples* and the voltage of each sample is the *quantized* or expressed by the nearest whole number. The accuracy to which this is done determines the sound quality. From then on, provided the discrete numbers which were output from the ADC are preserved, there is no further loss of sound quality whatsoever whilst the audio stays in the digital domain. To recover an analog waveform, for example to drive a loudspeaker, a digital-to-analog converter (DAC) is required. An ideal DAC would transfer all of the information in the digital domain to the analog output and thus cause no further quality loss. Practical DACs may fail to do this, often because of economic constraints.

Figure 1.8 shows a digital recorder in which an ADC provides data which are recorded. The same data are available at the DAC. Consequently, in a true digital recorder the recording medium has no sound quality and the quality is determined only by the converters. This is a great blessing because it means that all of one's efforts can go into designing good converters and the remaining parts of the system then do not need any special care.

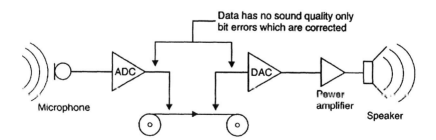

Figure 1.8 Digital recorder stores only discrete numbers which are recovered without error. Sound quality is determined by converters only and does not change if recorder is bypassed.

The first digital audio tape recorders became available in the 1970s. These were large and expensive and were only affordable by recording studios but their sound quality was very good and it was clear that this was the way of the future. This was underlined by the introduction of DAT (digital audio tape) in the mid-1980s. This low-cost stereo format using a small tape cassette failed as a consumer product for political and copyright reasons, but was adopted wholeheartedly by the audio industry as a professional device.

Digital video tape recorders became commercially available in the 1980s and these incorporated digital audio soundtracks, generally four. The audio and video data were multiplexed into the same data tracks on tape.

The Compact Disc (CD), launched in 1982, was the first consumer digital audio product and has proved extremely successful. The Compact Disc was based on the optical readout process developed for LaserVision video disks but adapted to record data. There is no physical contact between the pickup and the disk and hence no wear mechanism. The use of error correction resulted in a disk needing no special handling, which eliminates the scratches of vinyl disks. Track selection is via keypad rather than direct positioning of the pickup making the CD easy to use. The sound quality of a good CD player is determined by the DAC and unfortunately only a few CD players have DACs which are capable of accurately reproducing all of the data on the disk. However, even players containing DACs which are relatively poor still eclipse the quality of vinyl and are much more convenient to use.

In the early 1990s digital audio bit rate reduction or compression became economically viable. Compression allows a lower bit rate to be used to represent audio and is essential to contain the bandwidth required in DAB. It also reduces the data storage requirements in recorders, thus reducing cost. Two consumer formats using compression were launched: MiniDisc and DCC (Digital Compact Cassette). Although both could record as well as playing prerecorded media, these have not been successful because they are too expensive to compete with the Compact Cassette and cannot reach the sound quality of the Compact Disc.

1.2 Types of reproduction

The first audio systems were monophonic (from the Greek 'single sound'), invariably abbreviated to mono. Figure 1.9(a) shows that in mono there was one loudspeaker fed by a single signal. Whatever the position of the original sound sources, all of the sound would appear to come from one place, i.e. the loudspeaker. However accurate the reproduction of the sound waveform by a mono system, the spatial representation is clearly completely incorrect.

Stereophonic (from the Greek 'solid sound') systems attempt to reproduce spatial attributes of the sound. Early experiments involved large numbers of microphones (b) each feeding an independent speaker. Clearly this would be expensive and the technology of the day did not permit the recording or broadcasting of multiple channels.

Today's practical stereo systems are derived from the work of Alan Blumlein who showed that an infinity of virtual sound sources could be created at any points between only two loudspeakers fed by two separate signals as shown in (c). The signal representing each virtual source is fed simultaneously to both speakers and its location is determined by the relative intensity in the

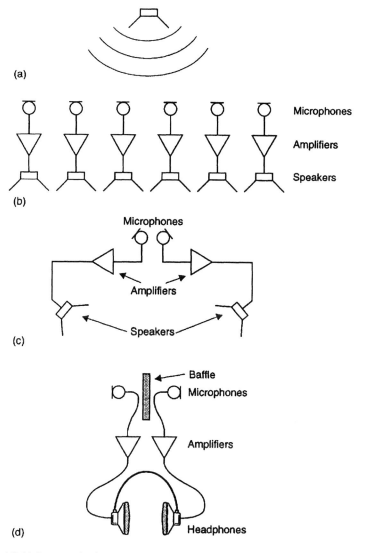

Figure 1.9 (a) In monophonic systems all sound appears to come from one place. (b) Early stereophonic systems relied on an array of microphones and speakers. (c) The stereo system invented by Blumlein created virtual sound sources at arbitrary positions between only two loudspeakers needing only two channels. (d) A dummy head connected to headphones gives realistic spatial effects but signal format is incompatible with (c).

two channels. This is known as intensity stereo and it is a fundamental requirement that coincident directional microphones are used so that only intensity differences are obtained. Loudspeakers must be used for reproduction as both ears must hear both speakers for the effect to work. Intensity stereo recordings cannot be reproduced accurately on headphones as the sound appears to be in the centre of the listener's head.

An alternative is shown in (d) where microphones are fitted in a dummy head and connected to headphones worn by the listener. This is known as binaural stereo. The realism obtained is very high because the listener's ears have been projected into the original sound field. However, as soon as the listener turns his head the effect is lost. Another problem is that binaural signals cannot be reproduced directly on loudspeakers because the signals to each ear cannot be kept separate.

The spatial coding of intensity stereo and binaural stereo is fundamentally different and incompatible with one another and with mono, and standards conversion is required to interchange program material between standards. Conversion to mono is achieved by adding the two stereo channels as shown in Figure 1.10(a). Intensity stereo gives good results as the spatial information simply collapses without affecting the tonality. As there are no phase differences in intensity stereo, adding the channels and dividing by two is always the correct procedure to obtain a mono equivalent without a level change.

Adding the two channels of a binaural signal gives very poor results as the spacing between the microphones produces timing and phase differences between them. Depending on the direction and frequency of the original sound, adding the channels may result in reinforcement or cancellation and this causes a serious change in tonality as well as loss of spatial information. It is not clear what gain factor should be used to avoid a level change. This poor mono compatibility is the main reason why broadcasters primarily use intensity stereo.

Mono signals can be reproduced on stereo systems simply by feeding both channels equally as in (b). The result is a central image with no spatial information as might be expected. Mono signals from single instruments can be positioned within the stereo image by varying the relative intensity of the signal fed to the two channels using a panoramic potentiometer or panpot (c). Today most pop records are made by panning a number of different mono signals into different locations in the stereo image (d).

Figure 1.11 shows that in order to listen to intensity stereo on headphones a device called a shuffler must be used. This electronically simulates the missing paths from each speaker to the opposite ear and moves the virtual sound sources to their correct forward location. One might be forgiven for thinking that personal stereo systems using headphones should incorporate such a device. In practice hardly any do and most give a disappointing inside-the-head image.

Transaural stereo[3] attempts to reproduce binaural stereo on loudspeakers by simulating the unwanted crosstalk between each speaker and the opposite ear and adding it in anti-phase to the signals. The listener then obtains, over a limited area, an approximation to the original binaural waveform at each ear. The transaural stereo processor is essentially an inverse shuffler.

The above stereophonic techniques give reasonably accurate spatial representation of the sound sources. Other techniques are used, such as spaced omnidirectional microphones, which produce both phase and intensity differences between the two channels. These are quite incapable of reproducing the spatial information in the original sound field, but instead a diffuse spacious effect is obtained. This is preferable to the point source of mono but it is only an effect.

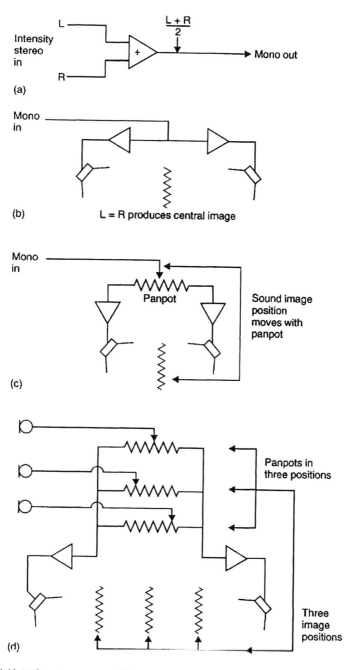

Figure 1.10 (a) Intensity stereo of Figure 1.9(c) can be converted to mono reasonably well by adding left and right signals. (b) Mono signals can be reproduced on intensity stereo speakers by supplying an identical signal to both channels. A central sound image results. (c) A sound image from a mono source can be moved or 'panned' by changing relative proportion of signals fed to left and right speakers. (d) Pop record production pans many different mono signals into an artificial stereo image.

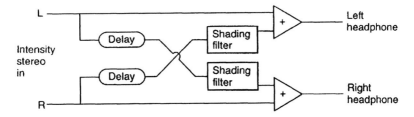

Figure 1.11 A standards converter known as a shuffler is necessary to listen to intensity stereo signals using headphones. This replicates electronically the fact that both ears hear both loudspeakers.

Stereophonic systems only give accurate spatial information in an arc between the two speakers. Surround sound systems attempt to encircle the listener with sound sources. Many of these use only two signal channels so that existing stereo recording and transmission equipment can be used. The two signals are encoded in such a way that the decoder can determine the direction from which the dominant sound source is coming. The sound is then steered to the appropriate speaker(s). Dolby Surround Sound and UHJ both work in this way.

In the late 1960s quadraphonic consumer equipment was made available in which four independent channels were conveyed. The technology of the day was not adequate and the system was a commercial failure. Recently with digital techniques it has become possible to multiplex an arbitrary number of audio channels into one transmission without crosstalk. This makes possible true surround sound.

1.3 Sound systems

There are many different uses for audio equipment and the performance criteria are quite different. Here some typical audio systems are outlined. A distinction commonly made is between professional and consumer equipment. In general professional equipment is a tool which someone uses to do his or her job. Failure or inconsistent performance is unacceptable as it could cause loss of income or a damaged reputation. Rugged construction is desirable for reliability. Before making a purchase most professionals insist on a demonstration or even a loan of the products under consideration so see how they function under the intended conditions. It is not unusual for manufacturers to take prototypes into the field to gauge reaction and obtain the views of skilled listeners. Figure 1.12 shows some of the applications of professional audio equipment.

A great deal of consumer equipment is purchased on its appearance without test. A number of low-cost loudspeakers are produced by injection moulding and many of these have imitation tweeters and reflex ports which are completely non-functional but give the appearance of a more sophisticated unit. No mass manufacturer displays prototypes as consumers would not understand why they were imperfect. Consumer equipment is for entertainment and is mostly produced at low cost to variable standards. The consumer will generally have little technical knowledge and the equipment installation will often be compromised by poor loudspeaker placement.

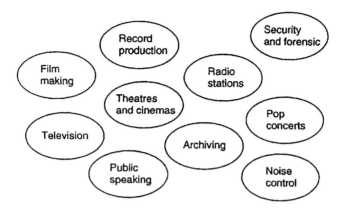

Figure 1.12 Some applications of professional audio equipment.

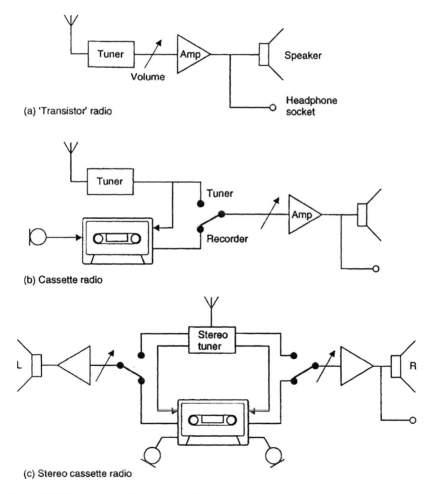

Figure 1.13 Consumer portable equipment types.

1.4 Portable consumer equipment

Figure 1.13 shows some typical portable consumer audio products. There is a signal source or sources, a source selector switch, a volume control, optional tone controls and an amplifier driving a loudspeaker. Portable mono products include the transistor radio and the radio cassette player, sometimes incorporating alarm clocks or timers which will switch the unit on at a predetermined time. Portable stereo products include the radio cassette player, more recently incorporating twin cassette decks and/or a CD player. Despite political correctness these portable units have come to be known almost universally as 'ghetto blasters'. Signal routing is extended to include the choice of the signal which is to be recorded by the cassette tape deck.

A constant problem with portable products is that battery power is extremely limited. Battery life is extended by failing to reproduce low frequencies and by having very light cones in the speakers to improve efficiency. These are then insufficiently rigid and the results are almost always disappointing. With the high cost of batteries, most of today's products contain an AC power supply or can be used with an external power supply. A mechanical switch disconnects the batteries when the external power connector is fitted.

The difficulties of portable loudspeaker reproduction are overcome in personal players which are designed only to drive headphones. The cassette player was implemented first, followed by Compact Disc players. Some models incorporate radio tuners and use the headphone lead as an antenna. Early personal CD players played for a short time and had to be handled like eggs to prevent skipping. Later units have lower power logic to extend playing time. A buffer memory is also installed in some players to provide an audio output whilst the pickup recovers the track after an impact.

1.5 Fixed consumer equipment

Fixed equipment gives much better quality than portable units. The traditional component hi-fi system is shown in Figure 1.14. A preamplifier acts as a signal selector and has volume and tone controls as well as determining which signals will be recorded. The preamplifier drives the power amplifier which is connected to a separate loudspeaker. Each signal source is in a separate box. Signal sources include radio tuners, cassette or open reel tape decks, vinyl disk players and CD players. In stereo systems the same units are present, but there are two loudspeakers and every circuit is duplicated. A balance control is added to adjust the relative level of the two channels.

The advance of microcircuitry has meant that the bulk and cost of domestic electronics has shrunk dramatically. A lot of hi-fi components are literally near-empty boxes with a few microchips in the corner. It is not unknown for the manufacturer to incorporate a thick steel plate to make the unit feel more substantial. An early result of circuit miniaturization was the receiver (US) or tuner amplifier (UK), containing a radio tuner, preamplifier and power amplifiers. As Figure 1.14 shows, this requires external speakers and has connections for external units such as tape decks, CD players and vinyl turntables.

The receiver approach was followed by the single unit hi-fi in which everything except the speakers is in a single box. In principle there is no reason why such a unit should not sound as good as a component system, provided

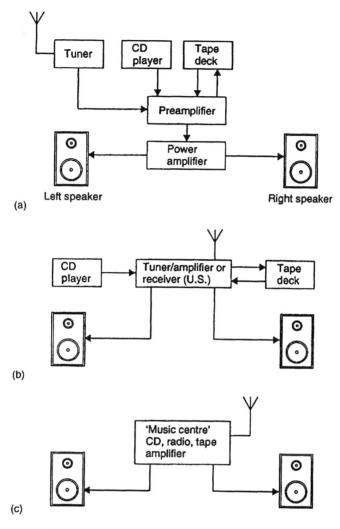

Figure 1.14 (a) Fixed hi-fi installation using individual components. (b) The receiver or tuner-amplifier reduces the number of boxes and interconnection. (c) Single unit hi-fi places every function in one enclosure. Performance is often compromised because the cost of performing every function well would make the unit cost high.

that the same circuitry is used. In practice few people will pay what appears to be a high price for one box and so these integrated units are normally built to a price rather than to a standard and the loudspeakers are usually marvels of penny pinching. A dramatic improvement can usually be had by discarding the standard speakers and using better models.

The traditional hi-fi sports a forest of control knobs and is difficult for non-technical users to wire up and operate. The appearance of a typical system is unimaginative, especially the loudspeakers which are usually box shaped. The lifestyle audio system has evolved for people who prefer equipment which is easier to use and which has a more imaginative appearance. Lifestyle systems usually have dramatic styling, few knobs, immaculate standards of surface finish and intelligent remote control systems.

1.6 High-end hi-fi

The high-end hi-fi market caters for the obsessive customer who often spends more time tinkering and comparing than actually listening for enjoyment. High-end products are frequently over-engineered, employing massive chunks of expensive materials for reasons which do not withstand scrutiny. Often a particular shortcoming is ruthlessly eliminated by some expensive technique. Whilst this improvement may be genuine, it is often at the expense of serious neglect in other areas. Consequently most high-end installations are unbalanced because too much attention is concentrated in a few places and because each unit is 'perfected' without considering how it will operate in a system. A common failing is that too much has been spent on the electronics and not enough on the loudspeakers. Precise signals are generated which the loudspeakers cannot reproduce. Despite the high cost of these systems their ergonomics are usually very poor and they represent extremely poor value for money although the high cost is usually sold as imparting exclusivity. A small industry has sprung up producing accessories for high-end hi-fi, many of which appear to defy the laws of physics and all of which operate in the area of diminishing returns. Naturally magazines are available to cater for the tastes of high-end hi-fi fans. In general these are difficult to read because the technical terms they contain often mean what the writer wants them to mean rather than what the skilled person would understand them to mean. The term pseudoscience has been coined to describe this kind of thing.

1.7 Public address

Many public buildings have combined background music and announcement systems. These can take on a very important role in the case of an emergency in giving evacuation instructions. Such systems must have back-up power and fire resistant wiring. Single failure points are eliminated as far as possible so that the system keeps working. Clearly the criteria are intelligibility before fidelity. Making a loudspeaker vandal- and water-proof and acceptable in appearance to an architect does not leave much chance that high quality will also be obtained.

Sound reinforcement is meant to make a live acoustic performance audible to a larger audience by augmenting the original sound. Clearly if the quality is poor it will be obvious that an amplification system is in use. With large venues, sound reinforcement gives way to public address (PA) systems which generate all of the sound. In PA systems efficiency is important and horn loudspeakers are often used. In sound reinforcement and PA the directivity of the loudspeakers is important so that sound is directed at the audience. In large indoor venues stray sound can excite reverberation and echoes which reduce intelligibility. In outdoor venues stray sound causes a nuisance.

Figure 1.15 shows the elements of a PA system. Conventional and radio microphones will feed a mixing console along with sources such as disks

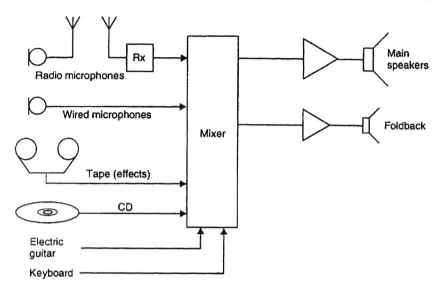

Figure 1.15 A public address (PA) system used for live concerts. Note the wide range of input types and the need for artiste foldback.

and tape, and electronic musical instruments such as synthesizers. The main console outputs will feed a bank of power amplifiers which drive the loud-speakers. In a large venue the main amplified sound will arrive back at the stage quite late and confuse the musicians. Further *foldback* loudspeakers will be required on stage so that the musicians can hear themselves. A different mix will be required for foldback than for the main speakers. Foldback speakers are often placed on the stage floor and angled up at the vocalists. Figure 1.16 shows that this will place them exactly in the response null of a directional microphone.

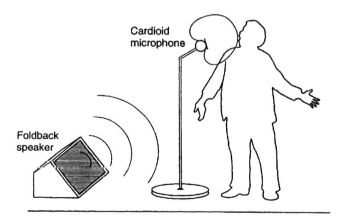

Figure 1.16 Foldback loudspeaker placed on stage floor is in response null of directional microphone reducing chances of feedback.

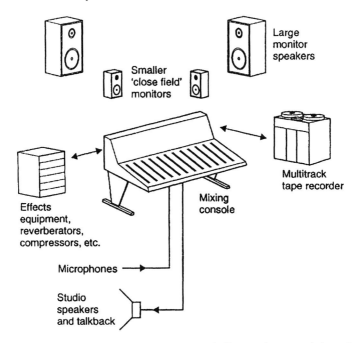

Figure 1.17 Multitrack recording studio. Note outboard effects equipment and alternative monitoring speakers.

1.8 Multitrack recording

Figure 1.17 shows the main components of a multitrack pop recording studio. The mixing console is designed so that it can be switched between two modes of operation. In recording, signals from the various microphones are controlled by the console and sent to the multitrack tape recorder. In mixing down, playback signals from the multitrack tape recorder are controlled to produce a two channel stereo mix which is sent to a stereo mastering recorder. The mixing console has switching so that it can route selected signals to outboard processing units such as reverberators, compressors, etc. and then accept the processed signal back into the mix. Such consoles are frequently fitted with automation systems which can move the faders in a predetermined manner as a function of time. The sound engineer can modify the fader motion until an ideal mix is obtained. The mix is monitored on a pair of loudspeakers and these are usually designed to be capable of extremely high levels. The high level is often obtained to the detriment of other qualities and such monitors are frequently quite disappointing. In many installations additional smaller loudspeakers are mounted closer to the console or even standing on the rear edge of it. These are often more accurate than the large units.

1.9 Sound radio

Figure 1.18 shows the basic elements of a small radio station. The mixer allows many different sound sources to be selected and mixed into the transmitter

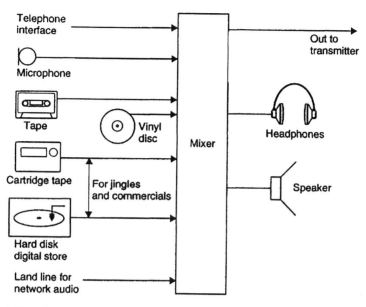

Figure 1.18 Radio station audio installation.

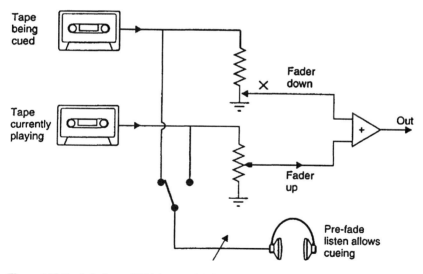

Figure 1.19 Pre-fade listen (PFL) is essential for broadcasting as it allows the operator to hear recording for cueing purposes.

feed. CDs, vinyl disks, cassette tapes, telephone lines, contribution lines and cartridge tapes may all be used in addition to live microphones. A useful function for cueing recordings to be aired is the *pre-fade listen (PFL)* technique shown in Figure 1.19 which routes signals from the sources to a monitoring point so that they can be heard whilst they are muted as far as transmission is concerned. If the announcer's microphone is faded down, the PFL signal may be monitored on loudspeakers, whereas if the microphone is in use headphones will be needed.

Frequently used recordings such as jingles and commercials were traditionally kept on endless tape cartridges. Once played these would keep running until they returned to the beginning of the recording so that they were self-cueing. The sound quality of cartridges is not very good and in many cases they have been replaced by a digital system based on hard disks.

In radio stations there are two important parameters: the time of day and the signal level. Both are prominently displayed near the mixer. In many radio stations items such as the news are not originated locally but are taken from a centrally originated network. Accurate timing is needed to switch to and from the networked news signals.

Signal level is important because transmitters must not be over-modulated by excess level. As many listeners use very low-cost radio sets, it is also important that very low audio levels are avoided as they would be lost in noise. Consequently, many radio stations use dynamic range compressors which boost low-level signals to ensure that nothing very quiet is transmitted. When listening on high-quality equipment the side effects of these compressors can be irritating.

1.10 Film and television sound

Clearly moving pictures without sound are unacceptable, but the presence of cameras, lighting and the other apparatus of image capture inevitably results in problems picking up sound. Television and film have solved these in different ways. Films usually have a larger budget and this allows the soundtrack to be completely recreated in post production using the live recording as a guide only. In most television programs this budget is not available and the solution is simply to allow microphones in shot. In both film and television accurate synchronization between the image and the sound, so-called 'lip-sync', is necessary. This was originally done mechanically, but is now done electronically using timecode signals which pass between different units. Traditionally the approach to television sound was that much of the material was speech and that small loudspeakers in the television set would be adequate. This view has had to change with the increased use of television sets to display rented movies on video cassettes. Proper reproduction of the full frequency range is then essential. The term home cinema has been coined to describe television sets having surround sound loudspeaker systems.

References

1. Rice, C.W. and Kellog, E.W., Notes on the development of a new type of hornless loud-speaker. *JAIEE*, **12**, 461–480 (1925)
2. Keller, A.C., Early hi-fi and stereo recording at Bell laboratories (1931–32). *J. Audio Eng. Soc.*, **29**, No. 4, 274–280 (1981)
3. Bauck, J. and Cooper, D.H., Generalised transaural stereo and applications. *J. Audio Eng. Soc.*, **44**, No. 9, 683–705 (1996)

Audio basics

In audio engineering there are many concepts which appear in almost every topic. The deciBel is a good example of a universal concept. Magnetism is important in microphones, motors, transformers, recording tape and loudspeakers. These and other common concepts are covered in this chapter.

2.1 Periodic and aperiodic signals

Sounds can be divided into these two categories and analysed both in the time domain in which the waveform is considered, or in the frequency domain in which the spectrum is considered. The time and frequency domains are linked by transforms of which the most well known is the Fourier transform.

Figure 2.1(a) shows that a periodic signal is one which repeats after some constant time has elapsed and goes on indefinitely in the time domain. In the frequency domain such a signal will be described as having a fundamental frequency and a series of harmonics or partials which are at integer multiples of the fundamental. The timbre of an instrument is determined by the harmonic structure. Where there are no harmonics at all, the simplest possible signal results which has only a single frequency in the spectrum. In the time domain this will be an endless sine wave.

Figure 2.1(b) shows an aperiodic signal known as white noise. The spectrum shows that there is equal level at all frequencies, hence the term 'white' which is analogous to white light containing all wavelengths. Transients or impulses may also be aperiodic. A spectral analysis of a transient (Figure 2.1(c)) will contain a range of frequencies, but these are not harmonics because they are not integer multiples of the lowest frequency. Generally the narrower an event in the time domain, the broader it will be in the frequency domain and vice versa.

2.2 Frequency response and linearity

The sine wave is extremely useful in testing audio equipment because of its spectral purity. It is a goal in sound reproduction that the timbre of the original sound shall not be changed by the reproduction process. There are two ways in which timbre can inadvertently be changed, as Figure 2.2 shows. In (a) the spectrum of the original sound shows a particular relationship between

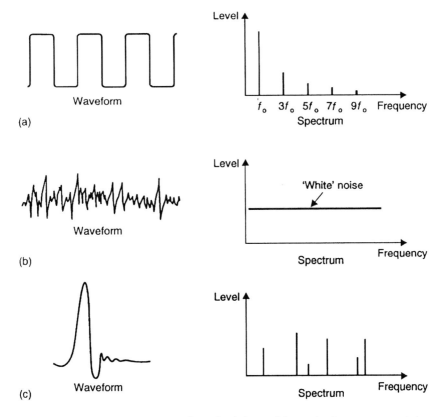

Figure 2.1 (a) Periodic signal repeats after a fixed time and has a simple spectrum consisting of fundamental plus harmonics. (b) Aperiodic signal such as noise does not repeat and has a continuous spectrum. (c) Transient contains an anharmonic spectrum.

harmonics. This signal is passed through a system (b) which has an unequal response at different frequencies. The result is that the harmonic structure (c) has changed, and with it the timbre. Clearly a fundamental requirement for quality sound reproduction is that the response to all frequencies should be equal.

Frequency response is easily tested using sine waves of constant amplitude at various frequencies as an input and noting the output level for each frequency.

Figure 2.3 shows that another way in which timbre can be changed is by non-linearity. All audio equipment has a transfer function between the input and the output which form the two axes of a graph. Unless the transfer function is exactly straight or *linear*, the output waveform will differ from the input. A non-linear transfer function will cause harmonic distortion which changes the distribution of harmonics and changes timbre.

As the sine wave is spectrally pure, the creation of harmonics by a non-linear process is at its most obvious. Consequently the use of a sine wave test signal and a spectrum analyser is a useful way of testing for harmonic distortion.

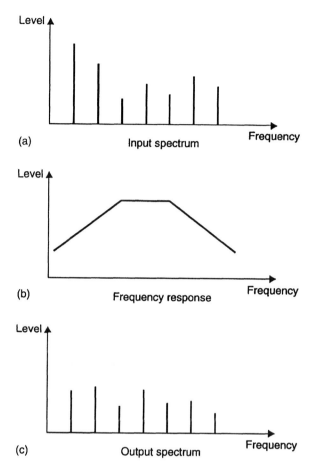

Figure 2.2 Why frequency response matters. Original spectrum at (a) determines timbre of sound. If original signal is passed through a system with deficient frequency response (b), the timbre will be changed (c).

At a real microphone placed before an orchestra a multiplicity of sounds may arrive simultaneously. The microphone diaphragm can only be in one place at a time, so the output waveform must be the sum of all the sounds. An ideal microphone connected by ideal amplification to an ideal loudspeaker will reproduce all of the sounds simultaneously by linear superimposition. However, should there be a lack of linearity anywhere in the system, the sounds will no longer have an independent existence, but will interfere with one another, changing one another's timbre and even creating new sounds which did not previously exist. This is known as *intermodulation*. Figure 2.4 shows that a linear system will pass two sine waves without interference. If there is any non-linearity, the two sine waves will intermodulate to produce sum and difference frequencies which are easily observed in the otherwise pure spectrum.

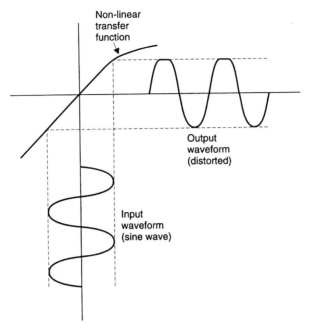

Figure 2.3 Non-linearity of the transfer function creates harmonics by distorting the waveform. Linearity is extremely important in audio equipment.

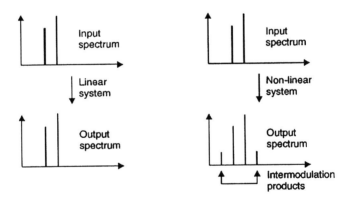

Figure 2.4 (a) A perfectly linear system will pass a number of superimposed waveforms without interference so that the output spectrum does not change. (b) A non-linear system causes intermodulation where the output spectrum contains sum and difference frequencies in addition to the originals.

2.3 The sine wave

As the sine wave is so useful it will be treated here in detail. Figure 2.5 shows a constant speed rotation viewed along the axis so that the motion is circular. Imagine, however, the view from one side in the plane of the rotation. From a distance only a vertical oscillation will be observed and if the position is

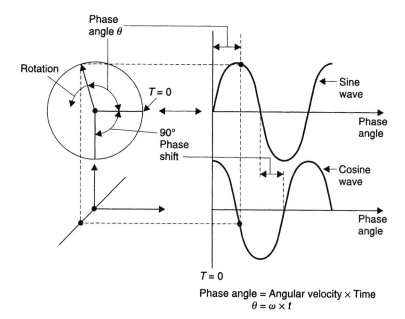

Figure 2.5 A sine wave is one component of a rotation. When a rotation is viewed from two places at right angles, one will see a sine wave and the other will see a cosine wave. The constant *phase shift* between sine and cosine is 90° and should not be confused with the time variant *phase angle* due to the rotation.

plotted against time the resultant waveform will be a sine wave. Geometrically it is possible to calculate the height or displacement because it is the radius multiplied by the sine of the phase angle.

The phase angle is obtained by multiplying the angular velocity ω by the time t. Note that the angular velocity is measured in radians per second whereas frequency f is measured in rotations per second or Hertz (Hz). As a radian is unit distance at unit radius (about 57°) then there are 2π radians in one rotation. Thus the waveform at a time t is given by $\sin \omega t$ or $\sin 2\pi f t$.

Imagine a second viewer who is at right angles to the first viewer. He will observe the same waveform, but at a different time. The displacement will be given by the radius multiplied by the cosine of the phase angle. When plotted on the same graph, the two waveforms are *phase shifted* with respect to one another. In this case the phase shift is 90° and the two waveforms are said to be *in quadrature*. Incidentally the motions on each side of a steam locomotive are in quadrature so that it can always get started (the term used is quartering). Note that the *phase angle* of a signal is constantly changing with time whereas the *phase shift* between two signals can be constant. It is important that these two are not confused.

The velocity of a moving component is often more important in audio than its displacement. The vertical component of velocity is obtained by differentiating the displacement. As the displacement is a sine wave, the velocity will be a cosine wave whose amplitude is proportional to frequency. In other words the displacement and velocity are in quadrature with the velocity lagging. This

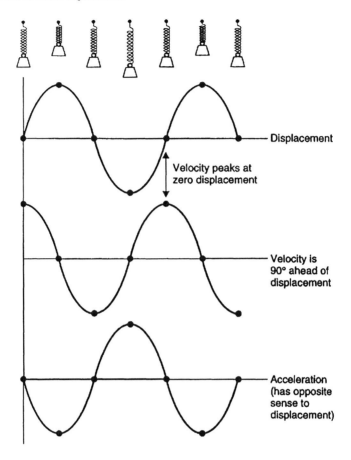

Figure 2.6 The displacement, velocity and acceleration of a body executing simple harmonic motion (SHM).

is consistent with the velocity reaching a minimum as the displacement reaches a maximum and vice versa. Figure 2.6 shows the displacement, velocity and acceleration waveforms of a body executing simple harmonic motion (SHM). Note that the acceleration and the displacement are always anti-phase.

2.4 Root mean square measurements

Figure 2.7(a) shows that according to Ohm's law, the power dissipated in a resistance is proportional to the square of the applied voltage. This causes no difficulty with direct current (DC), but with alternating signals such as audio it is harder to calculate the power. Consequently a unit of voltage for alternating signals was devised. Figure 2.7(b) shows that the average power delivered during a cycle must be proportional to the mean of the square of the applied voltage. Since power is proportional to the square of applied voltage, the same power would be dissipated by a DC voltage whose value was equal to the square root of the mean of the square of the AC voltage. Thus the volt

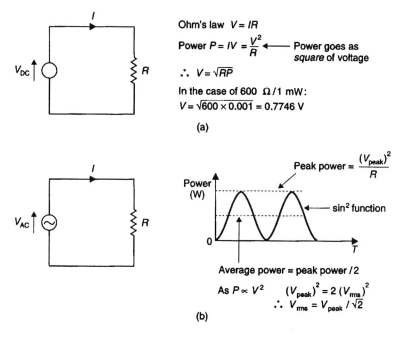

(a)

(b)

Figure 2.7 (a) Ohm's law: the power developed in a resistor is proportional to the square of the voltage. Consequently, 1 mW in 600 Ω requires 0.775 V. With a sinusoidal alternating input (b), the power is a sine square function which can be averaged over one cycle. A DC voltage which would deliver the same power has a value which is the square root of the mean of the square of the sinusoidal input.

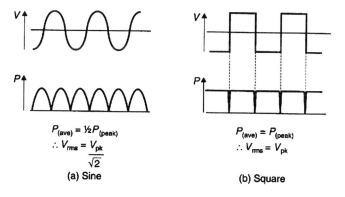

Figure 2.8 (a) For a sine wave the conversion factor from peak to rms is $\sqrt{2}$. (b) For a square wave the peak and rms voltage is the same.

rms (root mean square) was specified. An AC signal of a given number of volts rms will dissipate exactly the same amount of power in a given resistor as the same number of volts DC.

Figure 2.8(a) shows that for a sine wave the rms voltage V_{rms} is obtained by dividing the peak voltage V_{pk} by the square root of two. However, for a square wave (b) the rms voltage and the peak voltage are the same. Most moving coil AC voltmeters only read correctly on sine waves, whereas many electronic meters incorporate a true rms calculation.

On an oscilloscope it is often easier to measure the peak-to-peak voltage which is twice the peak voltage. The rms voltage cannot be measured directly on an oscilloscope since it depends on the waveform although the calculation is simple in the case of a sine wave.

2.5 The deciBel

The first audio signals to be transmitted were on telephone lines. Where the wiring is long compared to the electrical wavelength (not to be confused with the acoustic wavelength) of the signal, a transmission line exists in which the distributed series inductance and the parallel capacitance interact to give the line a characteristic impedance. In telephones this turned out to be about 600 Ω. In transmission lines the best power delivery occurs when the source and the load impedance are the same; this is the process of matching.

It was often required to measure the power in a telephone system, and one milliWatt (mW) was chosen as a suitable unit. Thus the reference against which signals could be compared was the dissipation of one milliWatt in 600 Ω. Figure 2.7 showed that the dissipation of 1 mW in 600 Ω was due to an applied voltage of 0.775 V rms. This voltage is the reference against which all audio levels are compared.

The deciBel is a logarithmic measuring system and has its origins in telephony[1] where the loss in a cable is a logarithmic function of the length. Human hearing also has a logarithmic response with respect to sound pressure level (SPL). In order to relate to the subjective response audio signal level measurements must also be logarithmic and so the deciBel was adopted for audio.

Figure 2.9 shows the principle of the logarithm. To give an example, if it is clear that 10^2 is 100 and 10^3 is 1000, then there must be a power between 2 and 3 to which 10 can be raised to give any value between 100 and 1000. That power is the logarithm to base 10 of the value, e.g. $\log_{10} 300 = 2.5$ approx. Note that 10^0 is 1.

Logarithms were developed by mathematicians before the availability of calculators or computers to ease calculations such as multiplication, squaring, division and extracting roots. The advantage is that armed with a set of log tables, multiplication can be performed by adding, division by subtracting. Figure 2.9 shows some examples. It will be clear that squaring a number is performed by adding two identical logs and the same result will be obtained by multiplying the log by 2.

The slide rule is an early calculator which consists of two logarithmically engraved scales in which the length along the scale is proportional to the log of the engraved number. By sliding the moving scale, two lengths can easily be added or subtracted and as a result multiplication and division is readily obtained.

The logarithmic unit of measurement in telephones was called the Bel after Alexander Graham Bell, the inventor. Figure 2.10(a) shows that the Bel was defined as the log of the *power* ratio between the power to be measured and

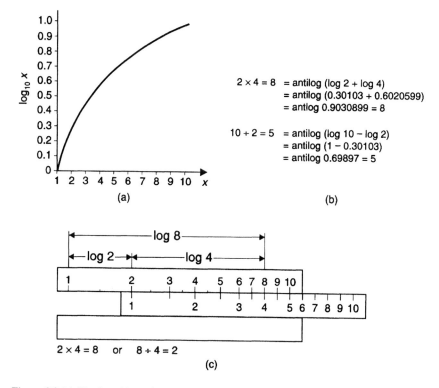

Figure 2.9 (a) The logarithm of a number is the power to which the base (in this case 10) must be raised to obtain the number. (b) Multiplication is obtained by adding logs, division by subtracting. (c) The slide rule has two logarithmic scales whose length can easily be added or subtracted.

some reference power. Clearly the reference power must have a level of 0 Bels since $\log_{10} 1$ is 0.

The Bel was found to be an excessively large unit for many purposes and so it was divided into 10 deciBels, abbreviated dB with a small d and a large B and pronounced deebee. Consequently the number of dB is 10 times the log of the power ratio. A device such as an amplifier can have a fixed power gain which is independent of signal level and this can be measured in dB. However, when measuring the power of a signal, it must be appreciated that the dB is a ratio and to quote the number of dBs without stating the reference is about as senseless as describing the height of a mountain as 2000 without specifying whether this is feet or metres. To show that the reference is one milliWatt into 600 Ω, the units will be dB(m). In radio engineering, the dB(W) will be found which is power relative to one Watt.

Although the dB(m) is defined as a power ratio, level measurements in audio are often done by measuring the signal voltage using 0.775 V as a reference in a circuit whose impedance is not necessarily 600 Ω. Figure 2.9(b) shows that as the power is proportional to the square of the voltage, the power ratio will be obtained by squaring the voltage ratio. As squaring in logs is performed by doubling, the squared term of the voltages can be replaced by multiplying

$$1 \text{ Bel} = \log_{10} \frac{P_1}{P_2} \quad 1 \text{ deciBel} = 1/10 \text{ Bel}$$

$$\text{Power ratio (dB)} = 10 \times \log_{10} \frac{P_1}{P_2}$$

(a)

As power $\propto V^2$, when using voltages:

$$\text{Power ratio (dB)} = 10 \log \frac{V_1^2}{V_2^2}$$

$$= 10 \times \log \frac{V_1}{V_2} \times 2$$

$$= 20 \log \frac{V_1}{V_2}$$

(b)

Figure 2.10 (a) The Bel is the log of the ratio between two powers, that to be measured and the reference. The Bel is too large so the deciBel is used in practice. (b) As the dB is defined as a power ratio, voltage ratios have to be squared. This is conveniently done by doubling the logs so the ratio is now multiplied by 20.

the log by a factor of two. To give a result in deciBels, the log of the voltage ratio now has to be multiplied by 20.

Whilst 600 Ω matched impedance working is essential for the long distances encountered with telephones, it is quite inappropriate for audio wiring in a studio. The wavelength of audio in wires at 20 kHz is 15 km. Most studios are built on a smaller scale than this and clearly analog audio cables are *not* transmission lines and they do not have a characteristic impedance. Consequently the reader is cautioned that anyone who attempts to sell exotic analog audio cables by stressing their transmission line characteristics is more of a salesman than a physicist.

In professional audio systems impedance matching is not only unnecessary it is also undesirable. Figure 2.11(a) shows that when impedance matching is required the output impedance of a signal source must be artificially raised so that a potential divider is formed with the load. The actual drive voltage must be twice that needed on the cable as the potential divider effect wastes 6 dB of signal level and requires unnecessarily high power supply rail voltages in equipment. A further problem is that cable capacitance can cause an undesirable high-frequency roll-off in conjunction with the high source impedance.

In modern professional audio equipment as shown in Figure 2.11(b), the source has the lowest output impedance practicable. This means that any ambient interference is attempting to drive what amounts to a short circuit and can only develop very small voltages. Furthermore shunt capacitance in the cable has very little effect. The destination has a somewhat higher impedance (generally a few k Ω) to avoid excessive currents flowing and to allow several loads to be placed across one driver.

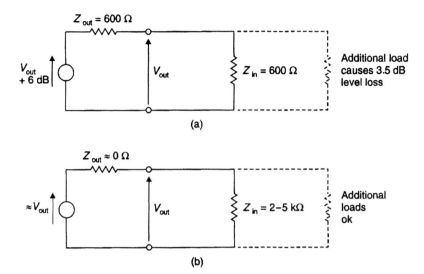

Figure 2.11 (a) Traditional impedance matched source wastes half the signal voltage in the potential divider due to the source impedance and the cable. (b) Modern practice is to use low-output impedance sources with high-impedance loads.

In the absence of a fixed impedance it is now meaningless to consider power. Consequently only signal voltages are measured. The reference remains at 0.775 V, but power and impedance are irrelevant. Voltages measured in this way are expressed in dB(u); the commonest unit of level in modern systems. Most installations boost the signals on interface cables by 4 dB. As the gain of receiving devices is reduced by 4 dB, the result is a useful noise advantage without risking distortion due to the drivers having to produce high voltages.

In order to make the difference between dB(m) and dB(u) clear, consider the lossless matching transformer shown in Figure 2.12. The turns ratio is 2:1 therefore the impedance matching ratio is 4:1. As there is no loss in the transformer, the power in is the same as the power out so that the transformer shows a gain of 0 dB(m). However, the turns ratio of 2:1 provides a voltage

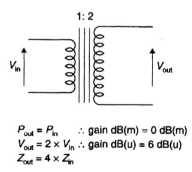

$$P_{out} = P_{in} \quad \therefore \text{ gain dB(m)} = 0 \text{ dB(m)}$$
$$V_{out} = 2 \times V_{in} \therefore \text{ gain dB(u)} = 6 \text{ dB(u)}$$
$$Z_{out} = 4 \times Z_{in}$$

Figure 2.12 A lossless transformer has no power gain so the level in dB(m) on input and output is the same. However, there is a voltage gain when measurements are made in dB(u).

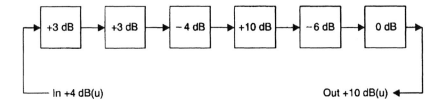

Figure 2.13 In complex systems each stage may have voltage gain measured in dB. By adding all of these gains together and adding to the input level in dB(u), the output level in dB(u) can be obtained.

gain of 6 dB(u). The doubled output voltage will develop the same power in to the quadrupled load impedance.

In a complex system signals may pass through a large number of processes, each of which may have a different gain. Figure 2.13 shows that if one stays in the linear domain and measures the input level in volts rms, the output level will be obtained by multiplying by the gains of all of the stages involved. This is a complex calculation.

The difference between the signal level with and without the presence of a device in a chain is called the *insertion loss* measured in dB. However, if the input is measured in dB(u), the output level of the first stage can be obtained by adding the insertion loss in dB. The output level of the second stage can be obtained by further adding the loss of the second stage in dB and so on. The final result is obtained by adding together all of the insertion losses in dB and adding them to the input level in dB(u) to give the output level in dB(u). As the dB is a pure ratio it can multiply anything (by addition of logs) without changing the units. Thus dB(u) of level added to dB of gain are still dB(u).

In acoustic measurements, the sound pressure level (SPL) is measured in deciBels relative to a reference pressure of 2×10^{-5} Pascals (Pa) rms. In order to make the reference clear the units are dB(SPL). In measurements which are intended to convey an impression of subjective loudness, a weighting filter is used prior to the level measurement which reproduces the frequency response of human hearing which is most sensitive in the mid-range. The most common standard frequency response is the so-called A-weighting filter, hence the term dB(A) used when a weighted level is being measured. At high or low frequencies, a lower reading will be obtained in dB(A) than in dB(SPL).

2.6 Audio level metering

There are two main reasons for having level meters in audio equipment: to line up or adjust the gain of equipment, and to assess the amplitude of the program material.

Line up is often done using a 1 kHz sine wave generated at an agreed level such as 0 dB(u). If a receiving device does not display the same level, then its input sensitivity must be adjusted. Tape recorders and other devices which pass signals through are usually lined up so that their input and output levels are identical, i.e. their insertion loss is 0 dB. Line up is important in large systems because it ensures that inadvertent level changes do not occur.

In measuring the level of a sine wave for the purposes of line up, the dynamics of the meter are of no consequence, whereas on program material the dynamics matter a great deal. The simplest (and cheapest) level meter is essentially an AC voltmeter with a logarithmic response. As the ear is logarithmic, the deflection of the meter is roughly proportional to the perceived volume, hence the term Volume Unit (VU) meter.

In audio recording and broadcasting, the worst sin is to overmodulate the tape or the transmitter by allowing a signal of excessive amplitude to pass. Real audio signals are rich in short transients which pass before the sluggish VU meter responds. Consequently the VU meter is also called the virtually useless meter in professional circles.

Broadcasters developed the Peak Program Meter (PPM) which is also logarithmic, but which is designed to respond to peaks as quickly as the ear responds to distortion. Consequently the attack time of the PPM is carefully specified. If a peak is so short that the PPM fails to indicate its true level, the resulting overload will also be so brief that the ear will not hear it. A further feature of the PPM is that the decay time of the meter is very slow, so that any peaks are visible for much longer and the meter is easier to read because the meter movement is less violent.

The original PPM as developed by the BBC was sparsely calibrated, but other users have adopted the same dynamics and added dB scales, Figure 2.14 shows some of the scales in use.

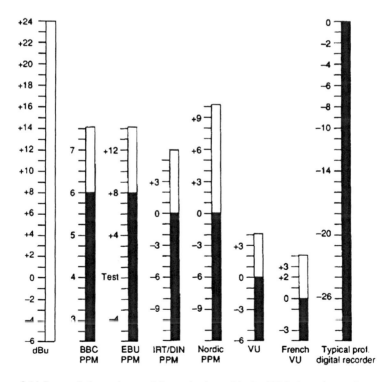

Figure 2.14 Some of the scales used in conjuction with the PPM dynamics. (After Francis Rumsey, with permission.).

In broadcasting, the use of level metering and line up procedures ensures that the level experienced by the listener does not change significantly from program to program. Consequently in a transmission suite, the goal would be to broadcast tapes at a level identical to that which was obtained during production. However, when making a recording prior to any production process, the goal would be to modulate the tape as fully as possible without clipping as this would then give the best signal-to-noise ratio. The level would then be reduced if necessary in the production process.

2.7 Vectors

Often several signals of the same frequency but with differing phases need to be added. When the two phases are identical, the amplitudes are simply added. When the two phases are 180° apart the amplitudes are subtracted. When there is an arbitrary phase relationship, vector addition is needed. A vector is simply an arrow whose length represents the amplitude and whose direction represents the phase shift. Figure 2.15 shows a vector diagram showing the phase relationship in the common three-phase electricity supply. The length of each vector represents the phase-to-neutral voltage which in many countries is about 230 V rms. As each phase is at 120° from the next, what will the phase-to-phase voltage be? Figure 2.15 shows that the answer can be found geometrically to be about 400 V rms. Consequently whilst a phase-to-neutral shock is not recommended, getting a phase-to-phase shock is recommended even less!

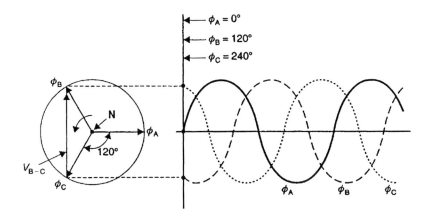

$$V_{B-C} = V_B \sin 120 - V_C \sin 240 \quad V_B = V_C = V_A$$

$$\therefore V_{B-C} = 2 \times V_A \times 0.866 = V_A \times 1.732$$

E.g. if phase voltage is 230 V rms,
phase–to–phase voltage = 398 V rms

Figure 2.15 Three-phase electricity uses three signals mutually at 120°. Thus the phase-to-phase voltage has to be calculated vectorially from the phase-to-neutral voltage as shown.

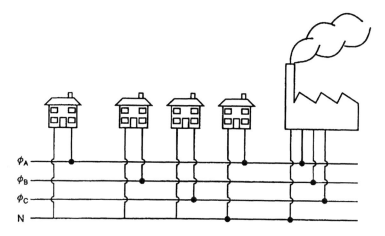

Figure 2.16 The possibility of a phase-to-phase shock is reduced in suburban housing by rotating phases between adjacent buildings. This also balances the loading on the three phases.

The three-phase electricity supply has the characteristic that although each phase passes through zero power twice a cycle, the total power is constant. This results in less vibration at the generator and in large motors. When a three-phase system is balanced (i.e. there is an equal load on each phase) there is no neutral current. Figure 2.16 shows that in most suburban power installations, each house only has a single phase supply for safety. The houses are connected in rotation to balance the load. Business premises such as recording studios and broadcasters will take a three-phase supply which should be reasonably balanced by connecting equal loading to each phase.

2.8 Phase angle and power factor

The power is only obtained by multiplying the voltage by the current when the load is resistive. Only with a resistive load will the voltage and the current be in the same phase. In both electrical and audio power distribution systems, the load may be *reactive* which means that the current and voltage waveforms have a relative phase shift. Mathematicians would describe the load as *complex*.

In a reactive load, the power in Watts, W, is given by multiplying the rms voltage, the rms current and the cosine of the relative phase angle ϕ. Clearly if the voltage and current are in quadrature there can be no power dissipated because cos ϕ is zero. cos ϕ is called the *power factor*. Figure 2.17 shows that this happens with perfect capacitors and perfect inductors connected to an AC supply. With a perfect capacitor, the current *leads* the voltage by 90°, whereas with a perfect inductor the current *lags* the voltage by 90°.

A power factor significantly less than one is bad news because it means that larger currents are flowing than are necessary to deliver the power. The losses in distribution are proportional to the current and so a reactive load is an inefficient load. Lightly loaded transformers and induction motors act as inductive loads with a poor power factor. In some industrial installations it is economic to install power factor correction units which are usually capacitor banks which balance the lagging inductive load with a capacitive lead.

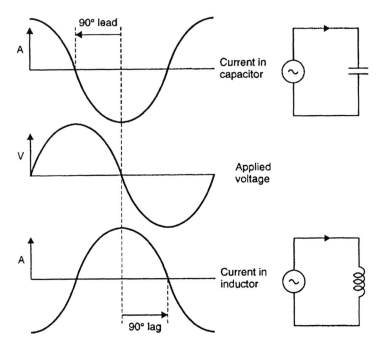

Figure 2.17 Ideal capacitors conduct current with a quadrature phase lead whereas inductors have a quadrature phase lag. In both cases the quadrature makes the product of current and voltage zero so no power is dissipated.

As the power factor of a load cannot be anticipated, AC distribution equipment is often rated in volt amps (VA) instead of Watts. With a resistive load, the two are identical, but with a reactive load the power which can be delivered falls. As loudspeakers are almost always reactive, audio amplifiers should be rated in VA. Instead amplifiers are rated in Watts leaving the unfortunate user to find out what reactive load can be driven.

2.9 Audio cabling

Balanced line working was developed for professional audio as a means to reject noise. This is particularly important for microphone signals because of the low levels, but is also important for line level signals where interference may be encountered from electrical and radio installations. Figure 2.18 shows how balanced audio should be connected. The receiver subtracts one input from the other which rejects any common mode noise or hum picked up on the wiring. Twisting the wires tightly together ensures that both pick up the same amount of interference.

The star-quad technique is possibly the ultimate interference rejecting cable construction. Figure 2.19 shows that in star-quad cable four conductors are twisted together. Diametrically opposite pairs are connected together at both ends of the cable and used as the two legs of a differential system. The interference pickup on the two legs is rendered as identical as possible by the construction so that it can be perfectly rejected at a well-engineered differential receiver.

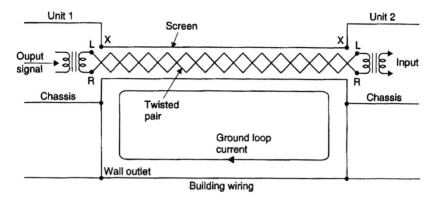

Figure 2.18 Balanced analog audio interface. Note that the braid plays no part in transmitting the audio signal, but bonds the two units together and acts as a screen. Loop currents flowing in the screen are harmless.

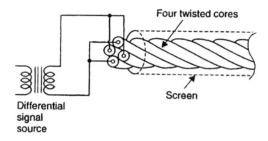

Figure 2.19 In star-quad cabling each leg of the balanced signal is connected to two conductors which are on opposite sides of a four-phase helix. The pickup of interference on the two legs is then as equal as possible so that the differential receiver can reject it.

The standard connector which has been used for professional audio for many years is the XLR which has three pins. It is easy to remember that pins 1, 2 and 3 connect to eXternal, Live and Return respectively. EXternal is the cable screen, Live is the in-phase leg of the balanced signal and Return is self-explanatory. The metal body shell of the XLR connector should be connected to both the cable screen and pin 1 although cheaper connectors do not provide a tag for the user to make this connection and rely on contact with the chassis socket to ground the shell. Oddly, the male connector (the one with pins) is used for equipment signal outputs, whereas the female (the one with receptacles) is used with signal inputs. This is so that when phantom power (see Chapter 5) is used, the live parts are insulated.

When making balanced cables it is important to ensure that the twisted pair is connected identically at both ends. If the two wires are inadvertently interchanged, the cable will still work, but a phase reversal will result, causing problems in stereo installations and preventing the continuity of absolute phase through any system.

In consumer equipment differential working is considered too expensive. Instead single-ended analog signals using coax cable are found using phono,

DIN and single pole jack connectors. Whilst these are acceptable in permanent installations, they will not stand repeated connection and disconnection and become unreliable. Very expensive gold-plated phono connectors are available which are advertised as being hand polished by Peruvian virgins or some such, but, however attractive they may look, they have absolutely no effect on the sound quality.

Effective unbalanced transmission over long distances is very difficult. When the signal return, the chassis ground and the safety ground are one and the same as in Figure 2.20(a), ground loop currents cannot be rejected. The only solution is to use equipment which is double insulated so that no safety ground is needed. Then each item can be grounded by the coax screen. As Figure 2.20(b) shows, there can then be no ground current as there is no loop. However, unbalanced working also uses higher impedances and lower signal levels and is more prone to interference. For these reasons some better quality consumer equipment will be found using balanced signals.

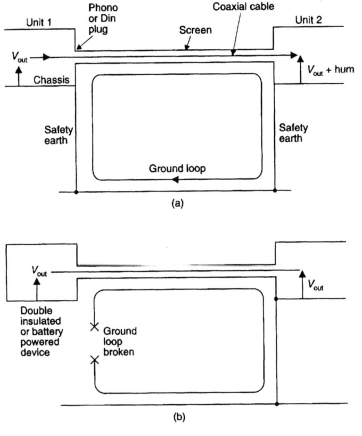

Figure 2.20 (a) Unbalanced consumer equipment cannot be protected from hum loops because the signal return and the screen are the same conductor. (b) With a floating signal source there will be no current in the screen. Source must be double insulated for safety.

2.10 Moving masses

Sound reproduction depends totally on microphones and loudspeakers which contain parts moving in sympathy with the audio waveform. The stylus of a vinyl disk player is another example. In CD players the objective lens is dynamically positioned to maintain accurate focus and tracking. In order to understand how these devices work, some basic knowledge of the physics of moving masses is essential.

Isaac Newton explained that a mass m would remain at rest or travel at constant velocity unless some net force acted upon it. If such a force F acts, the result is an acceleration a where $F = ma$.

Figure 2.21(a) shows a steady rotation which could be a mass tethered by a string. Tension in the string causes an inward force which accelerates the mass into a circular path. The tension can be described as a rotating vector. When viewed from the side, the displacement about the vertical axis appears sinusoidal. The vertical component of the tension is also sinusoidal, and out of phase with the displacement. As the two parameters have the same waveform, they must be proportional. In other words the restoring force on the mass is proportional to the displacement.

Figure 2.21(b) shows that the same characteristic is obtained if the mass is supported on a spring. The mass supported by a spring is found widely in audio because all diaphragms and styli need compliant support to keep them in place yet allow vibration. An ideal spring produces a restoring force which is proportional to the displacement. The constant of proportionality is called the *stiffness* which is the reciprocal of *compliance*. When such a system is displaced there is sustained *resonance*. Not surprisingly the displacement is

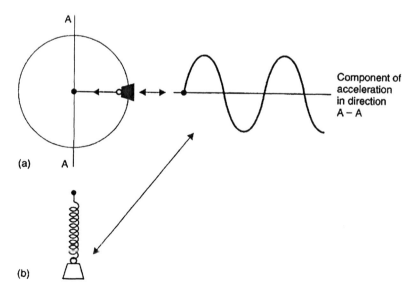

Figure 2.21 (a) Tethered rotating mass requires tension in the tether to accelerate it into a circular path. Component of acceleration in one axis is sinusoidal. (b) Mass supported on spring has identical characteristic.

sinusoidal and is called *simple harmonic motion* or SHM and has all of the characteristics of one dimension of a rotation as shown in Figure 2.6.

The only difference between the mass on a string and the mass on a spring is that when more energy is put into the system, the mass on a string goes faster because the displacement cannot increase but more tension can be produced. The mass on the spring oscillates at the same frequency but the amplitude has to increase so that the restoring force can be greater.

Eventually the resonance of a mass on a spring dies away. The faster energy is taken out of the system, the greater the rate of decay. Any mechanism which removes energy from a resonant system is called *damping*.

The motion of a rigid body can be completely determined by the mass, the stiffness and the damping factor. As audio signals contain a wide range of frequencies, it is important to consider what happens when resonant systems are excited by them.

Figure 2.22 shows the velocity and displacement of a mass-stiffness-damping system excited by a constant amplitude sinusoidal force acting on the mass at various frequencies. Below resonance, the frequency of excitation is low and

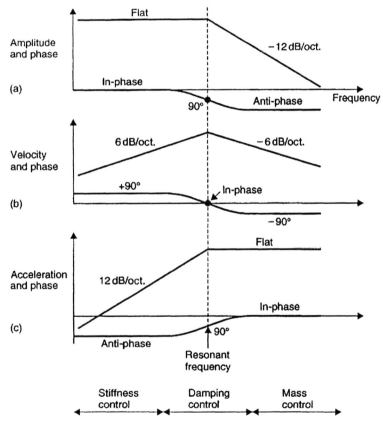

Figure 2.22 The behaviour of a mass-stiffness-damping system: (a) amplitude, (b) velocity, (c) acceleration.

little force is needed to accelerate the mass. The force needed to deflect the spring is greater and so the system is said to be *stiffness controlled*. The amplitude is independent of frequency, described as *constant amplitude* operation, and so the velocity rises at 6 dB/octave towards resonance. Above resonance the inertia of the mass is greater than the stiffness of the spring and the system is said to be *mass controlled*. With a constant force there is *constant acceleration*, yet as frequency rises there is less time for the acceleration to act. Thus velocity is inversely proportional to frequency which in engineering terminology is −6 dB/octave. As the displacement is the integral of the velocity the displacement curve is tilted by 6 dB so that below resonance the displacement is flat and in-phase with the force whereas above it falls at 12 dB/octave and is anti-phase with the force.

In the vicinity of resonance the amplitude is a function of the damping and is said to be *resistance controlled*. With no damping the Q factor is high and the amplitude at resonance tends to infinity, resulting in a sharp peak in the response. Increasing the damping lowers and broadens the peak so that with high damping the velocity is nearly independent of frequency. Figure 2.23 shows the effect of different damping factors on the impulse response, i.e. the response to a sudden shock. The underdamped system enters a decaying oscillation. The overdamped system takes a considerable time to return to rest. The critically damped system returns to rest in the shortest time possible subject to not overshooting.

Below resonance the displacement of the spring is proportional to the force. Here force and displacement are in-phase. Above resonance the acceleration of the mass is proportional to the force. Here force and acceleration are in-phase. It will be seen from Figure 2.22 and Figure 2.6 that the velocity leads the displacement but lags the acceleration. Consequently below resonance the velocity leads the applied force whereas above resonance it lags. Around

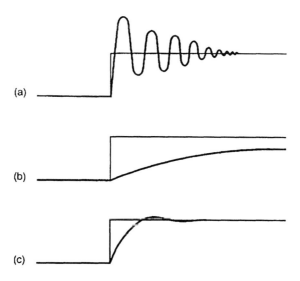

(a)

(b)

(c)

Figure 2.23 Response of a damped system to a step input: (a) underdamped, (b) overdamped, (c) critically damped.

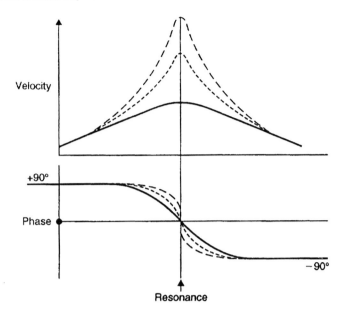

Figure 2.24 Phase passes through 0 ° at the resonant frequency. Rate of change of phase depends on the degree of damping.

resonance there is a phase reversal so that at the precise resonant frequency there is no phase shift at all. Figure 2.24 shows that the rate of phase change in the vicinity of resonance is a function of the damping.

It will be clear that the behaviour just noted has a direct parallel in the behaviour of an electronic damped tuned circuit consisting of an inductor, a capacitor and a resistor and the mathematics of both are one and the same. This is more than just a convenience because it means that an unwanted mechanical resonance or phase change can be suppressed by incorporating at some point a suitable electronic circuit designed to have the opposite characteristic. Additionally by converting mechanical parameters into electrical parameters the behaviour of a mechanism can be analysed as if it were an electronic circuit. This is particularly common in loudspeaker design.

2.11 Introduction to digital processes

However complex a digital process, it can be broken down into smaller stages until finally one finds that there are really only two basic types of element in use, and these can be combined in some way and supplied with a clock to implement virtually any process. Figure 2.25 shows that the first type is a *logical* element. This produces an output which is a logical function of the input with minimal delay. The second type is a *storage* element which samples the state of the input(s) when clocked and holds or delays that state. The strength of binary logic is that the signal has only two states, and considerable noise and distortion of the binary waveform can be tolerated before the state becomes uncertain. At every logical element, the signal is compared

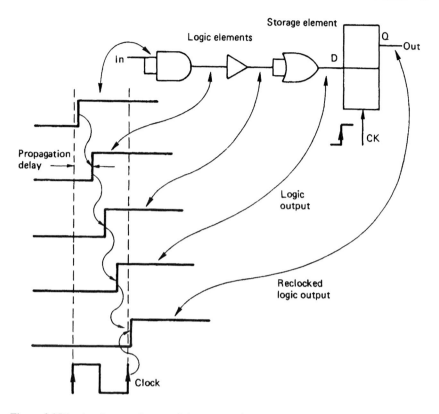

Figure 2.25 Logic elements have a finite propagation delay between input and output and cascading them delays the signal an arbitrary amount. Storage elements sample the input on a clock edge and can return a signal to near coincidence with the system clock. This is known as reclocking. Reclocking eliminates variations in propagation delay in logic elements.

with a threshold, and thus can pass through any number of stages without being degraded. In addition, the use of a storage element at regular locations throughout logic circuits eliminates time variations or jitter. Figure 2.25 shows that if the inputs to a logic element change, the output will not change until the *propagation delay* of the element has elapsed. However, if the output of the logic element forms the input to a storage element, the output of that element will not change until the input is sampled *at the next clock edge*. In this way the signal edge is aligned to the system clock and the propagation delay of the logic becomes irrelevant. The process is known as reclocking.

2.12 Logic elements

The two states of the signal when measured with an oscilloscope are simply two voltages, usually referred to as high and low. The actual voltage levels will depend on the type of logic family in use, and on the supply voltage used. Within logic, these levels are not of much consequence, and it is only necessary to know them when interfacing between different logic families or when driving external devices. The pure logic designer is not interested at all in these voltages, only in their meaning. Just as the electrical waveform from a microphone represents sound velocity, so the waveform in a logic circuit represents the truth of some statement. As there are only two states, there can only be *true* or *false* meanings. The true state of the signal can be assigned by the designer to either voltage state. When a high voltage represents a true logic condition and a low voltage represents a false condition, the system is known as *positive logic*, or *high true* logic. This is the usual system, but sometimes the low voltage represents the true condition and the high voltage represents the false condition. This is known as *negative logic* or *low true* logic. Provided that everyone is aware of the logic convention in use, both work equally well.

Positive logic name	Boolean expression	Positive logic symbol	Positive logic truth table	Plain English
Inverter or NOT gate	$Q = \overline{A}$	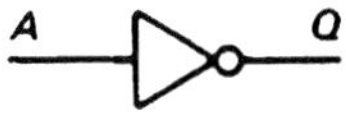	$\begin{array}{c\|c} A & Q \\ \hline 0 & 1 \\ 1 & 0 \end{array}$	Output is opposite of input
AND gate	$Q = A \cdot B$	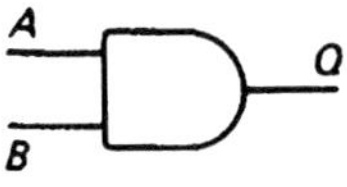	$\begin{array}{cc\|c} A & B & Q \\ \hline 0 & 0 & 0 \\ 0 & 1 & 0 \\ 1 & 0 & 0 \\ 1 & 1 & 1 \end{array}$	Output true when both inputs are true only
NAND (Not AND) gate	$Q = \overline{A \cdot B}$ $= \overline{A} + \overline{B}$	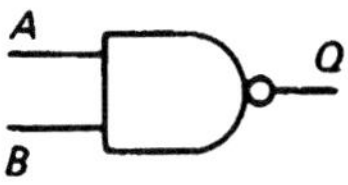	$\begin{array}{cc\|c} A & B & Q \\ \hline 0 & 0 & 1 \\ 0 & 1 & 1 \\ 1 & 0 & 1 \\ 1 & 1 & 0 \end{array}$	Output false when both inputs are true only
OR gate	$Q = A + B$	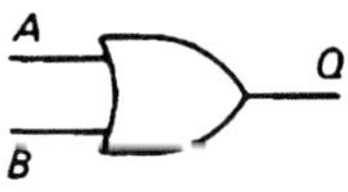	$\begin{array}{cc\|c} A & B & Q \\ \hline 0 & 0 & 0 \\ 0 & 1 & 1 \\ 1 & 0 & 1 \\ 1 & 1 & 1 \end{array}$	Output true if either or both inputs true
NOR (Not OR) gate	$Q = \overline{A + B}$ $= \overline{A} \cdot \overline{B}$	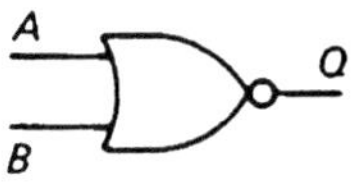	$\begin{array}{cc\|c} A & B & Q \\ \hline 0 & 0 & 1 \\ 0 & 1 & 0 \\ 1 & 0 & 0 \\ 1 & 1 & 0 \end{array}$	Output false if either or both inputs true
Exclusive OR (XOR) gate	$Q = A \oplus B$	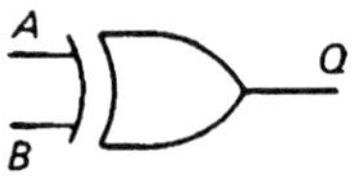	$\begin{array}{cc\|c} A & B & Q \\ \hline 0 & 0 & 0 \\ 0 & 1 & 1 \\ 1 & 0 & 1 \\ 1 & 1 & 0 \end{array}$	Output true if inputs are different

Figure 2.26 The basic logic gates compared.

In logic systems, all logical functions, however complex, can be configured from combinations of a few fundamental logic elements or *gates*. It is not profitable to spend too much time debating which are the truly fundamental ones, since most can be made from combinations of others. Figure 2.26 shows the important simple gates and their derivatives, and introduces the logical expressions to describe them, which can be compared with the truth-table notation. The figure also shows the important fact that when negative logic is used, the OR gate function interchanges with that of the AND gate.

If numerical quantities need to be conveyed down the two-state signal paths described here, then the only appropriate numbering system is binary, which has two symbols, 0 and 1. Just as positive or negative logic could be used for the truth of a logical binary signal, it can also be used for a numerical binary signal. Normally, a high voltage level will represent a binary 1 and a low voltage will represent a binary 0, described as a 'high for a one' system. Clearly a 'low for a one' system is just as feasible. Decimal numbers have several columns, each of which represents a different power of 10; in binary the column position specifies the power of two.

2.13 Storage elements

The basic memory element in logic circuits is the latch, which is constructed from two gates as shown in Figure 2.27(a), and which can be set or reset. A more useful variant is the D-type latch shown at Figure 2.27(b) which remembers the state of the input at the time a separate clock either changes state for an edge-triggered device, or after it goes false for a level-triggered device. D-type latches are commonly available with four or eight latches to the chip. A shift register can be made from a series of latches by connecting the Q output of one latch to the D input of the next and connecting all of the clock inputs in parallel. Data are delayed by the number of stages in the register. Shift registers are also useful for converting between serial and parallel data transmissions.

Where large numbers of bits are to be stored, cross-coupled latches are less suitable because they are more complicated to fabricate inside integrated circuits than dynamic memory, and consume more current.

In large random access memories (RAMs), the data bits are stored as the presence or absence of charge in a tiny capacitor as shown in Figure 2.27(c). The capacitor is formed by a metal electrode, insulated by a layer of silicon dioxide from a semiconductor substrate, hence the term MOS (metal oxide semiconductor). The charge will suffer leakage, and the value would become indeterminate after a few milliseconds. Where the delay needed is less than this, decay is of no consequence, as data will be read out before they have had a chance to decay. Where longer delays are necessary, such memories must be refreshed periodically by reading the bit value and writing it back to the same place. Most modern MOS RAM chips have suitable circuitry built in. Large RAMs store thousands of bits, and it is clearly impractical to have a connection to each one. Instead, the desired bit has to be addressed before it can be read or written. The size of the chip package restricts the number of pins available, so that large memories use the same address pins more than once. The bits are arranged internally as rows and columns, and the row address and the column address are specified sequentially on the same pins.

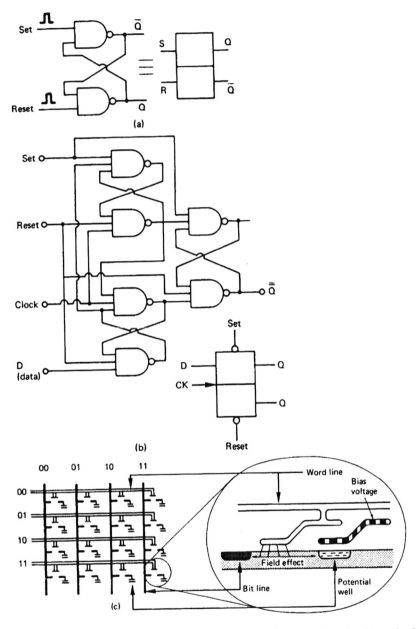

Figure 2.27 Digital semiconductor memory types. In (a) one data bit can be stored in a simple set–reset latch, which has little application because the D-type latch in (b) can store the state of the single data input when the clock occurs. These devices can be implemented with bipolar transistors or FETs, and are called static memories because they can store indefinitely. They consume a lot of power. In (c) a bit is stored as the charge in a potential well in the substrate of a chip. It is accessed by connecting the bit line with the field effect from the word line. The single well where the two lines cross can then be written or read. These devices are called dynamic RAMs because the charge decays, and they must be read and rewritten (refreshed) periodically.

2.14 Binary adding

The circuitry necessary for adding pure binary or two's complement (see Section 8.22) numbers is shown in Figure 2.28. Addition in binary requires two bits to be taken at a time from the same position in each word, starting at the least significant bit (LSB). Should both be ones, the output is zero, and there is a *carry-out* generated. Such a circuit is called a half adder, shown in (a), and is suitable for the least significant bit of the calculation. All higher stages will require a circuit which can accept a carry input as well as two data inputs. This is known as a full adder (b). Multi-bit full adders are available in chip form, and have carry-in and carry-out terminals to allow them to be cascaded to operate on long wordlengths. Such a device is also convenient for inverting a two's complement number, in conjunction with a set of inverters. The adder chip has one set of inputs grounded, and the carry-in permanently held true, such that it adds one to the one's complement number from the inverter.

When mixing by adding sample values, care has to be taken to ensure that if the sum of the two sample values exceeds the number range the result will be clipping rather than wraparound. In two's complement, the action necessary depends on the polarities of the two signals. Clearly if one positive and one negative number are added, the result cannot exceed the number range. If two positive numbers are added, the symptom of positive overflow is that the most significant bit sets, causing an erroneous negative result, whereas a negative overflow results in the most significant bit clearing. The overflow control circuit will be designed to detect these two conditions, and override the adder output. If the most significant bit (MSB) of both inputs is zero, the numbers are both positive, thus if the sum has the MSB set, the output is replaced with the maximum positive code (0111...). If the MSB of both inputs is set, the numbers are both negative, and if the sum has no MSB set, the output is replaced with the maximum negative code (1000...). These conditions can also be connected to warning indicators. Figure 2.28(c) shows this system in hardware. The resultant clipping on overload is sudden, and sometimes a PROM is included which translates values around and beyond maximum to soft-clipped values below or equal to maximum.

A storage element can be combined with an adder to obtain a number of useful functional blocks which will crop up frequently in digital equipment. Figure 2.29(a) shows that a latch is connected in a feedback loop around an adder. The latch contents are added to the input each time it is clocked. The configuration is known as an accumulator in computation because it adds up or accumulates values fed into it. In filtering, it is known as a discrete time integrator. If the input is held at some constant value, the output increases by that amount on each clock. The output is thus a sampled ramp.

Figure 2.29(b) shows that the addition of an inverter allows the difference between successive inputs to be obtained. This is digital differentiation. The output is proportional to the slope of the input.

2.15 Gain control by multiplication

Gain control is used extensively in audio systems. Digital filtering and transform calculations rely heavily on it, as do the processes in a mixer. Gain

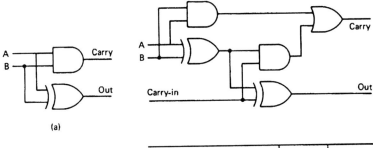

(a)

Data A	Bits B	Carry-in	Out	Carry-out
0	0	0	0	0
0	0	1	1	0
0	1	0	1	0
0	1	1	0	1
1	0	0	1	0
1	0	1	0	1
1	1	0	0	1
1	1	1	1	1

(b)

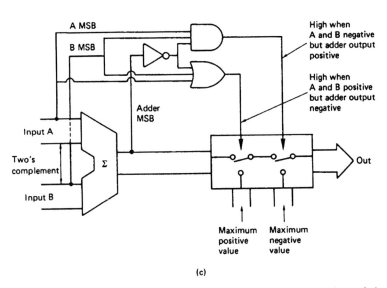

(c)

Figure 2.28 (a) Half adder; (b) full-adder circuit and truth table; (c) comparison of sign bits prevents wraparound on adder overflow by substituting clipping level.

is controlled in the digital domain by multiplying each sample value by a coefficient. If that coefficient is less than one, attenuation will result; if it is greater than one, amplification can be obtained.

Multiplication in binary circuits is difficult. It can be performed by repeated adding, but this is too slow to be of any use. In fast multiplication, one of the

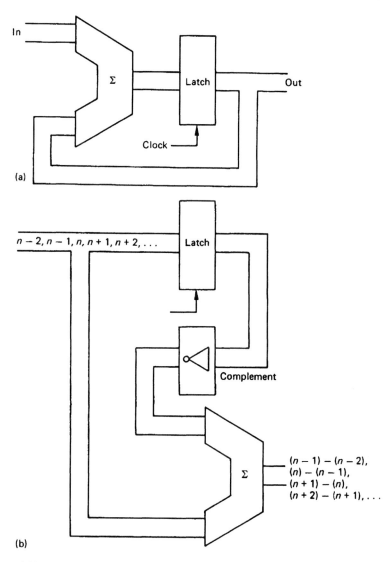

Figure 2.29 Two configurations which are common in processing. In (a) the feedback around the adder adds the previous sum to each input to perform accumulation or digital integration. In (b) an inverter allows the difference between successive inputs to be computed. This is differentiation.

inputs will be simultaneously multiplied by one, two, four, etc., by hard-wired bit shifting. Figure 2.30 shows that the other input bits will determine which of these powers will be added to produce the final sum, and which will be neglected. If multiplying by five, the process is the same as multiplying by four, multiplying by one, and adding the two products. This is achieved by adding the input to itself shifted two places. As the wordlength of such a device increases, the complexity increases greatly, so this is a natural application for an integrated circuit.

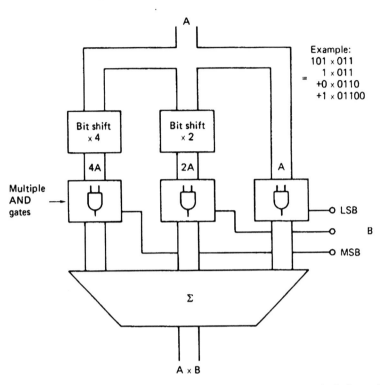

Figure 2.30 Structure of fast multiplier. The input A is multiplied by 1, 2, 4, 8, etc. by bit shifting. The digits of the B input then determine which multiples of A should be added together by enabling AND gates between the shifters and the adder. For long wordlengths, the number of gates required becomes enormous, and the device is best implemented in a chip.

2.16 Transforms

At its simplest a transform is a process which takes information in one domain and expresses it in another. Audio signals are in the time domain: their voltages (or sample values) change as a function of time. Such signals are often transformed to the frequency domain for the purposes of analysis or compression. In the frequency domain the signal has been converted to a spectrum; a table of the energy at different temporal or spatial frequencies. If the input is repeating, the spectrum will also be sampled, i.e. it will exist at discrete frequencies. In real program material, all frequencies are seldom present together and so those which are absent need not be transmitted and a coding gain is obtained. On reception an inverse transform or synthesis process converts back from the frequency domain to the time or space domains. To take a simple analogy, a frequency transform of piano music effectively works out what frequencies are present as a function of time; the transform works out which notes were played and so the information is not really any different from that contained in the original sheet music.

One frequently encountered way of entering the frequency domain from the time or spatial domains is the Fourier transform or its equivalent in sampled systems, the discrete Fourier transform (DFT). Fourier analysis holds that any periodic waveform can be reproduced by adding together an arbitrary number of harmonically related sinusoids of various amplitudes and phases. Figure 2.31 shows how a square wave can be built up of harmonics. The spectrum can be drawn by plotting the amplitude of the harmonics against frequency. It will be seen that this gives a spectrum which is a decaying wave. It passes through zero at all even multiples of the fundamental. The shape of the spectrum is a sin x/x curve. If a square wave has a sin x/x spectrum, it follows that a filter with a rectangular impulse response will have a sin x/x frequency response. It will be recalled that an ideal low-pass filter has a rectangular spectrum, and this has a sin x/x impulse response. These characteristics are known as a transform pair. In transform pairs, if one domain has one shape of the pair, the other domain will have the other shape. Thus a square wave has a sin x/x spectrum and a sin x/x impulse has a square spectrum. Figure 2.32 shows a number of transform pairs. Note the pulse pair. A time-domain pulse of infinitely short duration has a flat spectrum. Thus a flat waveform, i.e. a constant voltage, has only 0 Hz in its spectrum. Interestingly the transform of a Gaussian response in still Gaussian.

The Fourier transform specifies the amplitude and phase of the frequency components just once and such sine waves are endless. As a result the Fourier transform is only valid for periodic waveforms; i.e. those which repeat endlessly. Real program material is not like that and so it is necessary to break up the continuous time domain using windows. Figure 2.33(a) shows how a block of time is cut from the continuous input. By wrapping it into a ring it can be made to appear like a continuous periodic waveform for which a

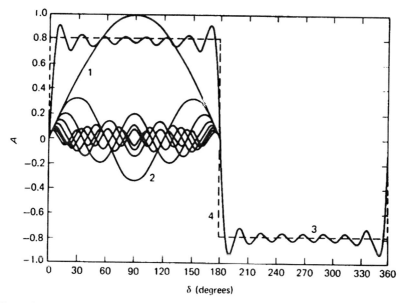

Figure 2.31 Fourier analysis of a square wave into fundamental and harmonics. A, amplitude; δ, phase of fundamental wave in degrees; 1, first harmonic (fundamental); 2, odd harmonics 3–15; 3, sum of harmonics 1–15; 4, ideal square wave.

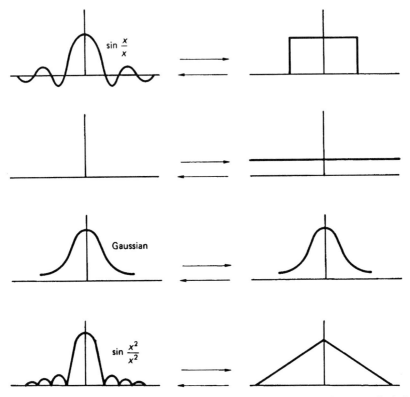

Figure 2.32 The concept of transform pairs illustrates the duality of the frequency (including spatial frequency) and time domains.

single transform, known as the short-time Fourier transform (STFT), can be computed. Note that in the Fourier transform of a periodic waveform the frequency bands have constant width. The inverse transform produces just such an endless waveform, but a window is taken from it and used as part of the output waveform. Rectangular windows are used in video compression, but are not generally adequate for audio because the discontinuities at the boundaries are audible. This can be overcome by shaping and overlapping the windows so that a crossfade occurs at the boundaries between them as in (b).

As has been mentioned, the theory of transforms assumes endless periodic waveforms. If an infinite length of waveform is available, spectral analysis can be performed to infinite resolution but as the size of the window reduces, so too does the resolution of the frequency analysis. Intuitively it is clear that discrimination between two adjacent frequencies is easier if more cycles of both are available. In sampled systems, reducing the window size reduces the number of samples and so must reduce the number of discrete frequencies in the transform. Thus for good frequency resolution the window should be as large as possible. However, with large windows the time between updates of the spectrum is longer and so it is harder to locate events on the time axis. Figure 2.34(a) shows the effect of two window sizes in a conventional

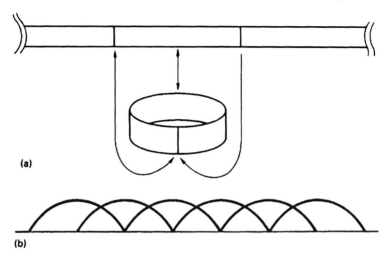

Figure 2.33 In (a) a continuous audio signal can be cut into blocks which are wrapped to make them appear periodic for the purposes of the Fourier transform. A better approach is to use overlapping windows to avoid discontinuities as in (b).

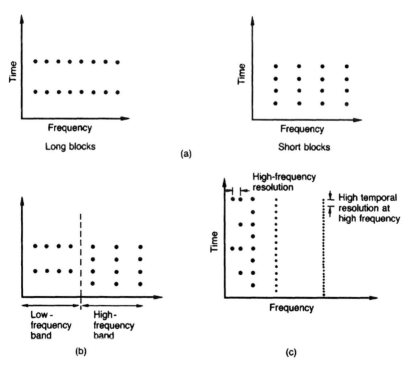

Figure 2.34 (a) In transforms greater certainty in the time domain leads to less certainty in the frequency domain and vice versa. Some transform coders split the spectrum as in (b) and use different window lengths in the two bands. In the recently developed wavelet transform the window length is inversely proportional to the frequency, giving the advantageous time/frequency characteristic shown in (c).

STFT and illustrates the principle of *uncertainty* also known as the Heisenberg inequality.

According to the uncertainty theory one can trade off time resolution against frequency resolution. In most program material, the time resolution required falls with frequency whereas the time (or spatial) resolution required rises with frequency. Fourier-based compression systems using transforms sometimes split the signal into a number of frequency bands in which different window sizes are available as in Figure 2.34(b). Some have variable length windows which are selected according to the program material. The Sony ATRAC system of the MiniDisc (see Chapter 11) uses these principles. Stationary material such as steady tones are transformed with long windows whereas transients are transformed with short windows.

The recently developed wavelet transform is one in which the window length is inversely proportional to the frequency. This automatically gives the advantageous time/frequency resolution characteristic shown in (c). The wavelet transform is considered in more detail in Section 2.19.

Although compression uses transforms, the transform itself does not result in any data reduction, as there are usually as many coefficients as input samples. Paradoxically the transform increases the amount of data because the coefficient multiplications result in wordlength extension. Thus it is incorrect to refer to transform compression; instead the term transform-*based* compression should be used.

2.17 The Fourier transform

Figure 2.35 shows that if the amplitude and phase of each frequency component is known, linearly adding the resultant components in an inverse transform results in the original waveform. In digital systems the waveform is expressed as a number of discrete samples. As a result the Fourier transform analyses the signal into an equal number of discrete frequencies. This is known as a discrete Fourier transform or DFT in which the number of frequency coefficients is equal to the number of input samples. The fast Fourier transform (FFT) is no more than an efficient way of computing the DFT.[2] As was seen in the previous section, practical systems must use windowing to create short-term transforms.

It will be evident from Figure 2.35 that the knowledge of the phase of the frequency component is vital, as changing the phase of any component will seriously alter the reconstructed waveform. Thus the DFT must accurately analyse the phase of the signal components.

There are a number of ways of expressing phase. Figure 2.36 shows a point which is rotating about a fixed axis at constant speed. Looked at from the side, the point oscillates up and down at constant frequency. The waveform of that motion is a sine wave, and that is what we would see if the rotating point were to translate along its axis whilst we continued to look from the side.

One way of defining the phase of a waveform is to specify the angle through which the point has rotated at time zero $(T = 0)$. If a second point is made to revolve at $90°$ to the first, it would produce a cosine wave when translated. It is possible to produce a waveform having arbitrary phase by adding together the sine and cosine wave in various proportions and polarities. For example,

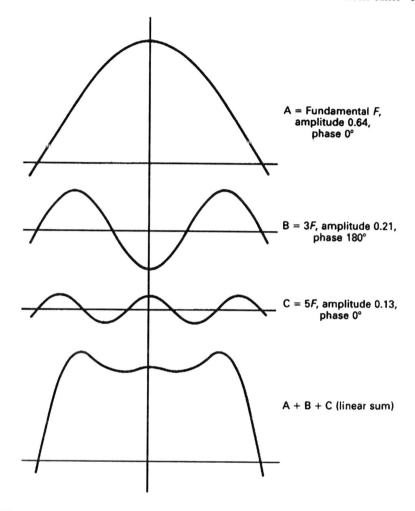

A = Fundamental F, amplitude 0.64, phase 0°

B = 3F, amplitude 0.21, phase 180°

C = 5F, amplitude 0.13, phase 0°

A + B + C (linear sum)

Figure 2.35 Fourier analysis allows the synthesis of any periodic waveform by the addition of discrete frequencies of appropriate amplitude and phase.

adding the sine and cosine waves in equal proportion results in a waveform lagging the sine wave by 45°.

Figure 2.36 shows that the proportions necessary are respectively the sine and the cosine of the phase angle. Thus the two methods of describing phase can be readily interchanged.

The discrete Fourier transform spectrum-analyses a string of samples by searching separately for each discrete target frequency. It does this by multiplying the input waveform by a sine wave, known as the basis function, having the target frequency and adding up or integrating the products. Figure 2.37(a) shows that multiplying by basis functions gives a non-zero integral when the input frequency is the same, whereas (b) shows that with a different input frequency (in fact all other different frequencies) the integral is zero showing

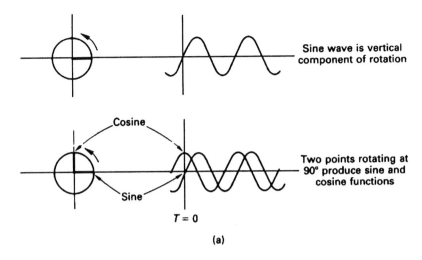

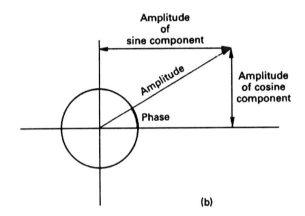

Figure 2.36 The origin of sine and cosine waves is to take a particular viewpoint of a rotation. Any phase can be synthesized by adding proportions of sine and cosine waves.

that no component of the target frequency exists. Thus from a real waveform containing many frequencies, all frequencies except the target frequency are excluded. The magnitude of the integral is proportional to the amplitude of the target component.

Figure 2.37(c) shows that the target frequency will not be detected if it is phase shifted 90° as the product of quadrature waveforms is always zero. Thus the discrete Fourier transform must make a further search for the target frequency using a cosine basis function. It follows from the arguments above that the relative proportions of the sine and cosine integrals reveal the phase of the input component. Thus each discrete frequency in the spectrum must be the result of a pair of quadrature searches.

Searching for one frequency at a time as above will result in a DFT, but only after considerable computation. However, a lot of the calculations are repeated

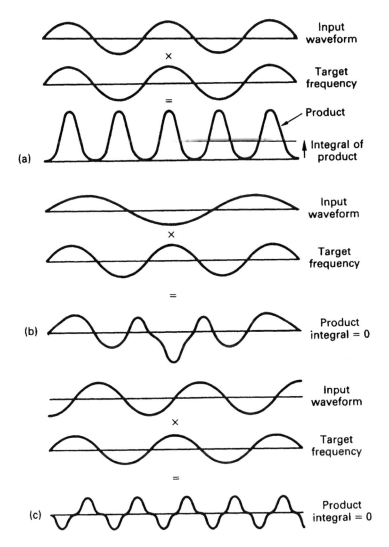

Input
waveform

×

Target
frequency

=

Product

Integral of
product

(a)

Input
waveform

×

Target
frequency

=

Product
integral = 0

(b)

Input
waveform

×

Target
frequency

=

Product
integral = 0

(c)

Figure 2.37 The input waveform is multiplied by the target frequency and the result is averaged or integrated. In (a) the target frequency is present and a large integral results. With another input frequency the integral is zero as in (b). The correct frequency will also result in a zero integral shown in (c) if it is at 90° to the phase of the search frequency. This is overcome by making two searches in quadrature.

many times over in different searches. The FFT gives the same result with less computation by logically gathering together all of the places where the same calculation is needed and making the calculation once.

The amount of computation can be reduced by performing the sine and cosine component searches together. Another saving is obtained by noting that every 180° the sine and cosine have the same magnitude but are simply inverted in sign. Instead of performing four multiplications on two samples 180° apart and adding the pairs of products it is more economical to subtract the sample values and multiply twice, once by a sine value and once by a cosine value.

The first coefficient is the arithmetic mean which is the sum of all of the sample values in the block divided by the number of samples. Figure 2.38 shows how the search for the lowest frequency in a block is performed. Pairs of samples are subtracted as shown, and each difference is then multiplied by the sine and the cosine of the search frequency. The process shifts one sample period, and a new sample pair are subtracted and multiplied by new sine and cosine factors. This is repeated until all of the sample pairs have been

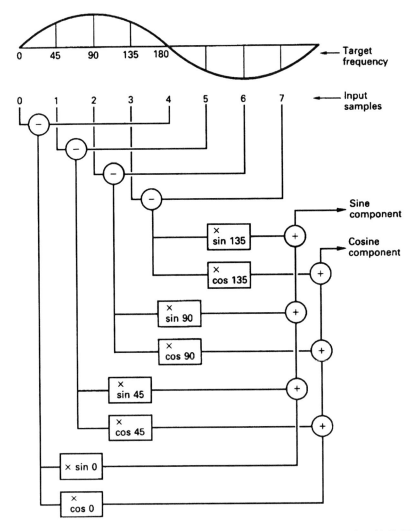

Figure 2.38 An example of a filtering search. Pairs of samples are subtracted and multiplied by sampled sine and cosine waves. The products are added to give the sine and cosine components of the search frequency.

multiplied. The sine and cosine products are then added to give the value of the sine and cosine coefficients, respectively.

It is possible to combine the calculation of the DC component which requires the sum of samples and the calculation of the fundamental which requires sample differences by combining stages shown in Figure 2.39(a) which take a pair of samples and add and subtract them. Such a stage is called a butterfly because of the shape of the schematic. Figure 2.39(b) shows how the first two components are calculated. The phase rotation boxes attribute the input to the sine or cosine component outputs according to the phase angle. As shown the box labelled 90° attributes nothing to the sine output, but unity gain to the cosine output. The 45° box attributes the input equally to both components.

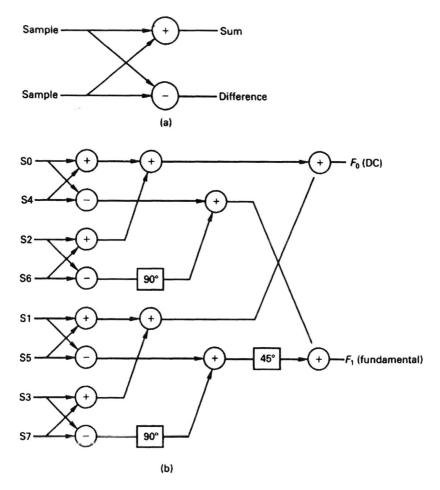

Figure 2.39 The basic element of an FFT is known as a butterfly as in (a) because of the shape of the signal paths in a sum and difference system. The use of butterflies to compute the first two coefficients is shown in (b).

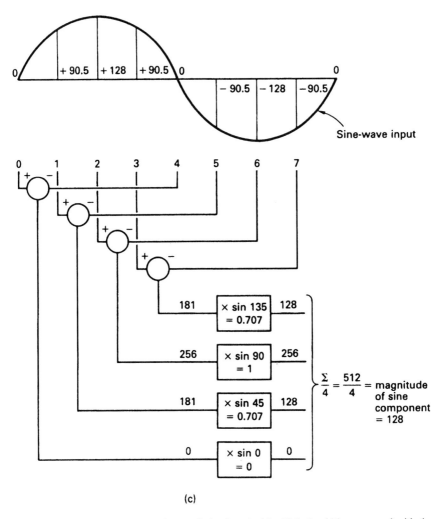

(c)

Figure 2.39 *(continued)* An actual example is given in (c) which should be compared with the result of (d) with a quadrature input.

Figure 2.39(c) shows a numerical example. If a sine-wave input is considered where zero degrees coincides with the first sample, this will produce a zero sine coefficient and non-zero cosine coefficient. Figure 2.39(d) shows the same input waveform shifted by 90°. Note how the coefficients change over.

Figure 2.39(e) shows how the next frequency coefficient is computed. Note that exactly the same first stage butterfly outputs are used, reducing the computation needed.

A similar process may be followed to obtain the sine and cosine coefficients of the remaining frequencies. The full FFT diagram for eight samples is shown in Figure 2.40(a). The spectrum this calculates is shown in Figure 2.40(b). Note that only half of the coefficients are useful in a real band-limited system

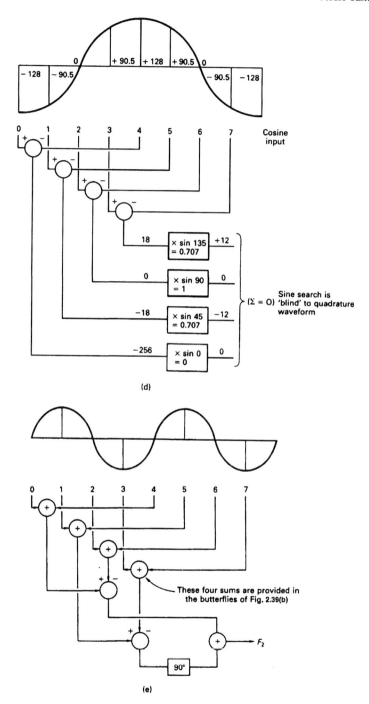

Figure 2.39 (*continued*) In (e) the butterflies for the first two coefficients form the basis of the computation of the third coefficient.

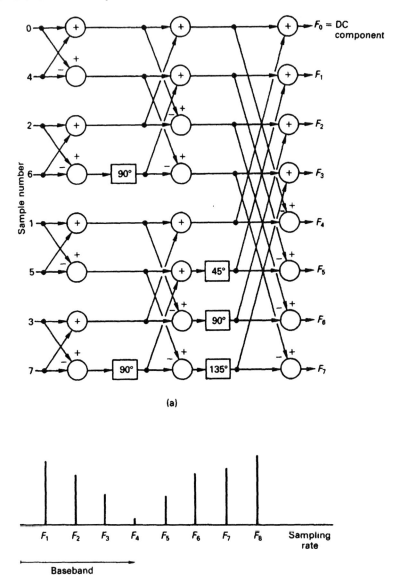

(a)

(b)

Figure 2.40 In (a) is the full butterfly diagram for an FFT. The spectrum this computes is shown in (b).

because the remaining coefficients represent frequencies above one half of the sampling rate.

In STFTs the overlapping input sample blocks must be multiplied by window functions. Figure 2.41 shows that multiplying the search frequency by the window has exactly the same result except that this need be done only once

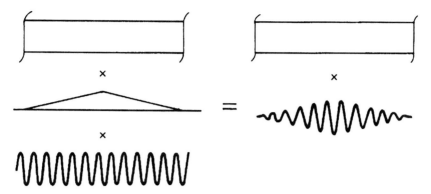

Figure 2.41 Multiplication of a windowed block by a sine wave basis function is the same as multiplying the raw data by a windowed basis function but requires less multiplication as the basis function is constant and can be pre-computed.

and much computation is saved. Thus in the STFT the basis function is a windowed sine or cosine wave.

The FFT is used extensively in such applications as phase correlation, where the accuracy with which the phase of signal components can be analysed is essential. It also forms the foundation of the discrete cosine transform.

2.18 The discrete cosine transform (DCT)

The DCT is a special case of a discrete Fourier transform in which the sine components of the coefficients have been eliminated leaving a single number. This is actually quite easy. Figure 2.42(a) shows a block of input samples to a transform process. By repeating the samples in a time-reversed order and performing a discrete Fourier transform on the double-length sample set a DCT is obtained. The effect of mirroring the input waveform is to turn it into an even function whose sine coefficients are all zero. The result can be understood by considering the effect of individually transforming the input block and the reversed block. Figure 2.42(b) shows that the phase of all the components of one block are in the opposite sense to those in the other. This means that when the components are added to give the transform of the double length block, all of the sine components cancel out, leaving only the cosine coefficients, hence the name of the transform.[3] In practice the sine component calculation is eliminated. Another advantage is that doubling the block length by mirroring doubles the frequency resolution, so that twice as many useful coefficients are produced. In fact a DCT produces as many useful coefficients as input samples. Clearly when the inverse transform is performed the reversed part of the waveform is discarded.

Figure 2.43 shows how a DCT is calculated by multiplying each sample in the input block by terms which represent sampled cosine waves of various frequencies. A given DCT coefficient is obtained when the result of multiplying every input sample in the block is summed. The DCT is primarily used in compression processing because it converts the input waveform into a form where redundancy can be easily detected and removed.

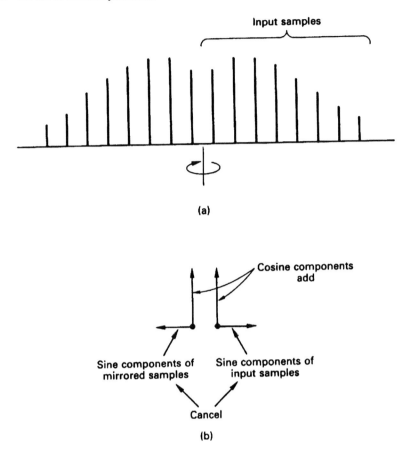

Figure 2.42 The DCT is obtained by mirroring the input block as shown in (a) prior to an FFT. The mirroring cancels out the sine components as in (b), leaving only cosine coefficients.

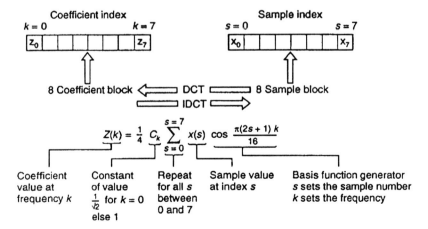

$$Z(k) = \frac{1}{4} \; C_k \sum_{s=0}^{s=7} x(s) \; \cos \frac{\pi(2s + 1) \, k}{16}$$

| Coefficient value at frequency k | Constant of value $\frac{1}{\sqrt{2}}$ for $k = 0$ else 1 | Repeat for all s between 0 and 7 | Sample value at index s | Basis function generator s sets the sample number k sets the frequency |

Figure 2.43 A DCT is calculated by multiplying the input sample block by various frequencies.

Fourier transform   Wavelet transform

Figure 2.44 Unlike discrete Fourier transforms, wavelet basis functions are scaled so that they contain the number of cycles irrespective of frequency. As a result their frequency discrimination ability is a constant proportion of the centre frequency.

2.19 The wavelet transform

The wavelet transform was not discovered by any one individual, but has evolved via a number of similar ideas and was only given a strong mathematical foundation relatively recently.[2,4,5] The wavelet transform is similar to the Fourier transform in that it has basis functions of various frequencies which are multiplied by the input waveform to identify the frequencies it contains. However, the Fourier transform is based on periodic signals and endless basis functions and requires windowing. The wavelet transform is fundamentally windowed, as the basis functions employed are not endless sine waves, but are finite on the time axis; hence the name. Wavelet transforms do not use a fixed window, but instead the window period is inversely proportional to the frequency being analysed. As a result a useful combination of time and frequency resolutions is obtained. High frequencies corresponding to transients in audio or edges in video are transformed with short basis functions and therefore are accurately located. Low frequencies are transformed with long basis functions which have good frequency resolution.

Figure 2.44 shows that a set of wavelets or basis functions can be obtained simply by scaling (stretching or shrinking) a single wavelet on the time axis. Each wavelet contains the same number of cycles such that as the frequency reduces the wavelet gets longer. Thus the frequency discrimination of the wavelet transform is a constant fraction of the signal frequency. In a filter bank such a characteristic would be described as 'constant Q'. Figure 2.45 shows the division of the frequency domain by a wavelet transform is logarithmic whereas with the Fourier transform the division is uniform. The logarithmic coverage is effectively dividing the frequency domain into octaves and as such parallels the frequency discrimination of human hearing. For a comprehensive treatment of wavelets the reader is referred to Strang and Nguyen.[6]

As it is relatively recent, the wavelet transform has yet to be widely used although it shows great promise as it is naturally a multi-resolution transform allowing scalable decoding. It has been successfully used in audio and in commercially available non-linear video editors and in other fields such as radiology and geology.

2.20 Magnetism

Magnetism is vital to sound reproduction as it used in so many different places. Microphones and loudspeakers rely on permanent magnets, recording

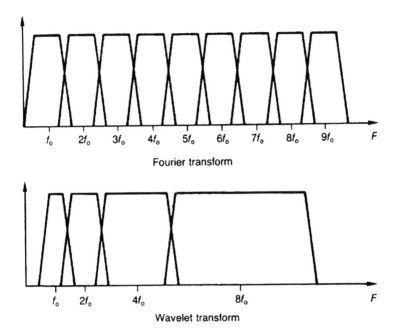

Figure 2.45 Wavelet transforms divide the frequency domain into octaves instead of the equal bands of the Fourier transform.

tape stores magnetic patterns and the tape is driven by motors which are driven by magnetism.

A magnetic field can be created by passing a current through a solenoid, which is no more than a coil of wire. When the current ceases, the magnetism disappears. However, many materials, some quite common, display a permanent magnetic field with no apparent power source. Magnetism of this kind results from the spin of electrons within atoms. Atomic theory describes atoms as having nuclei around which electrons orbit, spinning as they go. Different orbits can hold a different number of electrons. The distribution of electrons determines whether the element is diamagnetic (non-magnetic) or paramagnetic (magnetic characteristics are possible). Diamagnetic materials have an even number of electrons in each orbit, and according to the Pauli exclusion principle half of them spin in each direction. The opposed spins cancel any resultant magnetic moment. Fortunately there are certain elements, the transition elements, which have an odd number of electrons in certain orbits. The magnetic moment due to electronic spin is not cancelled out in these paramagnetic materials.

Figure 2.46 shows that paramagnetism materials can be classified as antiferromagnetic, ferrimagnetic and ferromagnetic. In some materials alternate atoms are antiparallel and so the magnetic moments are cancelled. In ferrimagnetic materials there is a certain amount of antiparallel cancellation, but a net magnetic moment remains. In ferromagnetic materials such as iron, cobalt or nickel, all of the electron spins can be aligned and as a result the most powerful magnetic behaviour is obtained.

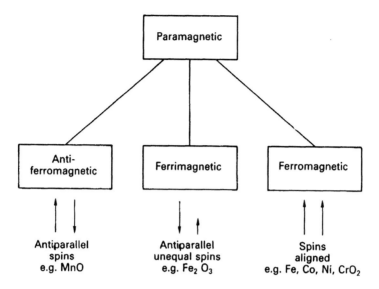

Figure 2.46 The classification of paramagnetic materials. The ferromagnetic materials exhibit the strongest magnetic behaviour.

It is not immediately clear how a material in which electron spins are parallel could ever exist in an unmagnetized state or how it could be partially magnetized by a relatively small external field. The theory of magnetic domains has been developed to explain what is observed in practice. Figure 2.47(a) shows a ferromagnetic bar which is demagnetized. It has no net magnetic moment because it is divided into domains or volumes which have equal and opposite moments. Ferromagnetic material divides into domains in order to reduce its magnetostatic energy. Figure 2.47(b) shows a domain wall which is around 0.1 µm thick. Within the wall the axis of spin gradually rotates from one state to another. An external field of quite small value is capable of disturbing the equilibrium of the domain wall by favouring one axis of spin over the other. The result is that the domain wall moves and one domain becomes larger at the expense of another. In this way the net magnetic moment of the bar is no longer zero as shown in Figure 2.47(c).

For small distances, the domain wall motion is linear and reversible if the change in the applied field is reversed. However, larger movements are irreversible because heat is dissipated as the wall jumps to reduce its energy. Following such a domain wall jump, the material remains magnetized after the external field is removed and an opposing external field must be applied which must do further work to bring the domain wall back again. This is a process of hysteresis where work must be done to move each way. Were it not for this non-linear mechanism magnetic recording would be impossible. If magnetic materials were linear, tapes would return to the demagnetized state immediately after leaving the field of the head and this book would be a good deal thinner.

Figure 2.48 shows a hysteresis loop which is obtained by plotting the magnetization M when the external field H is swept to and fro. On the

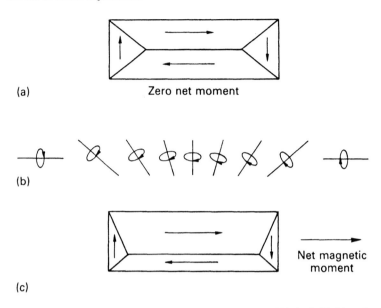

Figure 2.47 (a) A magnetic material can have a zero net moment if it is divided into domains as shown here. Domain walls (b) are areas in which the magnetic spin gradually changes from one domain to another. The stresses which result store energy. When some domains dominate, a net magnetic moment can exist as in (c).

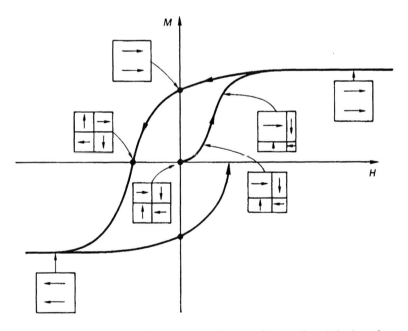

Figure 2.48 A hysteresis loop which comes about because of the non-linear behaviour of magnetic materials. If this characteristic were absent, magnetic recording would not exist.

macroscopic scale, the loop appears to be a smooth curve, whereas on a small scale it is in fact composed of a large number of small jumps. These were first discovered by Barkhausen. Starting from the unmagnetized state at the origin, as an external field is applied, the response is initially linear and the slope is given by the susceptibility. As the applied field is increased a point is reached where the magnetization ceases to increase. This is the saturation magnetization M_s. If the applied field is removed, the magnetization falls, not to zero, but the remanent magnetization M_r. This remanence is the magnetic memory mechanism which makes recording and permanent magnets possible. The ratio of M_r to M_s is called the squareness ratio. In recording, media squareness is beneficial as it increases the remanent magnetization.

If an increasing external field is applied in the opposite direction, the curve continues to the point where the magnetization is zero. The field required to achieve this is called the intrinsic coercive force $_mH_c$. A small increase in the reverse field reaches the point where, if the field where to be removed, the remanent magnetization would become zero. The field required to do this is the remanent coercive force, $_rH_c$.

As the external field H is swept to and fro, the magnetization describes a major hysteresis loop. Domain wall transit causes heat to be dissipated on every cycle around the loop and the dissipation is proportional to the loop area. For a recording medium, a large loop is beneficial because the replay signal is a function of the remanence and high coercivity resists erasure. The same is true for a permanent magnet. Heating is not an issue.

For a device such as a recording head, a small loop is beneficial. Figure 2.49(a) shows the large loop of a hard magnetic material used for recording media and for permanent magnets. Figure 2.49(b) shows the small loop of a soft magnetic material which is used for recording heads and transformers.

According to the Nyquist noise theorem, anything which dissipates energy when electrical power is supplied must generate a noise voltage when in

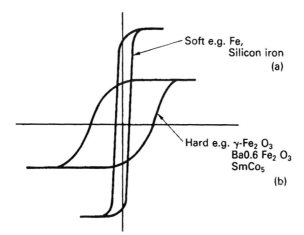

Figure 2.49 The recording medium requires a large loop area (a) whereas the head requires a small loop area (b) to cut losses.

thermal equilibrium. Thus magnetic recording heads have a noise mechanism which is due to their hysteretic behaviour. The smaller the loop, the less the hysteretic noise. In conventional heads, there are a large number of domains and many small domain wall jumps. In thin film heads there are fewer domains and the jumps must be larger. The noise this causes is known as Barkhausen noise, but as the same mechanism is responsible it is not possible to say at what point hysteresis noise should be called Barkhausen noise.

2.21 Electromagnetic compatibility (EMC)

EMC is a way of making electronic equipment more reliable by limiting both the amount of spurious energy radiated and the sensitivity to extraneous radiation. As electronic equipment becomes more common and more of our daily life depends upon its correct operation it becomes important to contain the unwanted effects of interference.

In audio equipment external interference can cause unwanted signals to be superimposed on the wanted audio signal. This is most likely to happen in sensitive stages handling small signals; e.g. microphone preamplifiers and tape replay stages. Interference can enter through any cables, including the power cable, or by radiation. Such stages must be designed from the outset with the idea that radio frequency (RF) energy may be present which must be rejected. Whilst one could argue that RF energy from an arcing switch should be suppressed at source one cannot argue that radio telephones should be banned as they rely on RF radiation. When designing from the outset, RF rejection is not too difficult. Putting it in afterwards is often impossible without an uneconomic redesign.

There have been some complaints from the high-end hi-fi community that the necessary RF suppression components will impair the sound quality of audio systems but this is nonsense. In fact good EMC design actually improves sound quality because by eliminating common impedances which pick up interference, distortion is also reduced.

In balanced signalling the screen does not carry the audio, but serves to extend the screened cabinets of the two pieces of equipment with what is effectively a metallic tunnel. For this to be effective against RF interference it has to be connected at both ends. This is also essential for electrical safety so that no dangerous potential difference can build up between the units. Figure 2.18 showed that connecting the screen at both ends causes an earth loop with the building ground wiring. Loop currents will circulate as shown but this is not a problem because by shunting loop currents into the screen, they are kept out of the audio wiring.

Some poorly designed equipment routes the X-pin of the XLR via the PCB instead of direct to the equipment frame. As Figure 2.50 shows, this effectively routes loop currents through the circuitry and is prone to interference. This approach does not comply with recent EMC regulations but there is a lot of old equipment still in service which should be put right. A simple track cut and a new chassis bonded XLR socket is often all that is necessary. Another false economy is the use of plastic XLR shells which cannot provide continuity of screening.

Differential working with twisted pairs is designed to reject hum and noise, but it only works properly if both signal legs have identical

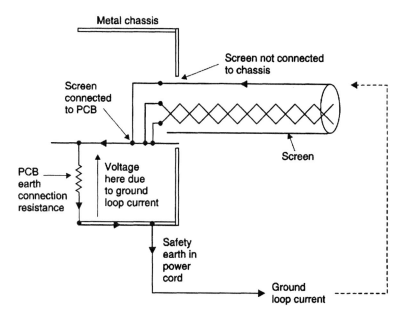

Figure 2.50 Poorly designed product in which screen currents pass to chassis via circuit board. Currents flowing in ground lead will raise voltages which interfere with the audio signal.

frequency/impedance characteristics at both ends. The easiest way of achieving this is to use transformers which give much better RF rejection than electronic balancing. Whilst an effective electronic differential receiver can be designed with care, a floating balanced electronic driver cannot compete with a transformer.

Analog audio equipment works at moderate frequencies and seldom has a radiation problem. However, any equipment controlled by a microprocessor or containing digital processing is a potential source of interference and once more steps must be taken in the design stage to ensure that radiation is minimized.

All AC powered audio devices contain some kind of power supply which rectifies the AC line to provide DC power. This rectification process is non-linear and can produce harmonics which leave the equipment via the power cable and cause interference elsewhere. Suitable power cord filtering must be provided to limit harmonic generation.

2.22 Electrical safety

Under fault conditions an excess of current can flow and the resulting heat can cause fire. Practical equipment must be fused so that excessive current causes the fuse element to melt, cutting off the supply. In many electronic devices the initial current exceeds the steady current because capacitors need to charge. Safe fusing requires the use of slow-blow fuses which have increased thermal mass. The switch-on surge will not blow the fuse, whereas a steady current of the same value will. Slow-blow fuses can be identified by the (T) after

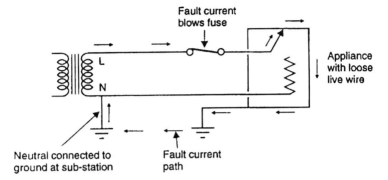

Figure 2.51 For electrical safety the metallic housing of equipment is connected to ground, preventing a dangerous potential existing in the case of a fault.

the rating, e.g. 3.15 A(T). Blown fuses should only be replaced with items of the same type and rating. Fuses do occasionally blow from old age, but any blown fuse should be considered an indicator of a potential problem. Replacing a fuse with one of a higher rating is the height of folly as no protection against fire is available. When dual-voltage 115/230 equipment is set to a different range a different fuse will often be necessary. In some small power supplies the power taken is small and the use of a fuse is not practicable. Instead a thermal switch is built in to the transformer. In the case of overheating this will melt. Generally these switches are designed to work once only after which the transformer must be replaced.

Except for low-voltage battery-powered devices, electrically powered equipment has to be considered a shock hazard to the user and steps must be taken to minimize the hazard. There are two main ways in which this is achieved. Figure 2.51 shows that the equipment is made with a conductive case which is connected to earth via a third conductor. In the event that a fault causes live wiring to contact the case, current is conducted to ground which will blow the fuse. Clearly disconnecting the earth for any purpose could allow the case to become live. The alternative is to construct the equipment in such a way that live wiring physically cannot cause the body to become live. In a double-insulated product all of the live components are encased in plastic so that even if a part such as a transformer or motor becomes live it is still insulated from the outside. Double-insulated devices need no safety earth.

Where there is a risk of cables being damaged or cut, earthing and double insulation are of no help because the live conductor in the cable may be exposed. Safety can be enhanced by the use of a residual current breaker (RCB) which detects any imbalance in live and neutral current. An imbalance means that current is flowing somewhere it shouldn't and this results in the breaker cutting off the power.

References

1. Martin, W.H., DeciBel – the new name for the transmission unit. *Bell System Tech. J.* January (1929)

2. Kraniauskas, P., *Transforms in Signals and Systems*, Chapter 6, Wokingham: Addison Wesley, (1992)
3. Ahmed, N., Natarajan, T. and Rao, K., Discrete cosine transform. *IEEE Trans. Computers*, **C-23**, 90–93 (1974)
4. Goupillaud, P., Grossman, A. and Morlet, J., Cycle-Octave and related transforms in seismic signal analysis. *Geoexploration*, **23**, 85–102 (1984/5)
5. Daubechies, I., The wavelet transform, time-frequency localization and signal analysis. *IEEE Trans. Info. Theory*, **36**, No. 5, 961–1005 (1990)
6. Strang, G. and Nguyen, T., *Wavelets and Filter Banks*, Wellesley, MA: Wellesley-Cambridge Press (1996)

Chapter 3

Sound and psychoacoustics

In this chapter the characteristics of sound as an airborne vibration and as a human sensation are tied together. The direction-sensing ability of the ear is not considered here as it will be treated in detail in Chapter 7.

3.1 What is sound?

There is a well-known philosophical riddle which goes 'If a tree falls in the forest and no one is there to hear it, does it make a sound?' This question can have a number of answers depending on the plane one chooses to consider. I believe that to understand what sound really is requires us to interpret this on many planes.

Physics can tell us the mechanism by which disturbances propagate through the air and if this is our definition of sound, then the falling tree needs no witness. We do, however, have the problem that accurately reproducing that sound is difficult because in physics there are no limits to the frequencies and levels which must be considered.

Biology can tell us that the ear only responds to a certain range of frequencies provided a threshold level is exceeded. If this is our definition of sound, then its reproduction is easier because it is only necessary to reproduce that range of levels and frequencies which the ear can detect.

Psychoacoustics can describe how our hearing has finite resolution in both time and frequency domains such that what we perceive is an inexact impression. Some aspects of the original disturbance are inaudible to us and are said to be masked. If our goal is the highest quality, we can design our imperfect equipment so that the shortcomings are masked. Conversely if our goal is economy we can use compression and hope that masking will disguise the inaccuracies it causes.

A study of the finite resolution of the ear shows how some combinations of tones sound pleasurable whereas others are irritating. Music has evolved empirically to emphasize primarily the former. Nevertheless we are still struggling to explain why we enjoy music and why certain sounds can make us happy and others can reduce us to tears. These characteristics must still be present in reproduced sound.

Whatever the audio technology we deal with, there is a common goal of delivering a satisfying experience to the listener. And we cannot neglect our own satisfaction. Possibly because it is a challenging and multidisciplinary subject, doing a good job in audio is particularly rewarding. However, some aspects of audio are emotive, some are technical. If we attempt to take an emotive view of a technical problem or vice versa we can look extremely foolish.

The frequency range of human hearing is extremely wide, covering some 10 octaves (an octave is a doubling of pitch or frequency) without interruption. There is hardly any other engineering discipline in which such a wide range is found. For example, in radio different wavebands are used so that the octave span of each is quite small. Whilst video signals have a wide octave span, the signal-to-noise and distortion criteria for video are extremely modest in comparison. Consequently audio is one of the most challenging subjects in engineering. Whilst the octave span required by audio can be met in electronic equipment, the design of mechanical transducers such as microphones and loudspeakers will always be difficult.

3.2 The ear

By definition, the sound quality of an audio system can only be assessed by human hearing. Many items of audio equipment can only be designed well with a good knowledge of the human hearing mechanism. The acuity of the human ear is astonishing. It can detect tiny amounts of distortion, and will accept an enormous dynamic range over a wide number of octaves. If the ear detects a different degree of impairment between two audio systems in properly conducted tests, we can say that one of them is superior. Thus quality is completely subjective and can only be checked by listening tests. However, any characteristic of a signal which can be heard can, in principle, also be measured by a suitable instrument although, in general, the availability of such instruments lags the requirement. The subjective tests will tell us how sensitive the instrument should be. Then the objective readings from the instrument give an indication of how acceptable a signal is in respect of that characteristic.

The sense we call hearing results from acoustic, mechanical, hydraulic, nervous and mental processes in the ear/brain combination, leading to the term psychoacoustics. It is only possible briefly to introduce the subject here. The interested reader is referred to Moore[1] for an excellent treatment.

Figure 3.1 shows that the structure of the ear is traditionally divided into the outer, middle and inner ears. The outer ear works at low impedance, the inner ear works at high impedance, and the middle ear is an impedance-matching device. The visible part of the outer ear is called the pinna which plays a subtle role in determining the direction of arrival of sound at high frequencies. It is too small to have any effect at low frequencies. Incident sound enters the auditory canal or meatus. The pipe-like meatus causes a small resonance at around 4 kHz. Sound vibrates the eardrum or tympanic membrane which seals the outer ear from the middle ear. The inner ear or cochlea works by sound travelling though a fluid. Sound enters the cochlea via a membrane called the oval window. If airborne sound were to be incident on the oval window directly, the serious impedance mismatch would cause most of the sound to be reflected. The middle ear remedies that mismatch by providing a mechanical advantage. The tympanic membrane is linked to the oval window by three bones known as ossicles which act as a lever system

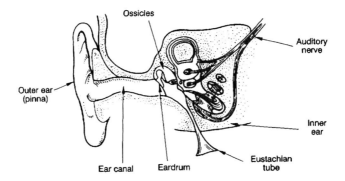

Figure 3.1 The structure of the human ear. See text for details.

such that a large displacement of the tympanic membrane results in a smaller displacement of the oval window but with greater force. Figure 3.2 shows that the malleus applies a tension to the tympanic membrane rendering it conical in shape. The malleus and the incus are firmly joined together to form a lever. The incus acts upon the stapes through a spherical joint. As the area of the tympanic membrane is greater than that of the oval window, there is a further multiplication of the available force. Consequently small pressures over the large area of the tympanic membrane are converted to high pressures over the small area of the oval window.

The middle ear is normally sealed, but ambient pressure changes will cause static pressure on the tympanic membrane which is painful. The pressure is relieved by the Eustachian tube which opens involuntarily whilst swallowing. The Eustachian tubes open into the cavities of the head and must normally be closed to avoid one's own speech appearing deafeningly loud.

The ossicles are located by minute muscles which are normally relaxed. However, the middle ear reflex is an involuntary tightening of the *tensor tympani* and *stapedius* muscles which heavily damp the ability of the tympanic membrane and the stapes to transmit sound by about 12 dB at frequencies below 1 kHz. The main function of this reflex is to reduce the audibility of one's own speech. However, loud sounds will also trigger this reflex which takes some 60–120 ms to occur, too late to protect against transients such as gunfire.

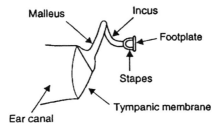

Figure 3.2 The malleus tensions the tympanic membrane into a conical shape. The ossicles provide an impedance-transforming lever system between the tympanic membrane and the oval window.

3.3 The cochlea

The cochlea, shown in Figure 3.3(a), is a tapering spiral cavity within bony walls which is filled with fluid. The widest part, near the oval window, is called the *base* and the distant end is the *apex*. Figure 3.3(b) shows that the cochlea is divided lengthwise into three volumes by Reissner's membrane and the basilar membrane. The *scala vestibuli* and the *scala tympani* are connected by a small aperture at the apex of the cochlea known as the *helicotrema*. Vibrations from the stapes are transferred to the oval window and become fluid-pressure variations which are relieved by the flexing of the round window. Effectively the basilar membrane is in series with the fluid motion and is driven by it except at very low frequencies where the fluid flows through the helicotrema, bypassing the basilar membrane.

Figure 3.3(c) shows that the basilar membrane is not uniform, but tapers in width and varies in thickness in the opposite sense to the taper of the cochlea. The part of the basilar membrane which resonates as a result of an applied sound is a function of the frequency. High frequencies cause resonance near to the oval window, whereas low frequencies cause resonances further away. More precisely the distance from the apex where the maximum resonance occurs is a logarithmic function of the frequency. Consequently tones spaced apart in octave steps will excite evenly spaced resonances in the basilar membrane. The prediction of resonance at a particular location on the membrane is called *place theory*. Essentially the basilar membrane is a mechanical frequency analyser. A knowledge of the way it operates is essential to an understanding of musical phenomena such as pitch discrimination, timbre, consonance and dissonance, and to auditory phenomena such as critical bands, masking and the precedence effect.

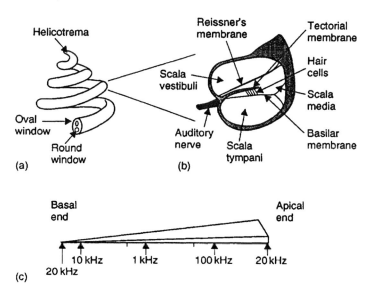

Figure 3.3 (a) The cochlea is a tapering spiral cavity. (b) The cross-section of the cavity is divided by Reissner's membrane and the basilar membrane. (c) The basilar membrane tapers so its resonant frequency changes along its length.

The vibration of the basilar membrane is sensed by the organ of Corti which runs along the centre of the cochlea. The organ of Corti is active in that it contains elements which can generate vibration as well as sense it. These are connected in a regenerative fashion so that the Q factor, or frequency selectivity of the ear is higher than it would otherwise be. The deflection of hair cells in the organ of Corti triggers nerve firings and these signals are conducted to the brain by the auditory nerve.

Nerve firings are not a perfect analog of the basilar membrane motion. A nerve firing appears to occur at a constant phase relationship to the basilar vibration; a phenomenon called phase locking, but firings do not necessarily occur on every cycle. At higher frequencies firings are intermittent, yet each is in the same phase relationship.

The resonant behaviour of the basilar membrane is not observed at the lowest audible frequencies below 50 Hz. The pattern of vibration does not appear to change with frequency and it is possible that the frequency is low enough to be measured directly from the rate of nerve firings.

3.4 Level and loudness

At 1kH₃, the ear can detect a sound pressure variation of only 2×10^{-5} Pa rms and so this figure is used as the reference against which sound pressure level (SPL) is measured. The sensation of loudness is a logarithmic function of SPL and consequently a logarithmic unit, the deciBel, is used in audio measurement. The deciBel was explained in detail in Section 2.5.

The dynamic range of the ear exceeds 130 dB, but at the extremes of this range, the ear is either straining to hear or is in pain. Neither of these cases can be described as pleasurable or entertaining, and it is hardly necessary to produce audio of this dynamic range since, amongst other things, the consumer is unlikely to have anywhere sufficiently quiet to listen to it. On the other hand extended listening to music whose dynamic range has been excessively compressed is fatiguing.

The frequency response of the ear is not at all uniform and it also changes with SPL. The subjective response to level is called loudness and is measured in *phons*. The phon scale and the SPL scale coincide at 1 kHz, but at other frequencies the phon scale deviates because it displays the actual SPLs judged by a human subject to be equally loud as a given level at 1 kHz. Figure 3.4 shows the so-called equal loudness contours which were originally measured by Fletcher and Munson and subsequently by Robinson and Dadson. Note the irregularities caused by resonances in the meatus at about 4 kHz and 13 kHz.

Usually, people's ears are at their most sensitive between about 2 and 5 kHz, and although some people can detect 20 kHz at high level, there is much evidence to suggest that most listeners cannot tell if the upper frequency limit of sound is 20 kHz or 16 kHz.[2,3] For a long time it was thought that frequencies below about 40 Hz were unimportant, but it is now clear that reproduction of frequencies down to 20 Hz improves reality and ambience.[4] The generally accepted frequency range for high-quality audio is 20 to 20 000 Hz, although for broadcasting an upper limit of 15 000 Hz is often applied.

The most dramatic effect of the curves of Figure 3.4 is that the bass content of reproduced sound is disproportionately reduced as the level is turned down.

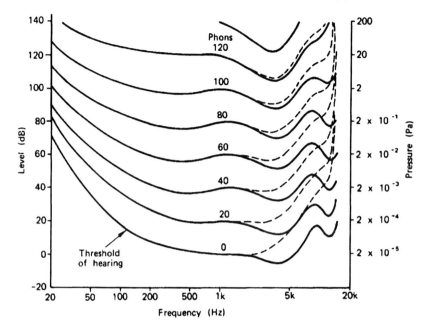

Figure 3.4 Contours of equal loudness showing that the frequency response of the ear is highly level dependent (solid line, age 20; dashed line, age 60).

This would suggest that if a powerful yet high quality reproduction system is available the correct tonal balance when playing a good recording can be obtained simply by setting the volume control to the correct level. This is indeed the case. A further consideration is that many musical instruments and the human voice change timbre with level and there is only one level which sounds correct for the timbre.

Audio systems with a more modest specification would have to resort to the use of tone controls to achieve a better tonal balance at lower SPL. A loudness control is one where the tone controls are automatically invoked as the volume is reduced. Although well meant, loudness controls seldom compensate accurately because they must know the original level at which the material was meant to be reproduced as well as the actual level in use. The equalization applied would have to be the difference between the equal loudness curves at the two levels.

There is no standard linking the signal level on a recording with the SPL at the microphone. The SPL resulting from a given signal level leaving a loudness control depends upon the sensitivity of the power amplifier and the loudspeakers and the acoustics of the listening room. Consequently loudness controls are doomed to be inaccurate and are eschewed on high-quality equipment.

A further consequence of level dependent hearing response is that recordings which are mixed at an excessively high level will appear bass light when played back at a normal level. Such recordings are more a product of self-indulgence than professionalism.

Loudness is a subjective reaction and is almost impossible to measure. In addition to the level dependent frequency response problem, the listener uses the sound not for its own sake but to draw some conclusion about the source. For example, most people hearing a distant motorcycle will describe it as being loud. Clearly at the source, it *is* loud, but the listener has compensated for the distance.

The best that can be done is to make some compensation for the level dependent response using *weighting curves*. Ideally there should be many, but in practice the A, B and C weightings were chosen where the A curve is based on the 40 phon response. The measured level after such a filter is in units of dBA. The A curve is almost always used because it most nearly relates to the annoyance factor of distant noise sources.

3.5 Frequency discrimination

Figure 3.5 shows an uncoiled basilar membrane with the apex on the left so that the usual logarithmic frequency scale can be applied. The envelope of displacement of the basilar membrane is shown for a single frequency at Figure 3.5(a). The vibration of the membrane in sympathy with a single frequency cannot be localized to an infinitely small area, and nearby areas are forced to vibrate at the same frequency with an amplitude that decreases with distance. Note that the envelope is asymmetrical because the membrane is tapering and because of frequency dependent losses in the propagation of vibrational energy down the cochlea. If the frequency is changed, as in Figure 3.5(b), the position of maximum displacement will also change. As the basilar membrane is continuous, the position of maximum displacement is infinitely variable allowing extremely good pitch discrimination of about one-twelfth of a semitone which is determined by the spacing of hair cells.

In the presence of a complex spectrum, the finite width of the vibration envelope means that the ear fails to register energy in some bands when there is more energy in a nearby band. Within those areas, other frequencies are mechanically excluded because their amplitude is insufficient to dominate the local vibration of the membrane. Thus the Q factor of the membrane

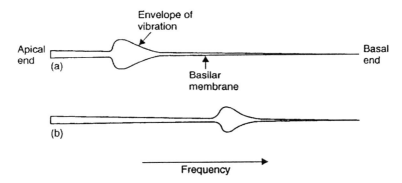

Figure 3.5 The basilar membrane symbolically uncoiled. (a) Single frequency causes the vibration envelope shown. (b) Changing the frequency moves the peak of the envelope.

is responsible for the degree of auditory masking, defined as the decreased audibility of one sound in the presence of another.

3.6 Critical bands

The term used in psychoacoustics to describe the finite width of the vibration envelope is *critical bandwidth*. Critical bands were first described by Fletcher.[5] The envelope of basilar vibration is a complicated function. It is clear from the mechanism that the area of the membrane involved will increase as the sound level rises. Figure 3.6 shows the bandwidth as a function of level.

As was shown in Chapter 2, transform theory teaches that the higher the frequency resolution of a transform the worse the time accuracy. As the basilar membrane has finite frequency resolution measured in the width of a critical band, it follows that it must have finite time resolution. This also follows from the fact that the membrane is resonant, taking time to start and stop vibrating in response to a stimulus. There are many examples of this. Figure 3.7 shows the impulse response. Figure 3.8 shows the perceived loudness of a tone burst increases with duration up to about 200 ms due to the finite response time.

The ear has evolved to offer intelligibility in reverberant environments which it does by averaging all received energy over a period of about 30 ms. Reflected sound which arrives within this time is integrated to produce a louder sensation, whereas reflected sound which arrives after that time can be temporally discriminated and is perceived as an echo. Our simple microphones

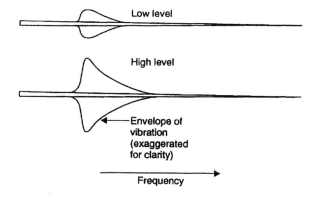

Figure 3.6 The critical bandwidth changes with SPL.

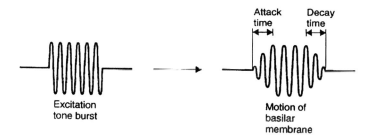

Figure 3.7 Impulse response of the ear showing slow attack and decay due to resonant behaviour.

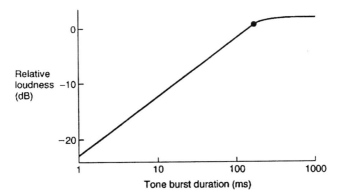

Figure 3.8 Perceived level of tone burst rises with duration as resonance builds up.

have no such ability which is why we often need to have acoustic treatment in areas where microphones are used.

A further example of the finite time discrimination of the ear is the fact that short interruptions to a continuous tone are difficult to detect. Finite time resolution means that masking can take place even when the masking tone begins after and ceases before the masked sound. This is referred to as forward and backward masking.[6]

As the vibration envelope is such a complicated shape, Moore and Glasberg have proposed the concept of equivalent rectangular bandwidth (ERB) to simplify matters. The ERB is the bandwidth of a rectangular filter which passes the same power as a critical band. Figure 3.9(a) shows the expression

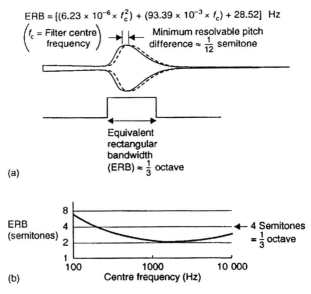

Figure 3.9 Effective rectangular bandwidth of critical band is much wider than the resolution of the pitch discrimination mechanism.

they have derived linking the ERB with frequency. This is plotted in (b) where it will be seen that one-third of an octave is a good approximation. This is about 30 times broader than the pitch discrimination also shown in (b).

Some treatments of human hearing liken the basilar membrane to a bank of fixed filters each of which is the width of a critical band. The frequency response of such a filter can be deduced from the envelope of basilar displacement as has been done in Figure 3.10. The fact that no agreement has been reached on the number of such filters should alert the suspicions of the reader. The fact that a third octave filter bank model cannot explain pitch discrimination some 30 times better is another cause for doubt. The response of the basilar membrane is centred upon the input frequency and no fixed filter can do this. However, the most worrying aspect of the fixed filter model is that according to Figure 3.10(b) a single tone would cause a response in several bands which would be interpreted as several tones. This is at variance with reality. Far from masking higher frequencies, we appear to be creating them!

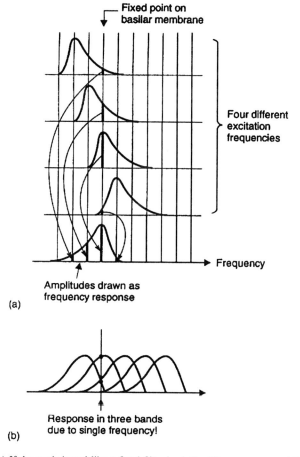

Figure 3.10 (a) If the ear behaved like a fixed filter bank the filter response could be derived as shown here. (b) This theory does not hold because a single tone would cause response in several bands.

This author prefers to keep in mind how the basilar membrane is actually vibrating in response to an input spectrum. If a mathematical model of the ear is required, then it has to be described as performing a finite resolution continuous frequency transform.

3.7 Beats

Figure 3.11(a) shows an electrical signal in which two equal sine waves of nearly the same frequency have been linearly added together. Note that the envelope of the signal varies as the two waves move in and out of phase. Clearly the frequency transform calculated to infinite accuracy is that shown at (b). The two amplitudes are constant and there is no evidence of the envelope modulation. However, such a measurement requires an infinite time. When a shorter time is available, the frequency discrimination of the transform falls and the bands in which energy is detected become broader.

When the frequency discrimination is too wide to distinguish the two tones as in (c), the result is that they are registered as a single tone. The amplitude of the single tone will change from one measurement to the next because the envelope is being measured. The rate at which the envelope amplitude changes is called a *beat* frequency which is not actually present in the input signal. Beats are an artefact of finite frequency resolution transforms. The fact that human hearing produces beats from pairs of tones proves that it has finite resolution.

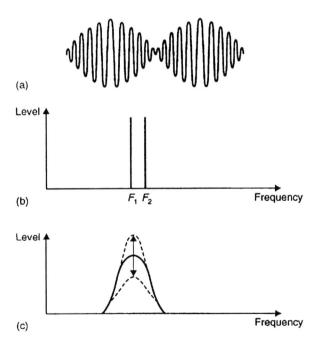

Figure 3.11 (a) Result of adding two sine waves of similar frequency. (b) Spectrum of (a) to infinite accuracy. (c) With finite accuracy only a single frequency is distinguished whose amplitude changes with the envelope of (a) giving rise to beats.

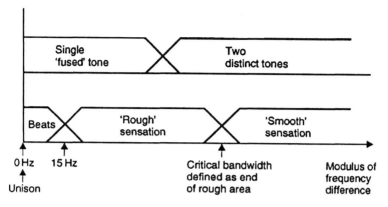

Figure 3.12 Perception of two-tone signal as frequency difference changes.

Measurement of when beats occur allows measurement of critical bandwidth. Figure 3.12 shows the results of human perception of a two-tone signal as the frequency difference, dF, changes. When dF is zero, described musically as *unison*, only a single note is heard. As dF increases, beats are heard, yet only a single note is perceived. The limited frequency resolution of the basilar membrane has *fused* the two tones together. As dF increases further, the sensation of beats ceases at $12-15\,Hz$ and is replaced by a sensation of roughness or *dissonance*. The roughness is due to parts of the basilar membrane being unable to decide the frequency at which to vibrate. The regenerative effect may well become confused under such conditions. The roughness persists until dF has reached the critical bandwidth beyond which two separate tones will be heard because there are now two discrete basilar resonances. In fact this is the definition of critical bandwidth.

3.8 Music and the ear

The characteristics of the ear, especially critical bandwidth, are responsible for the way music has evolved. Beats are used extensively in music. When tuning a pair of instruments together, a small tuning error will result in beats when both play the same nominal note. In certain pipe organs, pairs of pipes are sounded together with a carefully adjusted pitch error which results in a pleasing tremolo effect.

With certain exceptions, music is intended to be pleasing and so dissonance is avoided. Two notes which sound together in a pleasing manner are described as harmonious or *consonant*. Two sine waves appear consonant if they are separated by a critical bandwidth because the roughness of Figure 3.12 is avoided, but real musical instruments produce a series of harmonics in addition to the fundamental.

Figure 3.13 shows the spectrum of a harmonically rich instrument. The fundamental and the first few harmonics are separated by more than a critical band, but from the seventh harmonic more than one harmonic will be in one band and it is possible for dissonance to occur. Musical instruments have evolved to avoid the production of seventh and higher harmonics. It will be

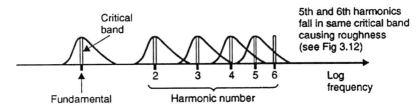

Figure 3.13 Spectrum of a real instrument with respect to critical bandwidth. High harmonics can fall in the same critical band and cause dissonance.

seen in Chapter 4 that violins and pianos are played or designed to excite the strings at a node of the seventh harmonic to suppress this dissonance.

Harmonic distortion in audio equipment is easily detected even in minute quantities because the first few harmonics fall in non-overlapping critical bands. The sensitivity of the ear to third harmonic distortion probably deserves more attention in audio equipment than the fidelity of the dynamic range or frequency response.

When two harmonically rich notes are sounded together, the harmonics will fall within the same critical band and cause dissonance unless the fundamentals have one of a limited number of simple relationships which makes the harmonics fuse. Clearly an octave relationship is perfect.

Figure 3.14 shows some examples. In (a) two notes with the ratio (interval) 3:2 are considered. The harmonics are either widely separated or fused and the combined result is highly consonant. The interval of 3:2 is known to musicians as a perfect fifth. In (b) the ratio is 4:3. All harmonics are either at least a third

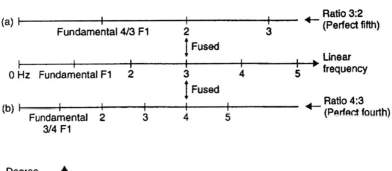

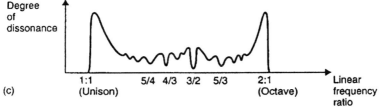

Figure 3.14 (a) Perfect fifth with a frequency ratio of 3:2 is consonant because harmonics are either in different critical bands or are fused. (b) Perfect fourth achieves the same result with 4:3 frequency ratio. (c) Degree of dissonance over range from 1:1 to 2:1.

of an octave apart or are fused. This relationship is known as a perfect fourth. The degree of dissonance over the range from 1:1 to 2:1 (unison to octave) was investigated by Helmholtz and is shown in Figure 3.14(c). Note that the dissonance rises at both ends where the fundamentals are within a critical bandwidth of one another. Dissonances in the centre of the scale are where some harmonics lie in within a critical bandwidth of one another. Troughs in the curve indicate areas of consonance. Many of the troughs are not very deep, indicating that the consonance is not perfect. This is because of the effect shown in Figure 3.13 in which high harmonics get closer together with respect to critical bandwidth. When the fundamentals are closer together, the harmonics will become dissonant at a lower frequency, reducing the consonance. Figure 3.14(c) also shows the musical terms used to describe the consonant intervals.

It is clear from Figure 3.14(c) that the notes of the musical scale have been arrived at empirically to allow the maximum consonance with pairs of notes and chords. Early instruments were tuned to the just diatonic scale in exactly this way. Unfortunately the just diatonic scale does not allow changes of key because the notes are not evenly spaced. A key change is where the frequency

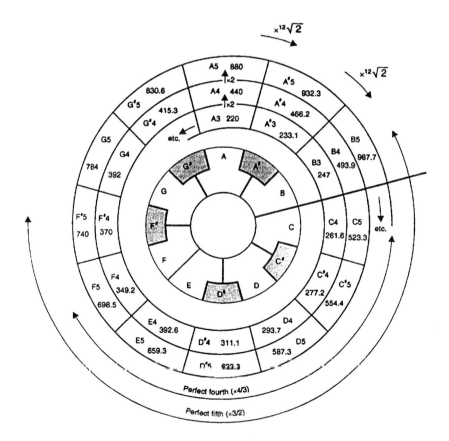

Figure 3.15 With a suitably tempered octave, scales can be played in different keys.

of every note in a piece of music is multiplied by a constant, often to bring the accompaniment within the range of a singer. In continuously tuned instruments such as the violin and the trombone this is easy, but with fretted or keyboard instruments such as the piano there is a problem.

The even tempered scale is a compromise between consonance and key changing. The octave is divided into 12 equal intervals called tempered semitones. On a keyboard, seven of the keys are white and produce notes very close to those of the just diatonic scale, and five of the keys are black. Music can be transposed in semitone steps by using the black keys.

Figure 3.15 shows an example of transposition where a scale is played in several keys.

3.9 The sensation of pitch

Frequency is an objective measure whereas *pitch* is the subjective near equivalent. Clearly frequency and level are independent, whereas pitch and level are not. Figure 3.16 shows the relationship between pitch and level. Place theory indicates that the hearing mechanism can sense a single frequency quite accurately as a function of the place or position of maximum basilar vibration. However, most periodic sounds and real musical instruments produce a series of harmonics in addition to the fundamental. When a harmonically rich sound is present the basilar membrane is excited at spaced locations. Figure 3.17(a) shows all harmonics, (b) shows even harmonics predominating and (c) shows odd harmonics predominating. It would appear that our hearing is accustomed to hearing harmonics in various amounts and the consequent regular pattern of excitation. It is the overall pattern which contributes to the sensation of pitch even if individual partials vary enormously in relative level.

Experimental signals in which the fundamental has been removed leaving only the harmonics result in unchanged pitch perception. The pattern in the remaining harmonics is enough uniquely to establish the missing fundamental. Imagine the fundamental in (b) to be absent. Neither the second harmonic nor

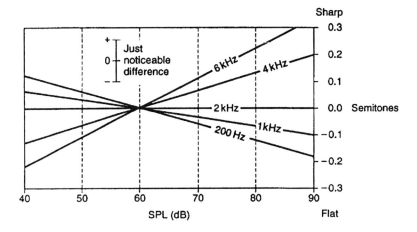

Figure 3.16 Pitch sensation is a function of level.

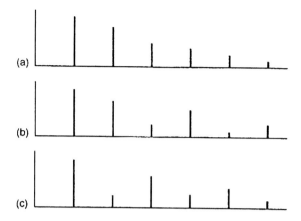

Figure 3.17 (a) Harmonic structure of rich sound. (b) Even harmonic predominance. (c) Odd harmonic predominance. Pitch perception appears independent of harmonic structure.

the third can be mistaken for the fundamental because if they were fundamentals a different pattern of harmonics would result. A similar argument can be put forward in the time domain, where the timing of phase-locked nerve firings responding to a harmonic will periodically coincide with the nerve firings of the fundamental. The ear is used to such time patterns and will use them in conjunction with the place patterns to determine the right pitch. At very low frequencies the place of maximum vibration does not move with frequency yet the pitch sensation is still present because the nerve firing frequency is used.

As the fundamental frequency rises it is difficult to obtain a full pattern of harmonics as most of them fall outside the range of hearing. The pitch

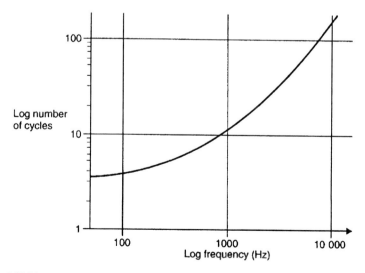

Figure 3.18 Pitch discrimination fails as frequency rises. The graph shows the number of cycles needed to distinguish pitch as a function of frequency.

discrimination ability is impaired and needs longer to operate. Figure 3.18 shows the number of cycles of excitation needed to discriminate pitch as a function of frequency. Clearly at around 5 kHz performance is failing because there are hardly any audible harmonics left. Phase locking also fails at about the same frequency. Musical instruments have evolved accordingly, with the highest notes of virtually all instruments found below 5 kHz.

3.10 The physics of sound

Sound is simply an airborne version of vibration which is why the two topics are inextricably linked. The air which carries sound is a mixture of gases, mostly nitrogen, some oxygen, a little carbon dioxide and so on. Gases are the highest energy state of matter, which is another way of saying that you have to heat ice to get water then heat it some more to get steam. The reason that a gas takes up so much more room than a liquid is that the molecules contain so much energy that they break free from their neighbours and rush around at high speed. As Figure 3.19(a) shows, the innumerable elastic collisions of these high speed molecules produce pressure on the walls of any gas container. In fact the distance a molecule can go without a collision, the mean free path, is quite short at atmospheric pressure. Consequently gas molecules also collide with each other elastically, so that if left undisturbed, in a container at a constant temperature, every molecule would end up with essentially the same energy and the pressure throughout would be constant and uniform.

Sound disturbs this simple picture. Figure 3.19(b) shows that a solid object which moves *against* gas pressure increases the velocity of the rebounding molecules, whereas in (c) one moving *with* gas pressure reduces that velocity. The average velocity and the displacement of all of the molecules in a layer of air near to a moving body is the same as the velocity and displacement of the body. Movement of the body results in a local increase or decrease in pressure of some kind. Thus sound is both a pressure and a velocity disturbance. Integration of the velocity disturbance gives the displacement.

Despite the fact that a gas contains endlessly rushing colliding molecules, a small mass or *particle* of gas can have stable characteristics because the molecules leaving are replaced by new ones with identical statistics. As a result acoustics seldom considers the molecular structure of air and the constant

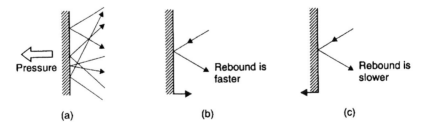

Figure 3.19 (a) The pressure exerted by a gas is due to countless elastic collisions between gas molecules and the walls of the container. (b) If the wall moves against the gas pressure, the rebound velocity increases. (c) Motion with the gas pressure reduces the particle velocity.

motion is neglected. Thus when particle velocity and displacement is considered in acoustics, this refers to the average values of a large number of molecules. The undisturbed container of gas referred to earlier will have a particle velocity and displacement of zero at all points.

When the volume of a fixed mass of gas is reduced, the pressure rises. The gas acts like a spring. However, a gas also has mass. Sound travels through air by an interaction between the mass and the springiness. Imagine pushing a mass via a spring. It would not move immediately because the spring would have to be compressed in order to transmit a force. If a second mass is connected to the first by another spring, it would start to move even later. Thus the speed of a disturbance in a mass/spring system depends on the mass and the stiffness.

After the disturbance had propagated the masses would return to their rest position. The mass/spring analogy is helpful for an early understanding, but is too simple to account for commonly encountered acoustic phenomena such as spherically expanding waves. It must be remembered that the mass and stiffness are distributed throughout the gas in the same way that inductance and capacitance are distributed in a transmission line. Sound travels through air without a net movement of the air.

3.11 The speed of sound

Unlike solids, the elasticity of gas is a complicated process. If a fixed mass of gas is compressed, work has to be done on it. This will create heat in the gas. If the heat is allowed to escape and the compression does not change the temperature, the process is said to be *isothermal*. However, if the heat cannot escape the temperature will rise and give a disproportionate increase in pressure. This process is said to be *adiabatic* and the diesel engine depends upon it. In most audio cases there is insufficient time for much heat transfer and so air is considered to act adiabatically. Figure 3.20 shows how the speed

$$\gamma = \text{adiabatic constant (1.4 for air)}$$

$$\text{Velocity } V = \sqrt{\frac{\gamma RT}{M}} \quad R = \text{gas constant } (8.31 \, \text{JK}^{-1} \, \text{mole}^{-1})$$

$$T = \text{absolute temp } ^\circ\text{K}$$

$$M = \text{molecular weight kg mole}^{-1}$$

Assume air is 21% O_2, 78% N_2, 1% Ar

Molecular weight $= 21\% \times 16 \times 2 + 78\% \times 14 \times 2 + 1\% \times 18 \times 1$

$$= 2.87 \times 10^{-2} \, \text{kg mole}^{-1}$$

$$V = \sqrt{\frac{1.4 \times 8.31 \, T}{2.87 \times 10^{-2}}} = 20.1 \sqrt{T}$$

at $20 \, ^\circ\text{C}$ $\quad T = 293 \, \text{K} \quad V = 20.1\sqrt{293} = 344 \, \text{ms}^{-1}$

Figure 3.20 Calculating the speed of sound from the elasticity of air.

of sound c in air can be derived by calculating its elasticity under adiabatic conditions.

If the volume allocated to a given mass of gas is reduced isothermally, the pressure and the density will rise by the same amount so that c does not change. If the temperature is raised at constant pressure, the density goes down and so the speed of sound goes up. Gases with lower density than air have a higher speed of sound. Divers who breath a mixture of oxygen and helium to prevent 'the bends' must accept that the pitch of their voices rises remarkably.

The speed of sound is proportional to the square root of the absolute temperature. Temperature changes with respect to absolute zero ($-273\,°C$) also amount to a few per cent except in extremely inhospitable places. The speed of sound experienced by most of us is about 1000 feet per second or 344 metres per second. Temperature falls with altitude in the atmosphere and with it the speed of sound. The local speed of sound is defined as Mach 1. Consequently supersonic aircraft are fitted with Mach meters.

As air acts adiabatically, a propagating sound wave causes cyclic temperature changes. The speed of sound is a function of temperature, yet sound causes a temperature variation. One might expect some effects because of this. Fortunately, sounds which are below the threshold of pain have such a small pressure variation compared with atmospheric pressure that the effect is negligible and air can be assumed to be linear. However, on any occasion where the pressures are higher, this is not a valid assumption. In such cases the positive half cycle significantly increases local temperature and the speed of sound, whereas the negative half cycle reduces temperature and velocity. Figure 3.21 shows that this results in significant distortion of a sine wave, ultimately causing a *shock wave* which can travel faster than the speed of sound until the pressure has dissipated with distance. This effect is responsible for the sharp sound of a handclap.

This behaviour means that the speed of sound changes slightly with frequency. High frequencies travel slightly faster than low because there is less time for heat conduction to take place. Figure 3.22 shows that a complex sound source produces harmonics whose phase relationship with the fundamental advances with the distance the sound propagates. This allows one mechanism

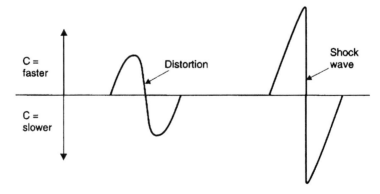

Figure 3.21 At high level, sound distorts itself by increasing the speed of propagation on positive half cycles. The result is a shock wave.

Figure 3.22 In a complex waveform, high frequencies travel slightly faster producing a relative phase change with distance.

(there are others) by which one can judge the distance from a known sound source. Clearly for realistic sound reproduction nothing in the audio chain must distort the phase relationship between frequencies. A system which accurately preserves such relationships is said to be phase linear.

3.12 Wavelength

Sound can be due to a one-off event known as percussion, or a periodic event such as the sinusoidal vibration of a tuning fork. The sound due to percussion is called transient whereas a periodic stimulus produces steady state sound having a frequency f.

Because sound travels at a finite speed, the fixed observer at some distance from the source will experience the disturbance at some later time. In the case of a transient, the observer will detect a single replica of the original as it passes at the speed of sound. In the case of the tuning fork, a periodic sound, the pressure peaks and dips follow one another away from the source at the speed of sound. For a given rate of vibration of the source, a given peak will have propagated a constant distance before the next peak occurs. This distance is called the wavelength, λ. Figure 3.23 shows that wavelength is defined as the distance between any two identical points on the whole cycle. If the source vibrates faster, successive peaks get closer together and the wavelength gets shorter. Figure 3.23 also shows that the wavelength is inversely proportional to the frequency. It is easy to remember that the wavelength of 1000 Hz is a foot (about 30 cm).

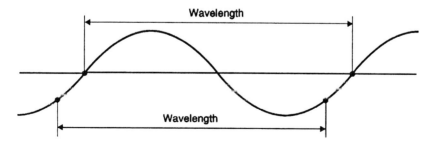

Figure 3.23 Wavelength is defined as the distance between two points at the same place on adjacent cycles. Wavelength is inversely proportional to frequency.

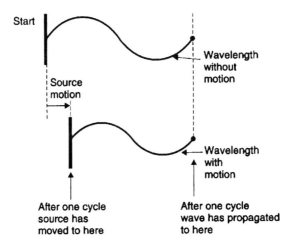

Figure 3.24 Periodic sounds are subject to Doppler shift if there is relative motion between the source and the observer.

3.13 The Doppler effect

If there is relative motion between the source and the observer, the frequency of a periodic sound will be changed. Figure 3.24 shows a sound source moving towards the observer. At the end of a cycle, the source will be nearer the observer than at the beginning of the cycle. As a result the wavelength radiated in the direction of the observer will be shortened so that the pitch rises. The wavelength of sounds radiated away from the observer will be lengthened. The same effect will occur if the observer moves. This is the Doppler effect which is most noticeable on passing motor vehicles whose engine notes appear to drop as they pass. Note that the effect always occurs, but it is only noticeable on a periodic sound. Where the sound is aperiodic, such as broadband noise, the Doppler shift will not be heard.

3.14 The wave number, k

Sound is a wave motion, and the way a wave interacts with any object depends upon the relative size of that object and the wavelength. The audible range of wavelengths is from around 17 mm to 17 m so dramatic changes in the behaviour of sound over the frequency range should be expected.

Figure 3.25(a) shows that when the wavelength of sound is large compared to the size of a solid body, the sound will pass around it almost as if it were not there. When the object is large compared to the wavelength, then simple reflection takes place as in (b). However, when the size of the object and the wavelength are comparable, the result can only be explained by diffraction theory.

The parameter which is used to describe this change of behaviour with wavelength is known as the wave number k and is defined as:

$$k = \frac{2\pi f}{c} = \frac{2\pi}{\lambda}$$

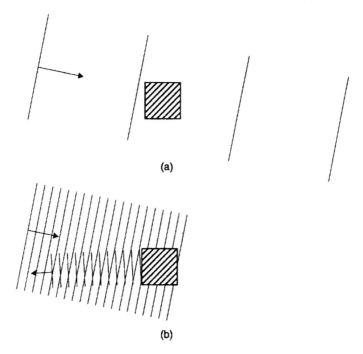

Figure 3.25 (a) Sound waves whose spacing is large compared to an obstacle simply pass round it. (b) When the relative size is reversed, an obstacle becomes a reflector.

Where f = frequency, c = the speed of sound and λ = wavelength. In practice the size of any object or distance a in metres is multiplied by k.

A good rule of thumb is that below $ka = 1$, sound tends to pass around as in Figure 3.25(a) whereas above $ka = 1$, sound tends to reflect as in Figure 3.25(b).

3.15 How sound is radiated

When sound propagates, there are changes in velocity v, displacement x, and pressure p. Figure 3.26 shows that the velocity and the displacement are always in quadrature. This is obvious because velocity is the differential of the displacement. When the displacement reaches its maximum value and is on the point of reversing direction, the velocity is zero. When the displacement is zero the velocity is maximum.

The pressure and the velocity are linked by the *acoustic impedance z* which is given by p/v. Just like electrical impedances which can be reactive, the acoustic impedance is complex and varies with acoustic conditions. Consequently the phase relationship between velocity and pressure also varies. When any vibrating body is in contact with the air, a thin layer of air must have the same velocity as the surface of the body. The pressure which results from that velocity depends upon the acoustic impedance.

The wave number can affect the way in which sound is radiated. Consider a hypothetical pulsating sphere as shown in Figure 3.27. The acoustic impedance

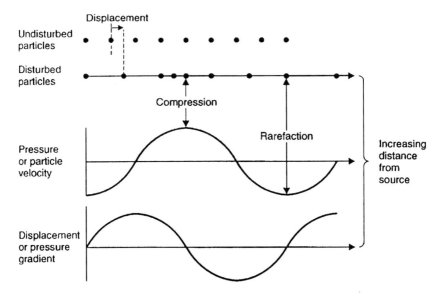

Figure 3.26 The pressure, velocity and displacement of particles as sound propagates.

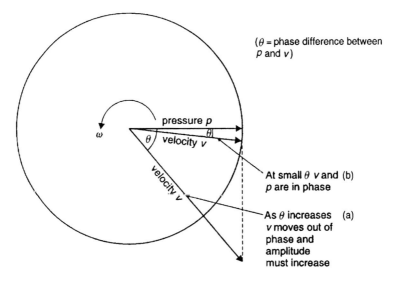

Figure 3.27 A pulsating sphere experiences an impedance which is a function of ka. With small ka pressure and velocity are in quadrature, but become coherent as ka rises.

changes with radius a. If the sphere pulsates very slowly, it will do work against air pressure as it expands and the air pressure will return the work as it contracts. There is negligible radiation because the impedance is reactive. Figure 3.27(a) shows that when ka is small there is a phase shift between the pressure and the velocity. As the frequency or the radius rises, as in (b), the phase angle reduces from 90° and the pressure increases. When *ka* is large, the phase angle approaches zero and the pressure reaches its maximum value compared to the velocity. The impedance has become resistive.

When *ka* is very large, the spherical radiator is at a distance and the spherical waves will have become plane waves. Figure 3.26 showed the relationships between pressure, velocity and displacement for a plane wave. A small air mass may have kinetic energy due to its motion and potential energy due to its compression. The total energy is constant, but the distribution of energy between kinetic and potential varies throughout the wave. This relationship will not hold when *ka* is small. This can easily occur especially at low frequencies where the wavelengths can be several metres.

3.16 Proximity effect

It will be seen in Chapter 5 that microphones can transduce either the pressure or the velocity component of sound. When *ka* is large, the pressure and velocity waveforms in a spherical wave are identical. However, it will be clear from Figure 3.27(a) and (b) that when *ka* is small the velocity exceeds the pressure component. This is the cause of the well-known proximity effect, also known as tip-up, which emphasizes low frequencies when velocity sensing

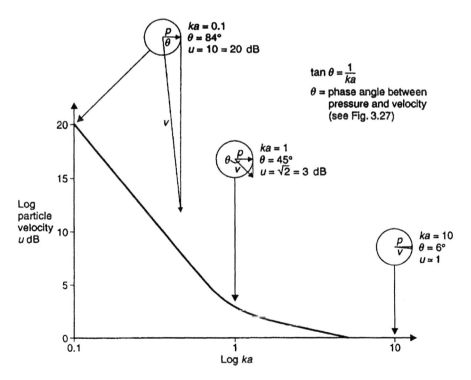

Figure 3.28 Proximity effect is due to raised velocity component of sound relative to pressure near to sound sources.

microphones are used close to a sound source. Figure 3.28 shows the response of a velocity microphone relative to that of a pressure microphone for different values of *ka*. Practical microphones often incorporate some form of bass cut filter to offset the effect.

3.17 Intensity and power

The sensation of sound is proportional to the average velocity. However, the displacement is the integral of the velocity. Figure 3.29 shows that to obtain an identical velocity or slope the amplitude must increase as the inverse of the frequency. Consequently for a given SPL low-frequency sounds result in much larger air movement than high frequency. The SPL is proportional to the *volume velocity U* of the source which is obtained by multiplying the vibrating area in m^2 by the velocity in ms^{-1}. As SPL is proportional to volume velocity, as frequency falls the volume or displacement must rise. This means that low frequency sound can only be radiated effectively by large objects, hence all of the bass instruments in the orchestra are much larger than their treble equivalents. This is also the reason why a loudspeaker cone is only seen to move at low frequencies.

The units of volume velocity are cubic metres per second (m^{3}s^{-1}) and so sound is literally an alternating current. The pressure *p* is linked to the current by the impedance just as it is in electrical theory. There are direct analogies between acoustic and electrical parameters and equations which are helpful. One small difficulty is that whereas alternating electrical parameters are measured in rms units, acoustic units are sometimes not. Thus when certain acoustic parameters are multiplied together the product has to be divided by two. This happens automatically with rms units. Figure 3.30 shows the analogous equations.

The intensity of a sound is the sound power passing through unit area. In the far field it is given by the product of the volume velocity and the pressure. In the near field the relative phase angle will have to be considered. Intensity is a vector quantity as it has direction which is considered to be perpendicular to the area in question. The total sound power is obtained by multiplying the intensity by the cross-sectional area through which it passes. Power is a scalar quantity because it can be radiated in all directions.

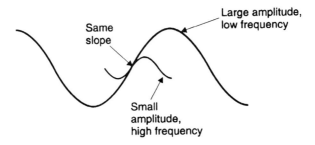

Figure 3.29 For a given velocity or slope, lower frequencies require greater amplitude.

Acoustic

U = volume velocity (m³/sec)

p = pressure (Pa)

Z = impedance = $\dfrac{p}{U}$

P = power = $\dfrac{|U|^2 \times \text{Real } (Z)}{2}$ (W)

↑

Note factor of 2 as U is not RMS

$|U|$ = amplitude of U Real (Z) = resistive part of impedance

Electrical

$\equiv I$ rms current (A)

$\equiv V$ rms voltage (V)

$\equiv \dfrac{V}{I} = Z$

$\equiv I^2 \times \text{Real } Z(\text{W})$

Figure 3.30 Electrical units are rms whereas many acoustic units are not, hence the factor of two difference in otherwise analogous equations.

3.18 The inverse square law

When a spherical sound wave is considered, there is negligible loss as it advances outwards. Consequently the sound power passing through the surface of an imaginary sphere surrounding the source is independent of the radius of that sphere. As the area of a sphere is proportional to the square of the radius, it will be clear that the intensity falls according to an inverse square law.

The inverse square law should be used with caution. There are a number of exceptions. As was seen in Figure 3.28 the proximity effect causes a deviation from the inverse square law for small *ka*. The area in which there is deviation from inverse square behaviour is called the *near field*.

In reverberant conditions a sound field is set up by reflections. As the distance from the source increases at some point the level no longer falls.

It is also important to remember that the inverse square law only applies to near-point sources. A line source radiates cylindrically and intensity is then inversely proportional to radius. Noise from a busy road approximates to a cylindrical source.

3.19 Wave acoustics

A proper understanding of the behaviour of sound requires familiarity with the principles of wave acoustics. Wave theory is used in many different disciplines including radar, sonar, antenna design and optics and the principles remain the same. Consequently the designer of a loudspeaker may obtain inspiration from studying a radar antenna or a CD pickup.

Figure 3.31 shows that when two sounds of equal amplitude and frequency add together, the result is completely dependent on the relative phase of the two. At (a) when the phases are identical, the result is the arithmetic sum. At (b) where there is a 180° relationship, the result is complete cancellation. This is constructive and destructive interference. At any other phase and/or amplitude relationship, the result can only be obtained by vector addition as shown in (c).

The wave theory of propagation of sound is based on interference and suggests that a wavefront advances because an infinite number of point sources

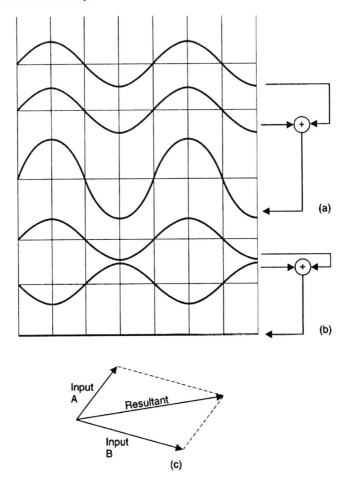

Figure 3.31 (a) Constructive interference between two in-phase signals. (b) Destructive interference between out-of-phase signals. (c) Vector addition is needed to find result of arbitrary phase relationship.

can be considered to emit spherical waves which will only add when they are all in the same phase. This can only occur in the plane of the wavefront. Figure 3.32(a) shows that at all other angles, interference between spherical waves is destructive. For any radiating body, such as a vibrating object, it is easy to see from Figure 3.32(b) that when ka is small, only weak spherical radiation is possible, whereas when ka is large, a directional plane wave can be propagated or beamed. Consequently high-frequency sound behaves far more directionally than low-frequency sound.

When a wavefront arrives at a solid body, it can be considered that the surface of the body acts as an infinite number of points which re-radiate the incident sound in all directions. It will be seen that when ka is large and the surface is flat, constructive interference only occurs when the wavefront is *reflected* such that the angle of reflection is the same as the angle of incidence.

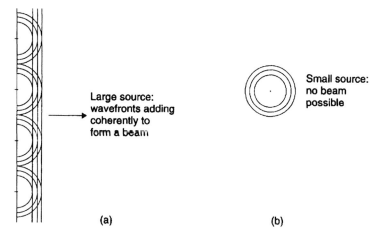

Large source:
wavefronts adding
coherently to
form a beam

Small source:
no beam
possible

(a)   (b)

Figure 3.32 (a) Plane waves can be considered to propagate as an infinity of spherical waves which cancel out in all directions other than forward to form a beam. (b) Where the sound source is small no beam can be formed.

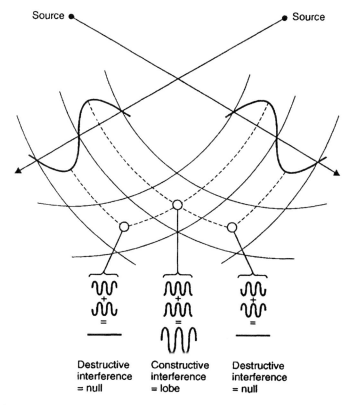

Destructive   Constructive   Destructive
interference   interference   interference
= null   = lobe   = null

Figure 3.33 Constructive and destructive interference between two identical sources.

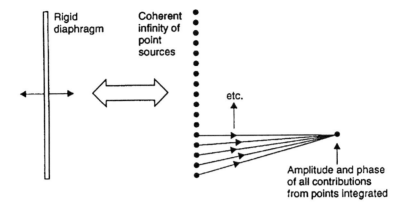

Figure 3.34 A rigid radiating surface can be considered as an infinity of coherent point sources. The result at a given location is obtained by integrating the radiation from each point.

When *ka* is small, the amount of re-radiation from the body compared to the radiation in the wavefront is very small. Constructive interference takes place beyond the body as if it were absent, thus it is correct to say that the sound diffracts around the body.

Figure 3.33 shows two identical sound sources which are spaced apart by a distance of several wavelengths and which vibrate in-phase. At all points equidistant from the sources the radiation adds constructively. The same is true where there are path length differences which are multiples of the wavelength. However, in certain directions the path length difference will result in relative phase reversal. Destructive interference means that sound cannot leave in those directions. The resultant diffraction pattern has a polar diagram which consists of repeating lobes with nulls between them.

Thus far this chapter has only considered the radiation of a pulsating sphere; a situation which is too simple to model many real-life sound radiators. The situation of Figure 3.33 can be extended to predict the results of vibrating bodies of arbitrary shape. Figure 3.34 shows a hypothetical rigid circular piston vibrating in an opening in a plane surface. This is apparently much more like a real loudspeaker. As it is rigid, all parts of it vibrate in the same phase. Following concepts advanced earlier, a rigid piston can be considered to be an infinite number of point sources. The result at an arbitrary point in space in front of the piston is obtained by integrating the waveform from every point source.

3.20 Radiation into smaller solid angles

A transducer can be affected dramatically by the presence of other objects, but the effect is highly frequency dependent. In Figure 3.35(a) a high frequency is radiated, and this simply reflects from the nearby object because the wavelength is short and the object is acoustically distant or in the *far field*. However, if the wavelength is made longer than the distance between the source and the object as in (b), the object is acoustically close or in the *near field* and becomes part of the source. The effect is that the object reduces the solid angle into which radiation can take place as well as raising the acoustic impedance

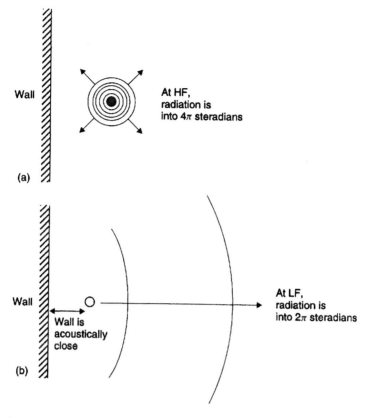

Figure 3.35 (a) At high frequencies an object is in the far field. (b) At low frequencies the same object is in the near field and increases velocity by constricting the radiation path.

the transducer sees. The volume velocity of the source is confined into a smaller cross-sectional area and consequently the velocity must rise in inverse proportion to the solid angle.

In Figure 3.36 the effect of positioning a loudspeaker is shown. In free space (a) the speaker might show a reduction in low frequencies which disappears when it is placed on the floor (b). In this case placing the speaker too close to a wall, or even worse, in a corner (c), will emphasize the low-frequency output. High-quality loudspeakers will have an adjustment to compensate for positioning. The technique can be useful in the case of small, cheap loudspeakers whose low-frequency response is generally inadequate. Some improvement can be had by corner mounting.

It will be evident that at low frequencies the long wavelengths make it impossible for two close-spaced radiators to get out of phase. Consequently when two radiators are working within one another's near field, they appear acoustically to be a single radiator. Each radiator will experience a doubled acoustic impedance because of the presence of the others. Thus the pressure for a given volume velocity will be doubled. As the intensity is proportional to the square of the pressure, it will be quadrupled.

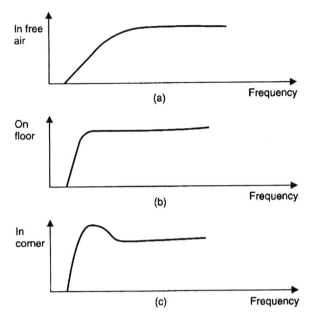

Figure 3.36 Loudspeaker positioning affects low-frequency response. (a) Speaker in free air appears bass deficient. (b) This effect disappears when floor mounted. (c) Bass is increased when mounted near a wall or corner.

The effect has to be taken into account when stereo loudspeakers are installed. At low frequencies the two speakers will be acoustically close and so will mutually raise their acoustic impedance causing a potential bass tip-up problem. When a pair of stereo speakers has been properly equalized, disconnecting one will result in the remaining speaker sounding bass light.

In Figure 3.37 the effect of positioning a microphone very close to a source is shown. The microphone body reduces the area through which sound can escape in the near field and raises the acoustic impedance, emphasizing the low frequencies. This effect will be observed even with pressure microphones as it is different in nature to and adds to the proximity effect described earlier. This is most noticeable in public address systems where the gain is limited to avoid howlround. The microphone must then be held close to obtain sufficient level and the plosive parts of speech are emphasized. The high signal levels generated often cause amplifier clipping, cutting intelligibility.

When inexperienced microphone users experience howlround they often misguidedly cover the microphone with a hand in order to prevent the sound from the speakers reaching it. This is quite the reverse of the correct action as the presence of the hand raises the local impedance and actually makes the howlround worse. The correct action is to move the microphone away from the body and (assuming a directional microphone) to point it away from the loudspeakers. In general this will mean pointing the microphone at the audience.

In Figure 3.38 a supra-aural headphone (one which sits above the ear rather than surrounding it) in free space has a very poor low-frequency response

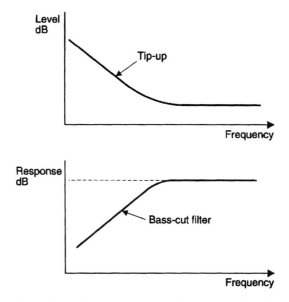

Figure 3.37 Bass tip-up due to close microphone positioning. A suitable filter will help intelligibility.

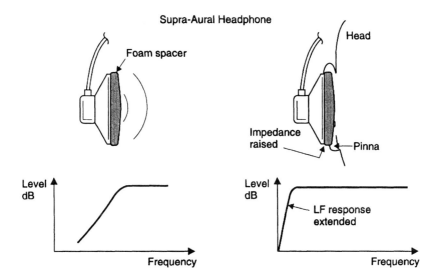

Figure 3.38 Supra-aural headphones rely on the bass tip-up in the near field to give a reasonable bass response.

because it is a dipole source and at low frequency air simply moves from front to back in a short circuit. However, the presence of the listener's head obstructs the short circuit and the bass tip-up effect gives a beneficial extension of frequency response to the intended listener, whilst those not wearing the headphones only hear high frequencies. Many personal stereo players incorporate a low-frequency boost to further equalize the losses. All practical headphones must be designed to take account of the presence of the user's head since headphones work primarily in the near field.

A dramatic example of bass tip-up is obtained by bringing the ear close to the edge of a cymbal shortly after it has been struck. The fundamental note which may only be a few tens of Hertz can clearly be heard. As the cymbal is such a poor radiator at this frequency there is very little damping of the fundamental which will continue for some time. At normal distances it is quite inaudible.

3.21 Refraction

If sound enters a medium in which the speed is different, the wavelength will change causing the wavefront to leave the interface at a different angle. This is known as refraction. The ratio of velocity in air to velocity in the medium is known as the refractive index of that medium; it determines the relationship between the angles of the incident and refracted wavefronts. This doesn't happen much in real life, it requires a thin membrane with different gases each side to demonstrate the effect. However, as was shown above in connection with the Doppler effect, wind has the ability to change the wavelength of sound. Figure 3.39 shows that when there is a wind blowing, friction with the earth's surface causes a velocity gradient. Sound radiated upwind will have its wavelength shortened more away from the ground than near it, whereas the reverse occurs downwind. Thus upwind it is difficult to hear a sound source because the radiation has been refracted upwards whereas downwind the radiation will be refracted towards the ground making the sound 'carry' better. Temperature gradients can have the same effect. As Figure 3.40(a) shows, the reduction in the speed of sound due to the normal fall in temperature with altitude acts to refract sound away from the earth. In the case of a temperature inversion (b) the opposite effect happens. Sometimes a layer of air forms in the atmosphere which is cooler than the air above and below it. Figure 3.40(c) shows that this acts as a waveguide because sound attempting to leave the layer is gently curved back in giving the acoustic equivalent of a mirage. In this way sound can travel hundreds of kilometres. Sometimes what appears to be thunder is heard on a clear sunny day. In fact it is the sound from a supersonic aircraft which may be a very long way away indeed

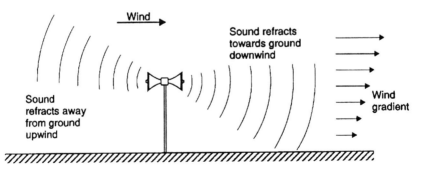

Figure 3.39 When there is a wind, the velocity gradient refracts sound downwards downwind of the source and upwards upwind of the source.

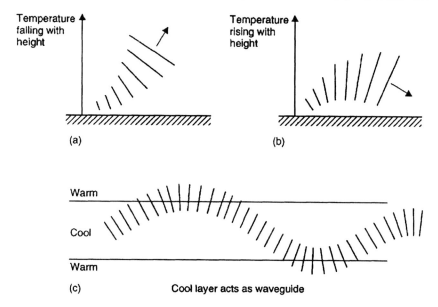

Figure 3.40 (a) Temperature fall with altitude refracts sound away from the earth. (b) Temperature inversion refracts sound back to earth. (c) Cool layer in the atmosphere can act as a waveguide.

3.22 Reflection, transmission and absorption

When two sounds of equal frequency and amplitude are travelling in opposite directions, the result is a *standing wave* where constructive interference occurs at fixed points one wavelength apart with nulls between. This effect can often be found between parallel hard walls, where the space will contain a whole number of wavelengths. As Figure 3.41 shows, reflection can also occur at

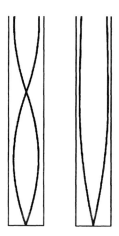

Figure 3.41 Standing waves in an organ pipe can exist at several different frequencies.

the impedance change at the end of a pipe. Here, a variety of different frequencies can excite standing waves at a given spacing. Wind instruments work on the principle of standing waves. The wind produces broadband noise, and the instrument resonates at the fundamental depending on the length of the pipe. The higher harmonics add to the richness or *timbre* of the sound produced.

In practice, many real materials do not reflect sound perfectly. As Figure 3.42 shows, some sound is reflected, some is transmitted and the remainder is absorbed. The proportions of each will generally vary with frequency. Porous materials are capable of being effective sound absorbers. The air movement is slowed by viscous friction among the fibres. Such materials include wood, foam, cloth and carpet. At low frequencies, large panels can move with the sound and dissipate energy in internal damping. Non-porous materials either reflect or transmit according to their mass. Thin, hard materials, such as glass, reflect high frequencies but transmit low frequencies. Substantial mass is required to prevent transmission of low frequencies, there being no substitute for masonry.

In real rooms with hard walls, standing waves can be set up in many dimensions, as Figure 3.43 shows. The frequencies at which the dominant standing waves occur are called eigentones. Any sound produced in such

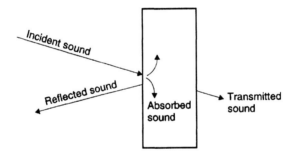

Figure 3.42 Incident sound is partially reflected, partially transmitted and partially absorbed. The proportions vary from one material to another and with frequency.

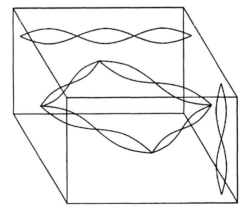

Figure 3.43 In a room, standing waves can be set up in three dimensions.

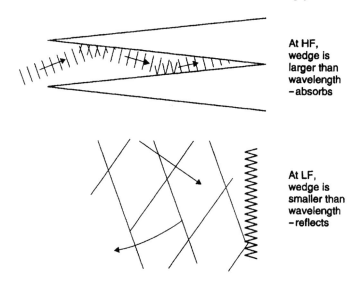

At HF,
wedge is
larger than
wavelength
– absorbs

At LF,
wedge is
smaller than
wavelength
– reflects

Figure 3.44 Anechoic wedges are effective until wavelength becomes too large to see them.

a room which coincides in frequency with an eigentone will be strongly emphasized or diminished (according to the listener's location) as a resonance which might take some time to decay. Clearly a cube would be the worst possible shape for a studio as it would have a small number of very powerful resonances.

At the opposite extreme, an *anechoic chamber* is a room treated with efficient absorption on every surface. Figure 3.44 shows that long wedges of foam absorb sound by repeated reflection and absorption down to a frequency determined by the length of the wedges (our friend *ka* again). Some people become distressed in anechoic rooms and musical instruments sound quiet, lifeless and boring. Sound of this kind is described as *dry*.

3.23 Reverberation

Reflected sound is needed in concert halls to amplify the instruments and add richness or *reverberation* to the sound. Since reflection cannot and should not be eliminated, practical studios, listening rooms and concert halls are designed so that resonances are made as numerous and close together as possible so that no single one appears dominant. Apart from choosing an irregular shape, this goal can be helped by the use of *diffusers* which are highly irregular reflectors. Figure 3.45 shows that if a two-plane stepped surface is made from a reflecting material, at some wavelengths there will be destructive interference between sound reflected from the upper surface and sound reflected from the lower. Consequently the sound cannot reflect back the way it came but must diffract off at any angle where constructive interference can occur. A diffuser made with steps of various dimensions will reflect sound in a complex manner. Diffusers are thus very good at preventing standing waves without the deadening effect that absorbing the sound would have.

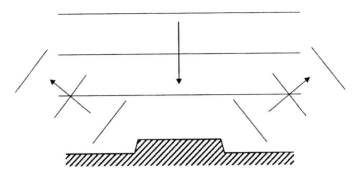

Figure 3.45 A diffuser prevents simple reflection of an incident wavefront by destructive interference. The diffracted sound must leave by another path.

In a hall having highly reflective walls, any sound will continue to reflect around for some time after the source has ceased. Clearly as more absorbent is introduced, this time will fall. The time taken for the sound to decay by 60 dB is known as the *reverberation time* of the room. The optimum reverberation time depends upon the kind of use to which the hall is put. Long reverberation times make orchestral music sound rich and full, but would result in intelligibility loss on speech. Consequently theatres and cinemas have short reverberation times, opera houses have medium times and concert halls have the longest. In some multi-purpose halls the reverberation can be modified by rotating wall panelling, although more recently this is done with electronic artificial reverberation using microphones, signal processors and loudspeakers.

Except at low frequencies, only porous materials make effective absorbers, but these cannot be used in areas which are prone to dampness or where frequent cleaning is required. This is why indoor swimming pools are so noisy.

References

1. Moore, B.C.J., *An Introduction to the Psychology of Hearing*, London: Academic Press (1989)
2. Muraoka, T., Iwahara, M. and Yamada, Y., Examination of audio bandwidth requirements for optimum sound signal transmission. *J. Audio Eng. Soc.*, **29**, 2–9 (1982)
3. Muraoka, T., Yamada, Y. and Yamazaki, M., Sampling frequency considerations in digital audio. *J. Audio Eng. Soc.*, **26**, 252–256 (1978)
4. Fincham, L.R., The subjective importance of uniform group delay at low frequencies. Presented at the 74th Audio Engineering Society Convention (New York, 1983), preprint 2056(H-1)
5. Fletcher, H., Auditory patterns. *Rev. Modern Physics*, **12**, 47–65 (1940)
6. Carterette, E.C. and Friedman, M.P., *Handbook of Perception*, 305–319. New York: Academic Press (1978)

Chapter 4

Sources of sound

If it is proposed to reproduce sound well, then it is a good idea to have a knowledge of how real sounds are produced in the first place. In this chapter the human voice, musical instruments and other common sound sources are considered. This chapter relies upon the principles outlined in Chapters 2 and 3.

4.1 Producing sounds

Before discussing specifics such as speech and musical instruments[1] it is useful to look at some of the mechanisms involved. Often the sound we hear is the result of two stages. First a raw sound or vibration is produced at the required pitch or frequency, then this is modified in some way to give a particular timbre or spectral content. In most cases the production of the vibration and the modification are two independent processes which do not affect one another. The exception is the flue pipe in which the two are intimately related.

It must be stressed that the sound made by the instrument is not the sound heard by the listener since this is strongly affected by the nature of the room in which the instrument is played. Equally the sound heard by the listener is frequently not that heard by the player since the player is often in the close field of the instrument and the directivity of most instruments is far from uniform.

4.2 Vibrating bars

Figure 4.1(a) shows a tuning fork; a simple mechanical vibration source. The prongs of the fork have mass and the flexing of the prongs apart or together provides a spring which tries to restore the prongs to their equilibrium position. A simplified version of the mechanical system is that shown in (b). If disturbed from equilibrium there is a constant interchange of energy between the spring and the masses which results in a sinusoidal oscillation called simple harmonic motion (SHM). As no external forces act, the centre of mass of the system must remain stationary. Consequently the masses must move in anti-phase. At the centre of the travel of the masses their velocity is a maximum. As the spring has its natural length all of the energy is kinetic. At the end of the travel the masses are stationary and the system contains only potential energy in the stretched spring.

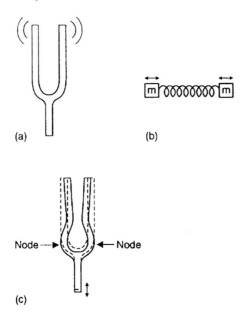

Figure 4.1 (a) The tuning fork is a stable pitch reference. (b) Simplified mechanical arrangement of two identical masses separated by a spring. (c) Two nodes exist in vibration pattern, causing handle to move axially.

The SHM of the tuning fork means that it radiates a pure tone with almost no harmonics. The resonance is so pure that the nature of the impact which caused the fork to ring only determines the amplitude. The tuning fork is also very stable, showing little change of frequency as time passes and so it is a useful reference for tuning instruments.

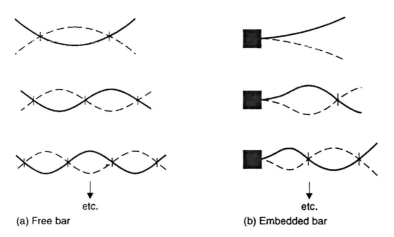

Figure 4.2 (a) Modes of a free bar. (b) Modes of an embedded bar.

The tuning fork has two nodes, one each side of the stem. As Figure 4.1(c) shows, this results in the handle moving lengthwise. If the stem is pressed against a flat surface it will radiate at the pitch of the fork.

Other mass/stiffness combinations exist. Figure 4.2(a) shows a free bar such as is used in a xylophone. When struck, this can vibrate in various modes, all of which allow its centre of mass to remain stationary. In general these modes are not harmonically related to the fundamental. Modes can exist simultaneously so that the bar has a timbre. The timbre will vary according to where the bar is struck. For example, if it is struck at a node of a particular mode, that mode will not be excited.

Vibrating bars can also be embedded in a heavy block so that one end is effectively immobilized. This approach is shown in Figure 4.2(b) and is used in certain chiming clocks and music boxes. Many modes are possible, although the music-box pin releases the end of the bar so that the vibration is mostly fundamental.

4.3 Vibrating panels

The vibrating bar can be extended into two dimensions to produce a vibrating panel. In a rectangular panel the modes in each dimension can be independent as shown in Figure 4.3(a). In a circular panel, such as a cymbal, the modes are radial and tangential. Figure 4.3(b) shows a radial mode in which the centre and the perimeter are moving in anti-phase, resulting in a circular node. Figure 4.3(c) shows a tangential mode in which there are two complete cycles around the perimeter. Consequently the nodes are radial. In a flat disk the radial and tangential modes can be independent, but in a conical disk they interact. This is why cymbals are slightly conical.

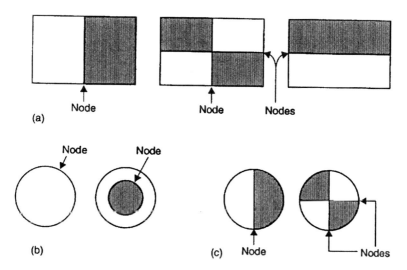

Figure 4.3 (a) Modes in the two dimensions of a rectangular panel can be independent. Circular panel has radial modes (b) and tangential modes (c).

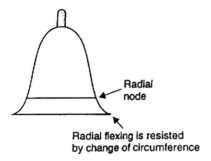

Figure 4.4 Church bell is a very deep cone. Radial modes cause stretching of perimeter making the timbre of bells slightly dissonant.

The church bell is a special case in which the radial and tangential modes interact strongly. Figure 4.4 shows that when the rim of the bell attempts to flex outwards in a radial mode, the perimeter must change in length. This requires the material of the bell to stretch and this is resisted strongly by the material stiffness. The result is that the modes of a bell are not harmonically related and this is evident in the somewhat dissonant sound made by bells.

4.4 Vibrating strings

The stretched string is another source of vibration. Figure 4.5 shows that when a taut string is deflected a restoring force is created by a component of the tension. This combines with the mass of the string to produce a vibration source. Strings, of necessity, must be fixed at the ends and so nodes are always found there. Further nodes can occur at other places as higher vibrational modes are excited. The timbre is determined by the mode structure which is

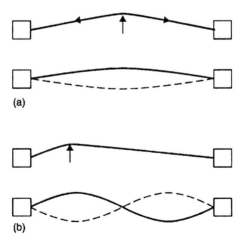

Figure 4.5 (a) Deflected taut string develops restoring force allowing SHM. (b) String can vibrate in a variety of modes, determined by where it is struck.

a function of where the excitation is applied along the string. In pianos, the hammers are arranged to strike at one-seventh the length of the string which will suppress the dissonant seventh harmonic. Violins and guitars are also generally bowed or plucked at this point.

Figure 4.5 also shows how the length, the tension and the mass per unit length affect the fundamental frequency. This expression is only true for strings which are not stiff. Stiff strings act to a degree like bars and an additional restoring force is available due to the bar being bent, raising the frequency. The curvature increases with the mode and consequently the modes are no longer integer multiples of the fundamental. This is called *inharmonicity*. In order to reduce inharmonicity, when a heavy string is required it is not made from solid thick material, but instead a thinner string is wrapped with wire. This can be seen in pianos and guitars. Figure 4.6 shows that a wrapped string is much more flexible than a solid one.

Stretched strings can be excited by striking as in the piano, plucking as in the harp and the guitar, or bowing as in the violin. Plucking and striking are both single acts and result in an initial transient followed by a decay. In contrast, bowing produces a sustained note which grows slowly. Bows are made of hair. Figure 4.7 shows that hair is covered with scales which lie down if it is stroked one way, but which rise up if it is stroked the other way. Bows need to be made with equal numbers of hairs facing in each direction. The gripping action of the scales and the additional friction from the rosin causes the bow and string to form a highly non-linear mechanism. As the bow is moved across the string, initially the string is gripped and deflected. At some point the grip will be broken by the restoring force of the string and the string will pull away as if it had been plucked. However, later in the cycle of oscillation the string will find that its velocity is almost the same as that

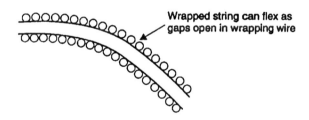

Figure 4.6 A solid string is stiffer than a wrapped string which suffers less inharmonicity.

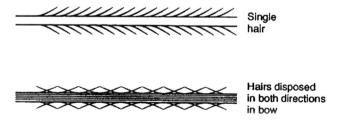

Figure 4.7 The bow of a stringed instrument contains equal numbers of hairs which are laid in opposite directions.

of the bow and it will be gripped once more. Effectively bowing consists of plucking the string at its own frequency. These plucks are transients which contain all frequencies. These broadband plucks are able to excite many modes of the string. The amplitudes of these modes are in inverse proportion to the harmonic number.

4.5 Vibrating diaphragms

The stretched flexible diaphragm of the drum is effectively a two-dimensional string. Traditionally animal skin was used, but synthetic diaphragms are also available. The tension in the skin provides a restoring force when deflected. The tension is quite high and the circular shape is used because it can best resist that tension without distorting. The diameter, the mass per unit area of the skin and the tension largely determine the fundamental frequency.

Like other vibrating systems, the drum skin can vibrate in various modes. Figure 4.8 shows that, like the cymbal, these modes can be radial or tangential. The drum differs from the cymbal in that the latter has a free edge whereas the edge of the drumskin is by definition a node.

Some drums have an open structure so that both sides of the diaphragm can radiate. In others, the rear of the diaphragm is enclosed. The air trapped beneath the diaphragm acts as a spring which provides a further restoring force when the diaphragm is deflected. The result is that the same fundamental frequency is obtained with less tension. However, the trapped air does not affect the higher modes so much because in these some of the diaphragm is moving out whilst another part is moving in. Consequently the timbre and the directivity of an enclosed drum differ from that of an open drum.

4.6 Using airflow

A flow of air can also be used to create vibration in a variety of ways. Figure 4.9(a) shows a flow of air through a constriction or *venturi*. The speed of the flow must increase in the venturi. As this increases the kinetic energy of the flow, the energy must come from somewhere and the result is that the potential energy due to the pressure must fall. The theory which explained this phenomenon was first put forward by Daniel Bernouilli and it is known as the Bernouilli effect in his honour. The effect is used in engine carburettors to draw fuel into the intake air. If the walls of the venturi are made flexible, an unstable condition results. The reduced pressure will pull the walls of the venturi together, shutting off the airflow. This terminates the pressure

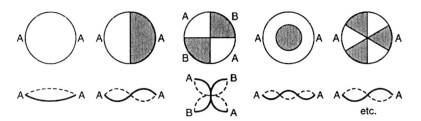

Figure 4.8 The modes of a taut skin. Note that the perimeter is a node.

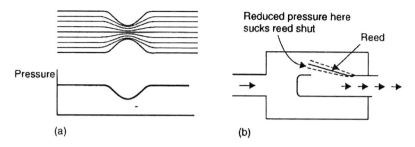

Reduced pressure here
sucks reed shut

Reed

Pressure

(a)   (b)

Figure 4.9 (a) Airflow through a constriction reduces pressure as described by Bernouilli. (b) Placing a reed in a constriction is unstable hence reed vibrates.

reduction and the walls then spring apart again, restarting the airflow and so on. The vocal cords work in this way, as do the trumpeter's lips. The resulting intermittent airflow is extremely rich in harmonics.

The Bernouilli effect is also present in the action of the reed. Figure 4.9(b) shows that air is blown through a gap between a flexible reed and a rigid base. As before, the increase in velocity in the gap causes a reduction in pressure which sucks the reed shut. This cuts off the airflow and the reed springs open again. If the reed overlaps the hole in the base it will form a good seal and interrupt the airflow completely. This produces a wide range of harmonics. However, if the reed is made a little smaller so that it fails to form a seal the higher harmonics will be less pronounced. This approach is used in the harmonium and the harmonica, also in the clarinet and oboe.

The airflow around bodies is seldom smooth and is frequently unstable. The fluttering of a flag is the result of instability. Air flowing over a cylinder breaks into turbulence part way round. However, the point at which it does so is chaotic and differs from one side to the other. As Figure 4.10 shows, the air flowing round the cylinder is being accelerated because its direction is changing. This requires a force and the reaction to the force is lift on the surface of the cylinder. If the flow is perfectly symmetrical, the lift cancels out. However, in the presence of instability there is no symmetry and there is a net alternating force.

This phenomenon can be observed in suspended telephone wires which 'sing' when a wind is blowing. The effect is subtle but can easily be heard by placing the ear against the supporting pole. It is a nuisance in suspension bridge wires and has caused factory chimneys to shake themselves to pieces. The solution which is frequently seen is to attach a flat strip or spoiler edgeways to the cylinder in a helix.

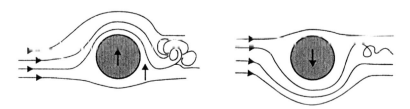

Figure 4.10 The airflow around a cylinder is unstable and causes an alternating force.

Figure 4.11 Airflow against an edge is also unstable.

A jet of air playing on an edge, as shown in Figure 4.11, is also unstable. The jet can be deflected by small pressure differences so that the flow is not evenly divided. However, an unbalanced flow raises the pressure on the side receiving more flow and this might be expected to apply a correction to restore an even balance. However, there is a lag in the system. It takes a finite time for the effect of the deflection at the jet to reach the edge. This time is a function of the distance from the jet to the edge and the airflow velocity (not the speed of sound). Thus the correction arrives too late and the pressure over-deflects the jet. The result is that the jet oscillates from one side of the edge to the other to produce what is called an *edge tone*.

As might be expected, the frequency is determined by the jet-to-edge distance (in organs this is called the *cut-up*) and the flow velocity. The process is sometimes called an air-reed but the mechanism of the reed is quite different. It must be recalled that most musical instruments evolved empirically long before acoustic theory came along to explain them. Two hundred years ago likening this effect to the action of a reed was quite plausible.

4.7 Resonators

Air has mass, but it is also compressible and can act as a spring. As a result it is possible to create a resonant system using only air. This was first described by Helmholtz and systems such as those in Figure 4.12 are called Helmholtz resonators. These resonators produce relatively pure tones as there is no mechanism for the production of modes. Consequently the sound lacks richness and their use in instruments is limited.

The pipe represents a more useful resonator because it allows a complex mode structure. Effectively a pipe is just a guide for a standing wave. Pipes can be open or closed at either end, but a pipe closed at both ends finds limited application. It is obvious that the end of a closed pipe will act as a reflector, sending a compression back the way it came, but an open end can also reflect because it acts as an impedance step. Some of the sound is radiated, but a compression is reflected back down the pipe as a rarefaction.

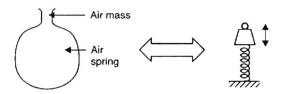

Figure 4.12 The Helmholtz resonator consists of an air mass and an air spring and has a single resonant frequency with no modes.

Figure 4.13(a) shows the first few modes of a pipe which is open at both ends. The ends form pressure nodes and displacement antinodes. The fundamental frequency is controlled by the length and the local value of c. The length will be a whole number of wavelengths. It is evident from (a) that at a point exactly half-way down the pipe there is no displacement for the fundamental mode. Only the pressure varies. Consequently if a rigid baffle as in (b) were slid into place after the resonance established, it would make no difference. Clearly a pipe closed at one end can be half the length of an open pipe for the same fundamental.

The same can be done for all of the odd-numbered modes of the open pipe as they have a central displacement node. However, halving the pipe length and closing one end eliminates the even modes of the open pipe. Thus the timbre of a closed pipe is different from that of an open pipe. Figure 4.13(c) shows that the spectrum from the open pipe contains all harmonics, whereas that from the closed pipe only contains odd harmonics.

If a closed-end pipe is made conical it will be able to resonate with both odd and even harmonics. The end of a pipe may also be flared to form a horn which produces better acoustic coupling and raises efficiency.

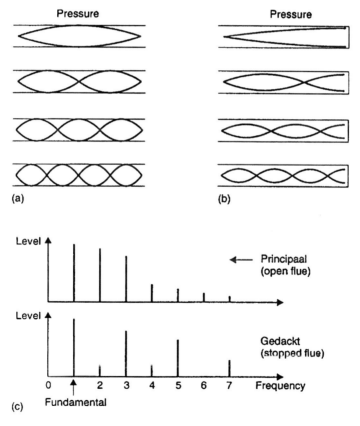

Figure 4.13 (a) Modes of a pipe open at both ends. (b) Modes of a pipe closed at one end. (c) Closed and open pipes have different harmonic structure hence different timbre.

4.8 The organ

The pipe organ has been described as the king of instruments and the range of sounds it can produce is remarkable.[2] Despite its brute power it is capable of great delicacy. The frequency range of the organ exceeds that of a whole orchestra, particularly at low frequencies. This is displayed very effectively in the organ symphony where the organ is integrated into the orchestra, sometimes taking the lead on the manuals, but often simply underscoring the rest of the orchestra with pedal notes which add great weight and warmth.

Organ pipes are many and varied, but all fall into one of two categories, the *flue* and the *reed*. Pipes may be of square section made from wood or of round section made from metal and may be open or stopped. Open pipes are tuned by a telescopic sliding section or by bending over the lip whereas stopped pipes use a sliding plug called a *tuning stopper*. As explained in Section 4.7, open pipes produce all harmonics whereas stopped pipes produce odd harmonics. The structure of some typical flue pipes is shown in Figure 4.14. The flue pipe

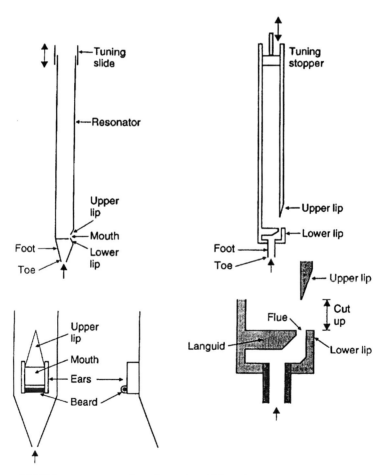

Figure 4.14 Major components of wooden and metal flue pipes.

works on the edge-tone principle. In the case of the wooden pipe, the air jet emerges from the flue which is the gap between the *languid* and the *lower lip*. In the metal pipe, the cylindrical shape transitions to flat at the mouth via the upper and lower lips. The flue is formed between the languid and the lower lip.

In both cases the air jet plays on the upper lip and generates an edge tone. However, the air jet passes across the *mouth* of the pipe and pressure within the pipe dominates the process. Thus the fundamental produced is controlled entirely by the length of the pipe and the timbre almost entirely by the shape and material of the pipe, although the cut-up height will have some effect. The harmonic structure is affected by the diameter of the pipe with respect to its length. Wide pipes favour lower harmonics whereas slim pipes favour higher harmonics. The cross-sectional area of a slim pipe is much smaller and sometimes *ears* are placed at the sides of the mouth to give the pipe better control over the flue jet. A further measure seen on bass pipes is a roller or *beard* which joins the ears across the bottom of the mouth to assist with prompt speech.

If the airflow velocity is dramatically increased the frequency of the edge tone may become so high that it cannot excite the fundamental. The resonance jumps to the second harmonic in an open pipe and the third harmonic in a stopped pipe. This is called *overblown* mode and never occurs in practical organs because the required pressure is not made available. It is easy to demonstrate with a recorder which is acoustically a mouth-blown flue pipe. By blowing hard the note shifts up in pitch and becomes squeaky.

The operation of a reed pipe is quite different. As Figure 4.15 shows, air is supplied to the *boot* which surrounds the reed mechanism. The reed is mounted on the side of the *shallot* above an opening which communicates with the hollow centre. The reed vibrates according to the Bernouilli effect described in Section 4.6. The fundamental frequency is set by the position of a tuning wire which alters the working length of the reed. The reeds used in pipe organs are generally metallic and have little internal damping. This gives them

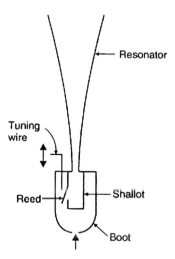

Figure 4.15 Components of a reed pipe.

a high Q factor which means that their resonant frequency is little affected by other influences. Unlike the flue pipe, where the pipe itself determines the fundamental frequency, in the reed pipe it is the reed which determines the fundamental and the resonances in the pipe determine only the harmonic structure.

The mouth of the flue pipe renders the bottom end of the pipe acoustically open and radiation occurs from both ends. In the reed pipe the bore of the shallot is so small in comparison with the cross-sectional area of the pipe that it is acoustically closed and radiation can only take place from the other end. Consequently a stopped reed pipe is an impossibility. The pipe itself is not just a resonator for the harmonics, but may also acts as an acoustic transformer coupling the radiation more efficiently to the surrounding air in the manner of a megaphone. Pipes may be conical and/or flared. The trompette looks like, and sounds like, its orchestral cousin.

The organ console consists of manuals, or keyboards used by the hands, and a pedal board. For a given fundamental note, there will be many different pipes which can all produce that note but with a wide range of timbre and level. The pipes which are to speak are selected by pulling out stops. These are knobs which are found at both sides of the keyboards. The coupling logic combines the key and stop inputs to determine which pipes receive air. In small organs this process may be completely mechanical. Recent organs may use completely electronic coupling implemented with logic gates. With electronic logic it is easy to provide stop combination memories. The organist can program these in advance and select them with a single press of a button on the front edge of the manual.

The control system of an organ is binary and the only control of the volume is by the number of stops selected. Some organs have a swell box which is a shuttered enclosure surrounding some of the pipes. The shutters can be opened and closed by the *swell pedal* which is mounted above the pedals.

Early organs were blown by manually operated bellows. The individual impulses from the bellows were smoothed out by a reservoir having flexible sides and a weighted top. The power required in a large organ may be tens of kiloWatts and later organs were blown by steam engines, then electrically. With kilometres of plumbing it is impossible to avoid leaks and a gentle hiss appears when the blowers are activated. The listener soon becomes unaware of the hiss, because its level and tonality is constant. However, any change in the hiss is extremely obvious. When recording an organ, moving a fader at the wrong time can be fatal. In organ symphonies fading down the organ microphones when the organ is not speaking may be good practice, but if the fader is brought up when the hiss is not masked by the orchestra the result is unsatisfactory. In Compact Discs which contain a number of different pieces it is generally better to allow the hiss to continue between tracks.

4.9 Wind instrument principles

Like the organ, wind instruments use resonant pipes which are excited either by edge tone, an actual reed or a reed mechanism formed by the player's lips. The mouthpiece of the recorder has all of the components of the organ flue pipe and the mouthpiece of a clarinet duplicates the structure of the organ reed pipe except that the reed will be softer with a lower Q factor. In the final category of wind instruments the mouthpiece is a simple bell which receives the player's lips. Unlike the organ, the edge tone instruments can be overblown to increase the octave span. Where the lips form the reed, the fundamental resonance can be changed by the player. Also unlike the organ, where each pipe only emits a single note, wind instruments must be capable of producing a range of notes from a single pipe.

Simple instruments such as bugles and posthorns have no means of changing their resonant behaviour which has a fixed fundamental and a range of harmonics. All the player can do is to excite different harmonics by changing the pitch of the lip reed. The resultant range of notes available is quite small and bugle music is somewhat restricted.

Figure 4.16 shows that in more sophisticated instruments resonant behaviour can be changed in a number of ways. Figure 4.16(a) shows the most obvious method is to change the length of the pipe telescopically. In the trombone this is done with the minimum of moving mass by sliding a U-shaped section between the fixed mouthpiece and trumpet. The trombone can vary its pitch infinitely to produce glissandi.

Figure 4.16(b) shows that variation in pipe length is obtained by bypassing discrete lengths of pipe using key-operated valves. In the trumpet there are three such sections of different length. Bypassing these sections raises the pitch by one, two and four semitones. The valves can be opened in seven binary

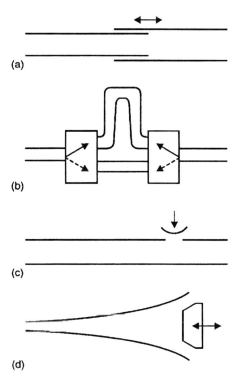

Figure 4.16 Fundamental of a pipe can be changed by telescoping (a), by discrete length changes introduced by valves (b), by covering side holes (c) or by obstructing the bell (d).

combinations to obtain nearly a musical scale. The scale is not perfect because the valves interact. When any single valve is pressed, the correct pitch can be obtained because the overall length of the pipe is changed by the correct amount. However, opening a second valve adds a different proportion of the overall length because the overall length was changed by the operation of the first valve. The French horn and the tuba employ a similar mechanism.

In Figure 4.16(c) a single pipe is made capable of playing a range of notes by placing holes in the side of the pipe which the player covers and uncovers. If the hole is large compared to the cross-sectional area of the pipe, the acoustic length of the pipe extends to the hole itself. However, if the hole is small the acoustic length is longer. This phenomenon is usefully employed on smaller instruments to move the finger holes within the hand span of the player. Where the instrument is too large the holes will be beyond the hand span and the holes will be uncovered by remote pads which are connected to keys by a linkage running down the instrument.

Figure 4.16(d) shows that the resonance can also be changed by placing a hand inside the bell of a flared pipe.

4.10 Wind instruments

Wind instruments are generally divided into woodwind and brass, although the distinction is based on general timbre rather than the actual material used. Consequently musicians do not see a contradiction in having in the woodwind section an instrument which is obviously made of metal. Brass instruments are excited by a lip reed whereas woodwind instruments are excited by an actual reed or edge tone.

The recorder is a miniature flue pipe whose fundamental is determined by the holes which are covered by the player's fingers. Some of the holes are small to put them within the player's hand span. As the recorder is acoustically open at both ends, it can be made to jump up by a number of whole octaves by overblowing so that the same finger holes are reused in each octave.

The flute is unusual in that it is played transversely rather than axially; i.e. one blows across it rather than down it to obtain a characteristic breathy sound. The flue is formed by the player's mouth and the mouthpiece consists of an aperture known as an *embouchure*. The flute is also played in overblown mode. In addition to controlling the air velocity, the flautist can also control the cut-up by rotating the instrument to move the sharp edge of the embouchure with respect to the lips. The flautist presses keys which mechanically operate pads. The piccolo is a smaller flute which plays an octave higher.

The oboe, the clarinet, the bassoon and the saxophone are all reed instruments which are played with keys to control pads which act on side holes. The oboe mouthpiece contains a double reed which produces a harmonically rich sawtooth-like waveform. The reed mouthpiece acoustically closes one end of the pipe, but because it is tapering or conical (with a slightly flared end) all harmonics are possible.

The clarinet has a single reed mouthpiece and is cylindrical which means it is acoustically similar to the organ reed pipe, only supporting odd harmonics. Because of this it can go lower in pitch than the oboe despite being a similar length. When the clarinet changes register the fundamental moves to the third harmonic instead of the second. This requires extra keys.

The bassoon is the lowest sounding woodwind instrument and its long conical pipe is folded to make it compact. It is blown through a thin pipe fitted with a double reed. Despite being made of brass, the saxophone is classified as a woodwind instrument because it has a single reed like the clarinet. It has a conical pipe which is folded and flared.

4.11 Brass instruments

Traditionally brass instruments are those which use the player's lips as a reed. The mouthpiece is a simple bell. The lip reed produces an acoustically closed end to the pipe and this would result in odd harmonics being emphasized in a parallel-sided pipe or tube. In most brass instruments at least some of the tube is tapered or conical to encourage even harmonics and to allow octave jumps by overblowing. The end of the tube is usually flared as this acts as an acoustic transformer which improves efficiency. As brass instruments were the first example of this technique, any acoustic device using a flared mouth is called a horn.

Brass instruments vary in the range covered, as Figure 4.17 shows, and therefore must have a wide range of acoustic lengths. In order to render the instruments more compact the tubing is coiled or folded.

The trumpet has a significant amount of cylindrical tubing before becoming conical and has a nominal length of about 2 m. Three piston valves allow extra pipe sections to be removed as was described in Section 4.9. Using overblowing, three octaves can be covered.

The trombone has 3 m or more of tubing, much of which is parallel sided to allow telescopic sliding, although there is a valved trombone which is uncommon. Like the violin, it is down to the player to pitch notes correctly.

Trumpets and trombones can be played with *mutes* which are obstructions designed to be placed in the bell. These reduce the level and change the timbre.

The tuba has 6 m of coiled pipework and is quite an impressive device which terminates in a substantial bell mouth. Four or five piston valves are provided as in the trumpet.

Unlike the previously mentioned brass instruments, the French horn faces backwards when played and this results in its timbre being quite mellow as

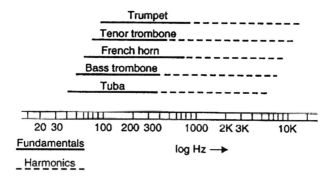

Figure 4.17 The range covered by various brass instruments.

far as the audience is concerned. It has the usual piston valves but the player
can also place a hand in the bell to raise the pitch by a semitone.

4.12 Stringed instruments

As was seen in Section 4.4, stringed instruments can be bowed or plucked.
Bowing produces a sound envelope which changes slowly whereas plucking
produces transient rich *pizzicato* sounds which decay rapidly.

The guitar is designed for plucking only. The sound made by the strings
is quite small and the majority of the sound is radiated by using the string
vibration to excite the body. Figure 4.18 shows the construction of a typical
acoustic guitar. The strings are attached to a *tailpiece* and pass over the *bridge*
and the *nut* before being wound around the tuning pegs.

The heights of the nut and the bridge are such that the strings are not quite
parallel to the fingerboard. The player determines the fundamental frequency
by pressing the string down onto *frets* which change its effective length. The
geometry of the string support is such that when the string is supported by a
given fret it can vibrate without touching the adjacent fret. If the *action* is not
correct, the result will be *fret buzz* which will require the bridge to be raised.

The frets are spaced such that the fundamental is raised in semitone intervals. As the strings are of different diameter, they have different stiffnesses.
Even if the *open tuning* of the strings is correct, as they are stopped, the string
stiffness drives the pitch up more for the thicker strings than it does for the
thinner ones. This can be compensated to some degree by turning the bridge
so that the thicker strings are slightly longer than the thinner ones. The further
the strings are stopped the greater this difference becomes.

The string tension causes a downthrust into the body and this is resisted by
a *soundpost* which connects the front and back of the body. Classical guitars
use gut or nylon strings which work with lower tension and put less thrust on
the instrument which is of relatively delicate construction. Greater level can
be obtained using steel strings, but these require more tension and a stronger
structure. The tailpiece is often taken to the end of the body for strength and
the tension is adjusted by *machine heads* which are worm gear driven shafts.

String vibrations pass down the bridge and the soundpost and excite the
thin wooden body. Sound is radiated directly by the flexing of the body and
also via the *soundhole*. The mass of air in the soundhole and the compliance
of the enclosed air form a resonant system which reinforces the bass.

The timbre of the guitar is determined by the vibration of the structure and
by the harmonic structure of the string vibration. If the guitar is played with

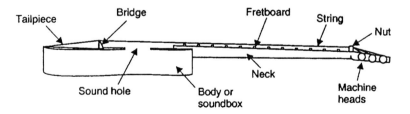

Figure 4.18 The components of an acoustic guitar.

a plectrum, the string will be bent to a sharper angle before it is released and this creates more high harmonics than fingering which bends the string to a radius.

The harp has no fretboard or stopping mechanism but has 48 vertically arranged strings which correspond over six octaves to the white keys of a piano. The bottom end of the strings terminates in a relatively small and inefficient soundboard which absorbs little string power and allows a lengthy sustain. Semitones are obtained by pedals in the base of the pillar which operate rods travelling up the pillar. These actuate length-changing mechanisms in the top of the harp.

The harpsichord is virtually a harp with a keyboard. The keys operate an ingenious mechanism which extends a peg to pluck the string as it rises, but retracts the peg as it falls so that the string is not fouled. The vibrations of the strings are amplified by a soundboard which is part of the cabinet.

The violin family (violin, viola, violincello and double bass) are all structurally similar and differ primarily in their range (see Figure 4.19). As the violin is designed to be bowed, the strings are arranged on a pronounced curve so that each one can be accessed by changing the angle of the bow. Practicality limits the number of strings to four as with any more it would be difficult to bow one without fouling those on either side. The body needs cut-outs to allow the bow to take up the appropriate angles. The cut-outs also raise the complexity of the vibrational response, richening the tone.

The fingerboard is unfretted so the player has to position the fingers by experience. The advantage of this arrangement over the use of frets is that by rocking the hand at typically 6 Hz, vibrato can be introduced. The bridge is supported by a soundpost as in the guitar, but the soundhole of the guitar is replaced by *f-holes* which are integral sign-shaped slots on either side of the bridge. These have two functions. First they allow the interior cavity of the instrument to act as a resonator, but most importantly they reduce the mass and stiffness of the top plate in the area of the bridge. The top plate can then respond to much higher frequencies allowing a brighter timbre. This is analogous to fitting a loudspeaker with a tweeter.

The violin family have very complicated vibrational responses causing the timbre to differ between notes. The 'cello and the double bass are floorstanding and the vibration picture is further complicated because floor vibrations are induced through the spike.

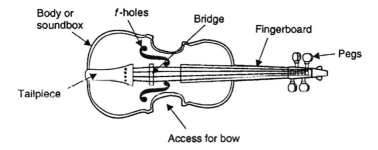

Figure 4.19 Structure of the violin family.

4.13 The pianoforte

Previous keyboard instruments such as the organ and the harpsichord were not touch sensitive. The piano e forte (Italian soft and loud) was a breakthrough because it produced a sound level which was a function of the effort with which the key was depressed. The pianoforte uses steel strings stretched across a substantial metal frame. There are three strings per note in the top five octaves, two in the next two octaves and one in the remainder. To reduce inharmonicity, the bass strings are wrapped. The key mechanism operates a hammer and a damper. The hammer flies forwards when the key is pressed, but the linkage is designed so that the hammer returns immediately it has struck. The damper only operates when the key is released. Strictly speaking the piano is a percussion instrument as it cannot sustain indefinitely.

The impact of the hammer depends upon the energy of the player. Heavy impacts clearly produce a higher level, but they also change the timbre. With a slow impact, the string starts to recoil immediately whereas with rapid impact there is no time for this to happen and the hammer puts a local deflection in the string which causes higher harmonics to be excited. Thus the timbre gets brighter as the instrument gets louder. Consequently it is important that pianoforte recordings are reproduced at the correct level. It is not possible to render the timbre forte simply by turning up the level.

The vibrations of the strings are modified and radiated by the soundboard which forms the base of the cabinet. Pedals are provided to modify the action. The *sustain* or *loud* pedal on the right retracts the dampers so that the string continues to sound after the key is released. The left pedal is the *soft* pedal which reduces the level. This can be done using the dampers, by using felt to soften the string impact, by reducing the hammer travel, or by moving the hammer bank laterally so that not all of the strings are struck. The grand has a centre pedal which acts to sustain only the bass strings. This is often absent in smaller upright and mini-pianos.

The pianoforte has an extremely wide frequency range and also a wide dynamic range so that it is one of the most demanding instruments as far as sound reproduction is concerned.

4.14 Percussion instruments

Percussion instruments are probably the oldest of all as they simply involve striking one object with another. The impact may result in a broadband noise which does not have a recognizable pitch. Instruments in which this happens are classified as *indefinite pitch* instruments. These are generally very simple as the concepts of tuning or producing a scale do not apply. However, resonant modification may be built in so that the impact simply excites a wide range of harmonics. These are called *definite pitch* instruments and clearly they must be more complex in order to offer a range of notes.

The triangle, the cymbal and the gong are all indefinite pitch devices whose mechanisms should be clear from the first sections of this chapter. The drum is available in various sizes, some of which are fitted with *snares* which are lengths of gut stretched across the membrane to give a rattling effect.

The drumkit consists of a selection of drums called tom-toms on stands and a pedal-operated bass drum or kick drum. Cymbals can be mounted on pedestals to be struck with the drumstick, or mounted in pairs so that a foot operated mechanism can clash them together. This is known as a hi-hat.

The xylophone and the larger marimba are definite pitch instruments which use tuned bars of wood or metal with resonating tubes beneath. They are played with hammers which may be hard or soft according to the timbre required. The celesta uses a keyboard to operate hammers. The bars are supported at the nodes of the fundamental resonance. As was seen in Section 4.2 bars do not naturally have harmonic modes and often the underside of the bar is curved to make the modes closer to harmonic. There are a number of instruments based on bells of various sizes.

The kettle drum or tympanum is a calfskin membrane stretched over a sealed chamber. Both the skin tension and the elasticity of the enclosed air determine the fundamental frequency. The pedal-timp has a mechanism to allow pre-adjusted shifts in tension to give a range of notes. The range is quite small and often a scaled pair of tympani are used. The player can determine the harmonic structure by striking the skin at different radii.

4.15 Electrical and electronic instruments

The electric guitar was originally developed to increase the sound level beyond the capabilities of an acoustic guitar so that larger venues could be used. Early electric guitars were no more than conventional steel-stringed guitars fitted with a pickup. The pickup consists of a permanent magnet with a coil of wire around it. Any ferrous object moving nearby will cause the magnetic field to vary and this will induce a signal in the coil. An amplifier and a loudspeaker are needed. Often the loudspeaker is deliberately designed to be non-linear and resonant in order to make the sound richer.

Later solid guitars dispensed with the hollow sound box and relied entirely on amplification. The six strings of the acoustic guitar were retained. Figure 4.20 shows an electric guitar. The guitar itself is a vibration source and its construction and materials have a large bearing on the sound quality and the sustain. The tension of steel strings puts a constant bending load on the neck and this is resisted by a truss rod which can be adjusted. The bridge is often fitted with individual height and position adjustments for each string. The bridge height is adjusted to set the action and the position of the bridge is adjusted to compensate for the increasing effect of string stiffness as the strings are stopped down. In practice the low-frequency strings will be longer than the high-frequency strings.

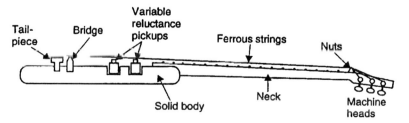

Figure 4.20 Structure of an electric guitar.

The bass guitar has four strings and is much heavier in order to give good sustain. The long neck and large machine heads can give a balance problem and some bass guitars fix the strings at the nut and the machine heads are fitted behind the bridge.

A large number of signal processors have been developed which modify the sound before the power amplification stage. Many of these are designed to be placed on the stage and the guitarist can operate them with a foot.

The Hammond organ was an early alternative to the pipe organ which used toothed tone wheels on a motor driven shaft. These operated pickups working on the same principle as the guitar pickup. Clearly an amplifier and loudspeaker system was required. The Leslie loudspeaker was specifically designed for electric organs and contains electrically driven rotating horn drive units which use the Doppler effect to obtain vibrato.

The vibraphone is a motorized version of the marimba in which rotating disks modulate the action of the resonator tubes behind the struck bars.

With developments in electronics it has been possible to completely synthesize sound signals. One of the first synthesizers was the Theremin which was a pair of high-frequency oscillators, one of which was tuned capacitively by moving the hand near an electrode. The beat frequency between the two oscillators was the audio output. Original Theremins are extremely rare and the most well-known recording to employ one is the Beach Boys' 'Good Vibrations'.

The Moog synthesizer was the first commercially available unit to electronically mimic the tone production and modification processes of a traditional instrument using oscillators and filters. Pioneering players such as Carlos and Tomita showed that synthesizers were not merely mimics of existing instruments but were instruments in their own right. Modern microelectronics has made it possible for synthesizers to be made at low cost. The size of the circuitry has shrunk so that it is all housed in the keyboard, which is what these units are now called. Synthesizers require some caution because they are capable of producing high levels at high frequencies far beyond the output of conventional instruments.

4.16 The human voice

The human voice is an extremely complex sound source because it is affected by so many factors. As Figure 4.21 shows, the primary power source is the expulsion of air from the lungs. The range of sounds which can be produced when inhaling is small and most people make no attempt to employ them. Air passing from the lungs to exit the nose and/or mouth must pass through the larynx which contains the vocal cords. These are one source of sound which is primarily periodic or *voiced*. If the vocal cords are not activated the throat can produce broadband or *unvoiced* sound in the lower part of the frequency range by a process called *glottal friction*. If the vocal cords are suddenly opened a pressure step is produced. This is known as a *glottal stop* and is common in London dialect as a replacement for the letter t.

Sounds produced in the larynx travel up the throat and are modified by the resonant cavities in the nose and mouth, both of which emit sound. This process is called *articulation*. Some sound is also radiated through vibration of the chest. The mouth resonances can be varied by the position of the tongue

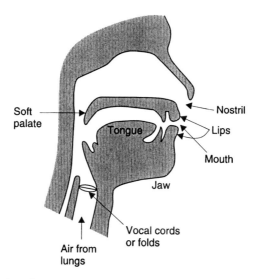

Figure 4.21 Production of the human voice.

which divides the mouth into two cavities. By almost blocking the airflow through the nose and mouth broadband noise can be produced in the upper part of the frequency range. With the use of the lips, tongue and teeth the flow of air can be started and stopped to produce transient sounds. For an excellent treatment of speech articulation the reader is recommended to consult Crystal.[3]

The vocal cords are two thin membranes which span the larynx. When placed under muscle tension they produce a slit-like opening which is unstable in the presence of airflow and which vibrates open and shut like a reed. The vibration produces a periodic pressure waveform which is roughly triangular. These voiced sounds are capable of sustained musical notes in singing and produce *vowels* in speech.

The fundamental frequency is determined by muscle action on the cords which can produce about a two-octave range. The muscles can change the tension in the cords or immobilize part of the cord. The centre frequency is determined by the size of the cords. In men they are about 25 mm long whereas in women they are about 15 mm long, accounting for the deeper voice of the male and the protuberance of the Adam's apple. As there is considerable variation in size from one individual to the next, singers are classified in one of six voices as in Figure 4.22. Classically trained singers can introduce frequency modulation into the vocal cords at a rate of 5.5–7.5 Hz. This is known as *vibrato* and causes the pitch to deviate by between 0.5 and 2 semitones.[4]

The triangular waveform from the vocal chords is rich in harmonics whose level falls at about 12 dB/octave. These harmonics are selectively reinforced by resonances, known as *formants*, in the mouth and nasal cavities. Clearly the mouth can be opened or closed and the nose can also be opened and closed by the soft palate. These two actions are quite independent. With the mouth open and the nasal cavity sealed by the soft palate, the shape of the mouth and the position of the tongue determines the timbre of the sound allowing a, e, i,

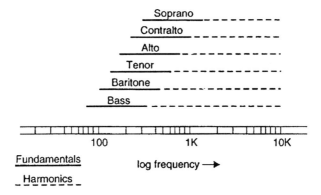

Figure 4.22 The vocal range is classified into six voices.

o, or u to be produced from the same fundamental. These sounds contain two or three formants; the definition of a vowel. With the mouth closed, the nose radiates sound and the nasal cavity and the mouth both act like low-pass filters. The position of the tongue determines the filtering action. If a sustained aaaa sound such as the a in father is produced and the mouth is closed or blocked the sound will change to mmmm. Pressing the tongue against the roof of the mouth changes this to nnnn. These sounds contain only one formant and so are not considered vowels.

Consonants are non-formant because the sounds are broadband. A broad spectrum can result from a sustained sound like a hiss or from a transient like a click. Some consonants result from the opening of the mouth and clearly cannot be sustained. When air pressure is built up in the mouth and released the result is a *plosive*. The p and b plosives are the audio engineer's nightmare as they can blow a microphone diaphragm all over the place and cause amplifier, tape or ADC overload. The p as in pot is formed by inflating the cheeks with unvoiced air and then suddenly releasing it by opening the lips. The b as in bus is formed by the same process except that the vocal cords are active while the cheeks inflate. The remaining plosives, k in kick, t as in top, d as in dog and g as in get, involve releasing pressure using the tongue against various parts of the mouth and teeth and the pressure step is much less pronounced.

Consonants which employ broadband sound such as f and s are called *fricatives*. They can be sustained and are formed by constricting the flow of air. The f as in fast is created by blowing air between the upper teeth and the lower lip whereas the s as in sat is created using upper and lower teeth in contact.

Normal conversation results in an average power of around 10 microWatts whereas shouting produces about a milliWatt. The spectrum of the voice changes with level and so it is not possible to simulate shouting using an electronic gain increase.

The singing voice at normal levels simply changes the fundamental frequency of the vocal cords to obtain the required notes but uses the same articulation and formant structure. Figure 4.23 shows that this results in an identical frequency envelope but filled by different harmonic structures. In this type of singing the intelligibility is as good as normal speech. However, when

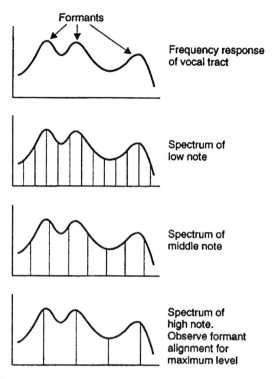

Formants

Frequency response
of vocal tract

Spectrum of
low note

Spectrum of
middle note

Spectrum of
high note.
Observe formant
alignment for
maximum level

Figure 4.23 In singing at moderate level the fundamental is changed by the vocal cords but the articulation and formant control remains the same.

an extremely high level is required, trained singers can slide the formants to frequencies which coincide with those from the vocal cords in order to produce a pronounced resonant peak. Sopranos are particularly good at this and can produce up to 120 dB corresponding to a peak power of around 1 Watt. The result of moving the formants is to reduce the intelligibility and the voice then has to be considered almost as a musical instrument.

The directivity of the human voice is remarkably wide and uniform sound can be picked up almost anywhere in a frontal hemisphere. The only one not to hear the human voice accurately is the speaker. Bone conduction causes the perceived timbre of one's own voice to be quite different from its actual timbre. Most people are quite surprised to hear recordings of their own voice and are, of course, incapable of judging the fidelity of such recordings.

Although the frequency range and dynamic range of the human voice is not very wide, enormous subtlety is present in the relative amplitude of the harmonic structure. A surprising number of defects in audio systems can be detected using speech. The majority of large studio monitor speakers give very poor results on female speech because of structural resonances which produce a timbral change. An audio system which can reproduce speech naturally is exceptional. On the other hand if it is only the spoken message which is important, speech remains intelligible after the waveform has been subject to any number of abuses.

4.17 Non-musical sound sources

Much of the soundtrack of television programmes and films is non-musical, yet such sounds are equally deserving of accurate reproduction. There is an infinity of non-musical sounds and the treatment here is necessarily selective. Forms of transport are commonplace noise sources which feature heavily in soundtracks and they will be considered in some detail.

The general principles of a noise source are not dissimilar to those of an instrument except that the noise is an accidental byproduct of a process rather than the sole requirement. There will be an excitation due to some vibration, a response due to the resonant characteristics, and further modification due to the acoustic environment.

4.18 The sound of a piston engine

The piston engine produces power by burning a fuel in air. The heat released expands the gases and the pressure pushes down on a piston which turns the crankshaft. The process is highly impulsive and the pressure changes are rapid, producing a wide harmonic structure.

Practical engines have several cylinders and a smoother delivery of power is achieved if the individual cylinders fire at different points around the rotation. The radial engine, used in aircraft, and the in-line engine can be designed to have precisely equal times between the firings by setting the crank pins at the appropriate angular positions on the crank shaft. In the four-stroke engine power is only delivered every two revolutions. In a four-cylinder, four-stroke engine two crank pins are at $0°$ and two are at $180°$. In a five-cylinder engine the pins are equally spaced $72°$ apart. In a six-cylinder engine there are two pins at $0°$, $120°$ and $240°$. The engine exhaust note frequency is determined by multiplying the crankshaft frequency by the number of cylinders and dividing by two.

In the vee engine, two cylinder blocks are placed at an angle to a common crankshaft. For economy, pairs of cylinders share the same crankpin. The result is that the firing of a vee engine is not uniform and the exhaust note has a characteristic irregularity as shown in Figure 4.24. The flat or boxer engine has horizontally opposed cylinders and fires regularly. However, the

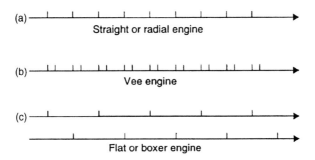

Figure 4.24 The sound of an engine is a function of the cylinder layout. The conventional in-line engine (a) produces evenly spaced exhaust pulses, whereas in the vee engine (b) they are irregular. The horizontally opposed engine (c) produces regular pulses but they appear on alternate sides of the engine.

two cylinder banks sometimes have separate exhaust systems which produce pulses alternately. This gives the VW 'beetle' its characteristic sound.

The opening of an exhaust valve produces a pressure step of hundreds of dB and some kind of silencing is a practical and a legal requirement. The silencer is an acoustic low-pass filter (Figure 4.25) which attempts to remove the higher harmonics from the step. An effective silencer can be quite large and expensive and may reduce the engine power by causing back pressure. Consequently small engines seldom have effective silencers, with lawnmowers and motorcycles being quite poor in this respect. Lawnmowers are noisy because of penny pinching, whereas many motorcyclists appear to *want* noisy machines as the noise level gives an illusion of power. One US pressure group has even gone so far as to suggest that noisy motorcycles are safer because people can hear them coming!

For racing purposes the exhaust pipe of an engine can be made to resonate so that a rarefaction occurs just as the exhaust valve opens, helping the exhaust gases to leave the engine. Clearly the length of the tuned pipes is critical. The sharply tuned engine will produce its maximum power at one speed and will need many gearbox ratios in order to use the greater power.

In addition to the exhaust note, engines produce other noises due to the air intake, vibrations of the structure, the valve mechanism and the cooling fan. Water-cooled engines are generally quieter than air-cooled engines. Not only is the water jacket an effective vibration damper, but the heat exchanger is very efficient, needing only a moderate airflow so the fan makes less noise. The air-cooled engine is less well damped and requires a powerful fan which is noisier.

The noise from a fan is often dominated by a discrete tone at the blade-passing frequency. This can be avoided by fitting the blades irregularly so that a broadband noise results.

The diesel engine compresses air only to increase its temperature. When the fuel is injected into the cylinder it burns instantly and produces a severe pressure step known as diesel knock. The shock waves travel around the engine and create a rattling noise.

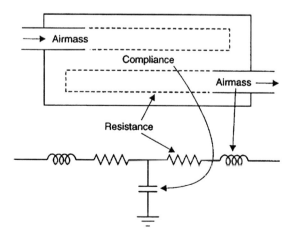

Figure 4.25 An engine silencer is an acoustic low-pass filter.

Both diesel and petrol engines can be turbocharged. A small turbine is driven by the exhaust gases and used to drive a compressor which packs more air into the cylinder. An intake silencer is needed to attenuate the blade-passing whine of the compressor. In certain diesel railway locomotives the turbocharger whine makes more noise than the rest of the engine as intake silencing is inadequate.

4.19 Vehicle noise

Road vehicles produce noise not only from the engine, but from the structure, the tyres and airflow. These all change as different functions of speed so the dominant sound will vary with speed.

Tyres have a tread in order to drain water from the contact patch. On wet roads the tyres produce broadband noise due to the squeezing of water into the tread and the subsequent shedding. The tread blocks are flexible and under braking or cornering forces they deflect and spring back when lifted from the road. This causes tyre squeal which only occurs in dry conditions on smooth surfaces. It is amusing to watch movies where the sound effects are put on so mechanically that the cars always squeal to a stop even when on gravel or wet roads! When all of the tread blocks are the same size the tyre sound is periodic. Military Land Rovers have a characteristic tone due to this effect. Some tyres have a pseudo-random variation in block size so that the tyre noise does not contain any dominant tones.

In forward motion, most modern vehicles have virtually silent transmissions because skew gears are employed which give absolutely smooth rotation. However, for economy, most vehicles use straight-cut reverse gears. These produce a modulation of the output shaft speed at the tooth-passing frequency. This excites the gearbox structure to give a pronounced whine.

Commercial vehicles often display a repertoire of sound effects. Structural rigidity is often lacking and driving over a bump will excite all manner of resonances. This is particularly noticeable on furniture vans which have large thin flat surfaces, and tipper trucks which are torsionally weak because the hinged-tipping body does not stiffen the chassis. Springs which are appropriate for full load are far too stiff for an unladen vehicle and act as if they were rigid. The result is tyre hop under braking and more body noise on bumps. Air brakes are often used and the exhausting air receives only token silencing. The brake mechanisms are often insufficiently rigid and squealing is quite common.

4.20 Aircraft noise

Aircraft come in many varieties and the range of sounds produced is equally wide. There are three contrasting types as far as sound radiation is concerned, piston engined, turboprop and turbojet. In piston-engined aircraft the level of silencing is quite poor and the engine exhaust noise masks virtually everything else. The geometry of the engine determines the timbre of the exhaust. Many light aircraft have flat four or six cylinder, air-cooled engines which produce a broad spectrum dominated by the exhaust periodicity. The classic Spitfire has a V-12 water-cooled engine. The vee construction produces an irregularity in the exhaust note and the water cooling damps the higher frequencies to produce an unmistakable sound which has become a legend.

The turboprop consists of a jet engine having a power turbine which turns the jet energy into shaft power to drive the propeller. The exhaust is relatively quiet because most of the energy has been removed by the power turbine. In contrast the compressor at the front of the engine turns at tens of thousands of rpm producing a high-level, high-frequency intake whine which is broadband noise containing a peak at the compressor blade-passing frequency. The intake duct is lined with acoustic absorbent to help attenuate the whine which is radiated directly forwards. In the turboprop the rotating propeller chops the intake whine to produce a high-frequency sound which is amplitude modulated by the propeller blade-passing frequency. When taxiing, reflections from the hard runway combine with the direct sound to produce comb filtering which has a phasing effect as the geometry changes. Vortices from other aircraft and wind gusts act as variable acoustic refractors so that the sound at the observer is amplitude modulated.

The turbojet engine produces thrust by accelerating gases backwards. The exhaust emerges at high velocity and there is a serious noise problem. To give an idea of the magnitude of the problem a powerful turbojet passes several tons of air a second. From the front the dominant sound is compressor whine whereas from the rear the exhaust noise dominates. Thus there is a distinct timbral change as the plane flies over. Military aircraft and Concorde use afterburning, a process which injects fuel into the exhaust to produce phenomenal amounts of flame, thrust and noise. The noise sounds like the world's largest blowlamp, which is in effect the case.

The pure turbojet is only needed for very high speeds and most airliners use the more efficient turbofan which is a cross between a turbojet and a turboprop. There is a power turbine which drives a large ducted fan which produces much of the thrust. The turbofan of the 747 is several metres across and turns quite slowly, consequently the intake noise of an approaching jumbo is more of a drone than a whine.

In aircraft with more than one engine the various noise sources interact. In some aircraft the engines are free running and turn at slightly different speeds. This results in beats at the frequency difference. In propeller-driven planes this can be very distracting and usually synchronizers are fitted which eliminate the beating.

Supersonic aircraft produce no audible warning of their approach. In subsonic aircraft the air moves out of the way because an area of high pressure precedes the craft. This cannot happen in a supersonic craft. Consequently the air is moved out of the way by brute force and the result is a shock wave which sweeps back from the nose like a bow wave. At the rear of the craft a vacuum is left after its passage and the air rushes to fill it causing another shock wave. The pressure with respect to the craft is relatively constant at a given place, but on the ground the pressure profile between the two shock waves sweeps past the listener. The result is known as an N-wave because of its waveform. Another way of considering the origin of the N wave is the mechanism shown in Section 3.4 for extremely high amplitude sounds.

4.21 Helicopter noise

The helicopter is an extremely complex sound source because significant parts of it are rotating. The helicopter rotor blades are long and thin and act as line

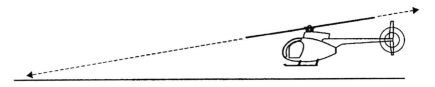

Figure 4.26 Forward-flying helicopter tilts the rotor disk which directs blade noise to the ground ahead.

sources of noise. The noise is broadband and is due to the airflow past them. Line sources are extremely directional and the blades radiate strongly at right angles to their length but very little axially. In the plane of the rotor the sound is emitted in rotating beams rather like those of a lighthouse. To the stationary listener in the plane of the rotor the beams of sound sweep past at the blade-passing frequency to create the characteristic chopping sound.

Figure 4.26 shows that a helicopter has to tilt the rotor forwards in order to fly along. This puts listeners on the ground in the plane of the blades, so the first warning of the approach of a helicopter is the chopping sound. However, as the machine passes overhead the listener moves onto the blade axis where the chopping effect disappears and the noise of the engines will be heard. The tail rotor makes more noise than the main rotor because it is turning much faster and it is in a confused airflow. The noise from the tail is emitted laterally and so will not be heard as the machine approaches but it may dominate the sound as it passes.

Helicopters have to compensate for the effect of forward speed adding to and subtracting from the speed of the blades as they turn. If nothing is done the retreating blade produces less lift than the advancing blade and this generates a rolling couple. However, the rotor system acts like a huge gyroscope and there is a $90°$ phase lag in the effect which actually pitches the machine back having the effect of slowing it down. In order to maintain speed the angle of the blades has to alter sinusoidally as they rotate; the angle of the retreating blade becomes greater and that of the advancing blade becomes less.

The retreating blade has slower relative airflow and a greater pitch angle and so it is more likely to stall. However, the period of time for which the maximum pitch is attained is quite small and in practice the pitch can be increased briefly beyond the normal stall limit. This results in what is called dynamic lift. The pressure differential across the blade can be very high and a powerful pulse of low-frequency sound is emitted. This is heard on the ground as blade barking which is normally only audible during tight manoeuvres.

References

1. Fletcher, N.H. and Rossing, T.D., *The Physics of Musical Instruments*, New York: Springer Verlag (1991)
2. Sumner, W.L. *The Organ* (5th ed.), London: Macdonald and Company (1975)
3. Crystal, D., *The Cambridge Encyclopaedia of Language*, Cambridge: Cambridge University Press (1987)
4. Sundberg, J., *The Science of the Singing Voice*, DeKalb: Northern Illinois University Press (1987)

Chapter 5

Microphones

In this chapter the principles and practice of microphones are considered. The polar or directional characteristics of a microphone are most important and are treated here before consideration of the operating principle. Polar characteristics assume even greater importance if the illusion of stereophony is to be made realistic. The microphone is a measuring device and its output consists of *intelligence* rather than power. Consequently it is possible to conceive of an ideal microphone and the best practice approaches this quite closely.

5.1 Introduction

We use electrical signalling to carry audio because it allows ease of processing and simple transmission over great distances using cable or radio. It is essential to have transducers such as microphones and loudspeakers which can convert between real sound and an electrical equivalent. Figure 5.1(a) shows that even if the electrical system is ideal, the overall quality of a sound system is limited by the quality of both microphone and loudspeaker. Figure 5.1(b) shows that in a broadcast sound system or when selling prerecorded material, the quality of the final loudspeaker is variable, dependent upon what the consumer can afford. However, it must be assumed that at least a number of consumers will have high-quality systems. The quality of the sound should exceed or at least equal that of the consumer's equipment even after all of the recording, production and distribution processes have been carried out. Consequently the microphone used in production must be of high quality. The loudspeakers used for monitoring the production process must also be of high quality so that any defect in the microphone or elsewhere can be identified.

5.2 Microphone principles

The job of the microphone is to convert sound into an electrical signal. As was seen in Chapter 3, sound consists of both pressure and velocity variations and microphones can use either or both in order to obtain various directional characteristics.[1]

Figure 5.2(a) shows a true pressure microphone which consists of a diaphragm stretched across an otherwise sealed chamber. In practice a small pinhole is provided to allow changes in atmospheric pressure to take place

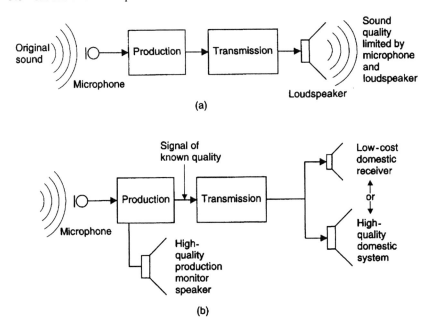

Figure 5.1 (a) The final sound quality of an audio system is limited by both microphones and loudspeakers. (b) Sound production must be performed using high-quality loudspeakers on the assumption that the transmitted quality should be limited only by the listener's equipment.

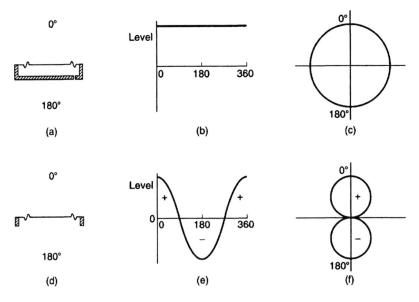

Figure 5.2 (a) Pressure microphone only allows sound to reach one side of the diaphragm. (b) Pressure microphone is omnidirectional for small ka. (c) Directional characteristic is more intuitive when displayed in polar form. (d) Velocity or pressure gradient microphone exposes both sides of diaphragm. (e) Output of velocity microphone is a sinusoidal function of direction. (f) In polar coordinates velocity microphone shows characteristic figure-of-eight shape for small ka.

without causing damage. Some means is provided to sense the diaphragm motion and convert it into an electrical output signal. This can be done in several ways which will be considered in Section 5.4. The output of such a microphone for small values of *ka* is completely independent of direction as (b) shows.

Unlike human hearing, which is selective, microphones reproduce every sound which reaches them. Figure 5.3(a) shows the result of placing a microphone near to a hard wall. The microphone receives a combination of direct and reflected sound between which there is a path-length difference. At frequencies where this amounts to a multiple of a wavelength, the reflection will reinforce the direct sound, but at intermediate frequencies cancellation will occur, giving a comb-filtering effect. Clearly a conventional microphone should not be positioned near a reflecting object.

The path-length difference is zero at the wall itself. The pressure zone microphone (PZM) of (b) is designed to be placed on flat surfaces where it will not suffer from reflections. A pressure capsule is placed facing and parallel to a flat surface at a distance which is small compared to the shortest wavelength of interest. The acoustic impedance rises at a boundary because only half-space can be seen and the output of a PZM is beneficially doubled.

Figure 5.2(d) shows the *pressure gradient* (PG) microphone in which the diaphragm is suspended in free air from a symmetrical perimeter frame. The maximum excursion of the diaphragm will occur when it faces squarely across the incident sound. As Figure 5.2(e) shows, the output will fall as the sound moves away from this axis, reaching a null at 90°. If the diaphragm were truly weightless then it would follow the variations in air velocity perfectly, hence the term *velocity microphone*. However, as the diaphragm has finite mass then

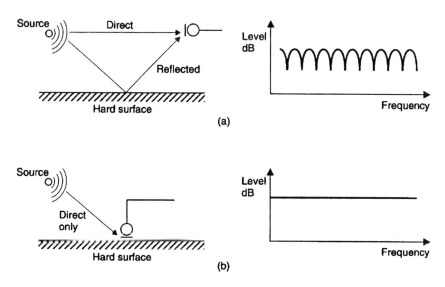

Figure 5.3 (a) Microphone placed several wavelengths from reflective object suffers comb filtering due to path-length difference. (b) Pressure zone microphone is designed to be placed at a boundary where there is no path-length difference.

a pressure difference is required to make it move, hence the more accurate term, pressure gradient microphone.

The pressure gradient microphone works by sampling the passing sound wave at two places separated by the front-to-back distance. Figure 5.4 shows that the pressure difference rises with frequency as the front-to-back distance becomes a greater part of the cycle. The force on the diaphragm rises at 6 dB/octave. Eventually the distance exceeds half the wavelength at the critical frequency where the pressure gradient effect falls rapidly. Fortunately the rear of the diaphragm will be starting to experience shading at this frequency so that the drive is only from the front. This has the beneficial effect of transferring to pressure operation so that the loss of output is not as severe as the figure suggests. The pressure gradient signal is in phase with the particle displacement and is in quadrature with the particle velocity.

In practice the directional characteristics shown in Figure 5.2(b) and (e) are redrawn in *polar coordinates* such that the magnitude of the response of the microphone corresponds to the distance from the centre point at any angle. The pressure microphone (c) has a circular polar diagram as it is omnidirectional or *omni* for short. Omni microphones are good at picking up ambience and reverberation which makes them attractive for music and sound-effects recordings in good locations. In acoustically poor locations they cannot be used because they are unable to discriminate between wanted and unwanted sound. Directional microphones are used instead.

The PG microphone has a polar diagram (f) which is the shape of a figure-of-eight. Note the null at 90° and that the polarity of the output reverses beyond 90° giving rise to the term *dipole*. The figure-of-eight microphone (sometimes just called an *eight*) responds in two directions giving a degree of ambience pickup, although the sound will be a little drier than that of an

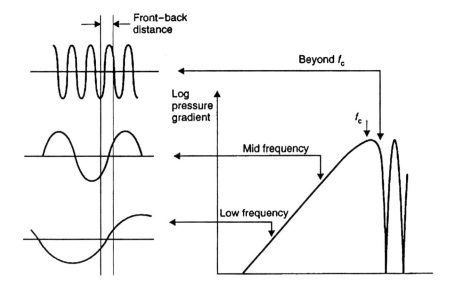

Figure 5.4 The pressure gradient microphone diaphragm experiences a pressure difference which rises with frequency up to the critical frequency, f_c.

omni. A great advantage of the figure-of-eight microphone over the omni is that it can reject an unwanted sound. Rather than point the microphone at the wanted sound, a better result will be obtained by pointing the null or dip in the polar diagram at the source of the unwanted sound.

Unfortunately the PG microphone cannot distinguish between gradients due to sound and those due to gusts of wind. Consequently PG microphones are more sensitive to wind noise than omnis.

If omni and eight characteristics are combined, various useful results can be obtained. Figure 5.5(a) shows that if the omni and eight characteristics are added equally, the result is a heart-shaped polar diagram called a *cardioid*. This response is obtained because at the back of the eight the output is anti-phase and has to be subtracted from the output of the omni. With equal signals this results in a null at the rear and a doubling at the front. This useful polar response will naturally sound drier than an eight, but will have the advantage of rejecting more unwanted sound under poor conditions. In public address applications, use of a cardioid will help to prevent *feedback* or *howl-round* which occurs when the microphone picks up too much of the signal from the loudspeakers. Virtually all hand-held microphones have a cardioid response where the major lobe faces axially so that the microphone is pointed

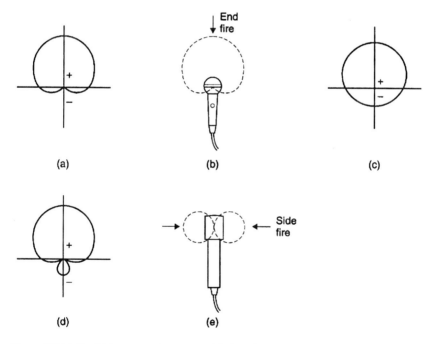

Figure 5.5 (a) Combining an omni response with that of an eight in equal amounts produces the useful cardioid directivity pattern. (b) Hand-held fixed cardioid response microphones are usually built in the end-fire configuration where the body is placed in the null. (c) Sub-cardioid obtained by having more omni in the mix gives better ambience pickup than cardioid. (d) Hyper-cardioid obtained by having more eight in the mix is more directional than cardioid but the presence of the rear lobe must be considered in practice. (e) Microphones with variable polar diagram are generally built in the side-fire configuration.

at the sound source. This is known as an *end-fire* configuration shown in Figure 5.5(b).

Where a fixed cardioid-only response is required, this can be obtained using a single diaphragm where the chamber behind it is not sealed, but open to the air via an acoustic labyrinth. Figure 5.6(a) shows that the asymmetry of the labyrinth means that sound which is incident from the front reaches the rear of the diaphragm after a path difference allowing pressure gradient operation. Sound from the rear arrives at both sides of the diaphragm simultaneously, nulling the pressure gradient effect. Sound incident at 90° experiences half the path-length difference, giving a reduced output in comparison with the on-axis case. The overall response has a cardioid polar diagram. This approach is almost universal in hand-held cardioid microphones.

In variable directivity microphones there are two such cardioid mechanisms facing in opposite directions as shown in (b). The system was first devised by the Neumann company.[2] The central baffle block contains a pattern of tiny holes, some of which are drilled right through and some of which are blind. The blind holes increase the volume behind the diaphragms, reducing the resonant frequency in pressure operation when the diaphragms move in anti-phase. The holes add damping because the viscosity of air is significant in such small cross-sections.

The through-drilled holes allow the two diaphragms to move in tandem so that pressure gradient operation is allowed along with further damping. Sound incident from one side (c) acts on the *outside* of the diaphragm on that side directly, but passes *through* the other diaphragm and then through the cross-drilled holes to act on the *inside* of the first diaphragm. The path-length difference creates the pressure gradient condition. Sound from the 'wrong' side (d) arrives at both sides of the far diaphragm without a path-length difference.

The relative polarity and amplitude of signals from the two diaphragms can be varied by a control. By disabling one or other signal, a cardioid response can

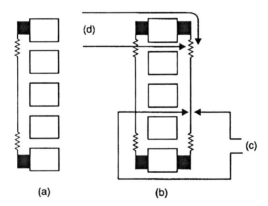

Figure 5.6 (a) Fixed cardioid response is obtained with a labyrinth delaying sound reaching the rear of the diaphragm. (b) Double cardioid capsule is the basis of the variable directivity microphone. (c) Sound arriving from the same side experiences a path-length difference to create a pressure gradient. (d) Sound arriving from the opposite side sees no path-length difference and fails to excite diaphragm.

be obtained. Combining them equally results in an omni, whereas combining them with an inversion results in a figure-of-eight response. Unequal combination can obtain the sub-cardioid shown in Figure 5.5(c) or a hyper-cardioid shown in Figure 5.5(d).

Where a flexible polar response is required, the end-fire configuration cannot be used as the microphone body would then block the rearward access to the diaphragm. The *side-fire* configuration is shown in Figure 5.5(e) where the microphone is positioned *across* the approaching sound, usually in a vertical position. For television applications where the microphone has to be out of shot such microphones are often slung from above pointing vertically downwards.

In most applications the polar diagrams noted above are adequate, but on occasions it proves to be quite impossible to approach the subject close enough and then a highly directional microphone is needed. Picking up ball sounds in sport is one application. Figure 5.7 shows the use of a conventional cardioid microphone fitted with a parabolic reflector. Only wavefronts arriving directly on-axis are focused on the microphone which is rendered highly directional. Parabolic units are popular for recording wildlife, but they are large and clumsy.

Figure 5.8(a) shows that the *shotgun* microphone consists of a conventional microphone capsule which is mounted at one end of a slotted tube. Sound wavefronts approaching from an angle will be diffracted by the slots such that each slot becomes a re-radiator launching sound down the inside of the tube. However, Figure 5.8(b) shows that the radiation from the slots travelling down the tube will not add coherently and will be largely cancelled. A wavefront approaching directly on axis as in Figure 5.8(c) will pass directly down the outside and the inside of the tube as if the tube were not there and consequently will give a maximum output.

Contact microphones are designed to sense the vibrations of the body of a musical instrument or the throat of a speaking human to produce an electrical signal. In fact these are not microphones at all but are *accelerometers*. Figure 5.9 shows that in an accelerometer a mass is supported on a compliance. When the base of the accelerometer is vibrated, the mass tends to lag

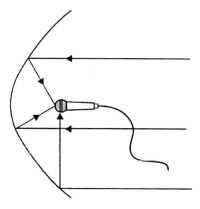

Figure 5.7 Parabolic reflector with cardioid microphone is very directional but fails at low frequencies.

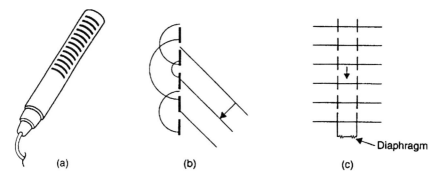

Figure 5.8 (a) Shotgun microphone has slotted tube. (b) Off-axis sound enters slots to produce multiple incoherent sources which cancel. (c) On-axis sound is unaware of tube.

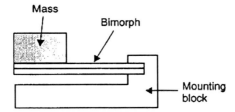

Figure 5.9 Contact microphone is actually an accelerometer. Suspended mass lags body vibration actuating sensor.

behind, causing a relative movement. Any convenient transducing mechanism can be used to sense the movement.

5.3 Microphone limitations

The directivity patterns given in Section 5.2 are only true where ka is small and are thus ideal. In practice at high frequencies, ka will not be small and the actual polar diagram will differ due to diffraction becoming significant.

Figure 5.10(a) shows the result of a high-frequency sound arriving off-axis at a large diaphragm. It will be clear that at different parts of the diaphragm the sound has a different phase and that in an extreme case cancellation will occur, reducing the output significantly.

When the sound is even further off-axis, shading will occur. Consequently at high frequency the polar diagram of a nominally omni microphone may look something like that shown in Figure 5.10(b). The high-frequency polar diagram of an eight may resemble Figure 5.10(c). Note the narrowing of the response such that proper reproduction of high frequencies is only achieved when the source is close to the axis.

The frequency response of a microphone should ideally be flat and this is often tested on-axis in anechoic conditions. However, in practical use the surroundings will often be reverberant and this will change the response at high frequencies because the directivity is not independent of

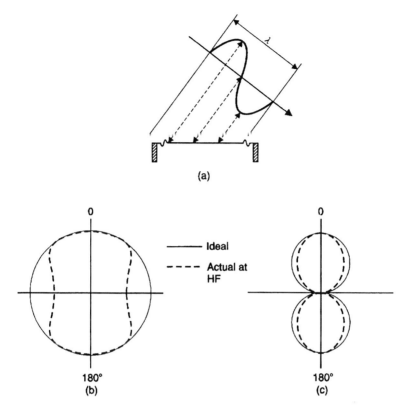

Figure 5.10 (a) Off-axis response is impaired when *ka* is not small because the wavefront reaches different parts of diaphragm at different times causing an aperture effect. (b) Polar diagram of practical omni microphone at high frequency shows narrowing of frontal response due to aperture effect and rear loss due to shading. (c) Practical eight microphone has narrowing response at high frequency.

frequency. Consequently a microphone which is flat on axis but which has a directivity pattern which narrows with frequency will sound dull in practical use. Conversely a microphone which has been equalized flat in reverberant surroundings may appear too bright to an on-axis source.

Pressure microphones being omnidirectional have the most difficulty in this respect because shading makes it almost impossible to maintain the omnidirectional response at high frequencies. Clearly an omni based on two opposing cardioids will be better at high frequencies than a single pressure capsule.

It is possible to reduce *ka* by making the microphone diaphragm smaller but this results in smaller signals making low noise difficult to achieve. However, developments in low-noise circuitry will allow diaphragm size beneficially to be reduced.

In the case of the shotgun microphone, the tube will become acoustically small at low frequencies and will become ineffective causing the polar diagram to widen. If high directivity is required down to low frequencies the assembly must be made extremely long. The parabolic reflector microphone has the same characteristics. As such microphones are not normally used for full frequency range sources the addition of a high-pass filter is beneficial as it removes low frequencies without affecting the quality of, for example, speech.

The electrical output from a microphone can suffer from distortion with very loud signals or from noise with very quiet signals. In passive microphones distortion will be due to the linear travel of the diaphragm being exceeded whereas in active microphones there is the additional possibility of the circuitry being unable to handle large amplitude signals. Generally a maximum SPL will be quoted at which a microphone produces more than 0.5% THD.

Noise will be due to thermal effects in the transducer itself and in the circuitry. Microphone noise is generally quoted as the SPL which would produce the same level as the noise. The figure is usually derived for the noise after A weighting (see Section 3.4). The difference between the 0.5% distortion SPL and the self-noise SPL is the dynamic range of the microphone. 110 dB is considered good but some units reach an exceptional 130 dB.

In addition to thermal noise, microphones may also pick up unwanted signals and hum fields from video monitors, lighting dimmers and radio transmissions. Considering the low signal levels involved, microphones have to be well designed to reject this kind of interference. The use of metal bodies and grilles is common to provide good RF screening.

The output voltage for a given SPL is called the sensitivity. The specification of sensitivity is subject to as much variation as the mounting screw thread. Some data sheets quote output voltage for 1 Pa, some for 0.1 Pa. Sometimes the output level is quoted in dBV, sometimes dBu (see Section 2.7). The outcome is that in practice preamplifier manufacturers provide a phenomenal range of gain on microphone inputs and the user simply turns up the gain to get a reasonable level.

It should be noted that in reverberant conditions pressure and pressure gradient microphones can give widely differing results. For example, where standing waves are encountered, a pressure microphone positioned at a pressure node would give an increased output whereas a pressure gradient microphone in the same place would give a reduced output. The effect plays havoc with the polar diagram of a cardioid microphone.

The proximity effect should always be considered when placing microphones. As explained in Section 3.16, proximity effect causes an emphasis of low frequencies when a PG microphone is too close to a source. A true PG microphone such as an eight will suffer the most, whereas a cardioid will have 6 dB less proximity effect because half of the signal comes from an omni response.

5.4 Microphone mechanisms

There are two basic mechanisms upon which microphone operation is based: electrostatic and electrodynamic. The electrodynamic microphone operates on the principle that a conductor moving through a magnetic field will generate a voltage proportional to the rate of change of flux. As the magnetic flux is constant, then this results in an output proportional to the velocity of the conductor, i.e. it is a *constant velocity* transducer. The most common type has a moving coil driven by a diaphragm. In the ribbon microphone the diaphragm itself is the conductor.

The electrostatic microphone works on the variation in capacitance between a moving diaphragm and a fixed plate. As the capacitance varies directly with the spacing the electrostatic microphone is a *constant amplitude* transducer. There are two forms of electrostatic microphone: the condenser, or capacitor, microphone and the electret microphone.

The ribbon and electrostatic microphones have the advantage that there is extremely direct coupling between the sound waveform and the electrical output and so very high quality can be achieved. The moving coil microphone is considered to be of lower quality but is cheaper and more robust.

Microphones can be made using other techniques but none can offer high quality and use is restricted to consumer or communications purposes or where something unbreakable is required.

Figure 5.11(a) shows that the vibrations of a diaphragm can alter the pressure on a quantity of carbon granules, altering their resistance to current flow. This construction was used for telephone mouthpieces for many years. Whilst adequate for speech, the noise level and distortion are high.

Figure 5.11(b) shows the moving iron or variable reluctance microphone in which a ferrous diaphragm vibrates adjacent to a permanent magnet. The variation of the air gap causes the flux of the magnet to vary and this induces a signal in a coil.

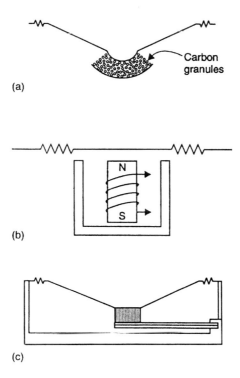

Figure 5.11 (a) Carbon microphone relies on variations of pressure affecting resistance of carbon granules. (b) Ferrous diaphragm of variable reluctance microphone changes flux as it moves. Coil responds to flux changes. (c) Piezo-electric microphone relies on voltage produced by bending a crystal.

Figure 5.11(c) shows a piezo-electric or crystal microphone in which the diaphragm vibrations deform a bimorph crystal of Rochelle salt or barium titanate. This material produces a voltage when under strain. A high input impedance is required.

5.5 Electrodynamic microphones

There are two basic implementations of the electrodynamic microphone; the ribbon and the moving coil. Figure 5.12(a) shows that in the ribbon microphone the diaphragm is a very light metallic foil which is suspended under light tension between the poles of a powerful magnet. The incident sound causes the diaphragm to move and the velocity of the motion results in an EMF being generated across the ends of the ribbon.

The most common form which the ribbon microphone takes is the figure-of-eight response although cardioid and pressure variants are possible using the techniques outlined above. The output voltage of the ribbon is very small but the source impedance of the ribbon is also very low and so it is possible

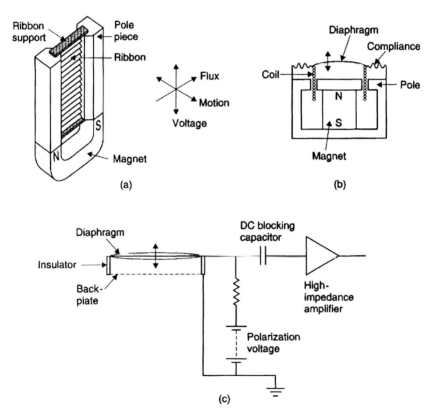

Figure 5.12 (a) Now obsolete ribbon microphone suspends conductive diaphragm in magnetic field. Low output impedance requires matching transformer. (b) The moving coil microphone is robust but indirect coupling impairs quality. (c) Capacitor microphone has very tight coupling but requires high-impedance electronics and needs to be kept dry.

to use a transformer to produce a higher output voltage at a more convenient impedance.

As the driving force on a pressure gradient transducer is proportional to frequency, the resonant frequency of the ribbon is set below the audio band at only a few Hertz. Consequently the ribbon works in the mass-controlled region (see Section 5.8) where for constant drive force the velocity falls at 6 dB/octave. This balances the 6 dB/octave rise of the pressure gradient effect to produce a flat response. The very high compliance needed to set the resonance below the audio band means that the ribbon microphone is shock sensitive. If the body of the microphone is moved, the diaphragm will lag behind causing relative motion. Good mounting isolation is required.

The advantage of the ribbon microphone is that the motion of the ribbon is directly converted to an electrical signal. This is potentially very accurate. However, unless the transformer is of extremely high quality the inherent accuracy will be lost. A further problem is that to obtain reasonable sensitivity the diaphragm must be relatively large giving a low cut-off frequency and leading to the directivity problems mentioned in Section 5.3. Traditionally the magnet was also large, leading to a heavy construction. A further problem is that the diaphragm is extremely delicate and a single exposure to wind might destroy it.

Although the ribbon microphone was at one time the best available it has been overshadowed by the capacitor microphone and is little used today. The ribbon principle deserves to be revisited with modern techniques. A smaller diaphragm and a physically smaller rare-earth magnet would push up the cut-off frequency by reducing the path-length difference. The smaller output could be offset by incorporating an amplifier in an active design.

The most common version of the electrodynamic microphone is the moving coil system shown in Figure 5.12(b). The diaphragm is connected to a cylindrical former upon which is wound a light coil of wire. The coil operates in the radial flux pattern of a cylindrical magnet. As the output is proportional to velocity, a moving coil pressure microphone has to work in the resistance-controlled domain (see Section 2.8) using a mid-band resonance which is heavily damped. The range is often extended by building in additional damped resonances. A moving coil pressure gradient microphone would need to operate in mass control.

As it is possible to wind many turns of wire on the coil, the output of such a microphone is relatively high. The structure is quite robust and can easily withstand wind and handling abuse. However, the indirect conversion, whereby the sound moves the diaphragm and the diaphragm moves the coil, gives impaired performance because the coil increases the moving mass and the mechanical coupling between the coil and diaphragm is never ideal. Consequently the moving coil microphone, generally in a cardioid response form, finds common application in outdoor use, for speech or for public address, but is considered inadequate for accurate music work.

5.6 Electrostatic microphones

In the capacitor (or condenser) microphone the diaphragm is highly insulated from the body of the microphone and is fed with a high polarizing voltage via a large resistance. Figure 5.12(c) shows that a fixed metallic grid forms a capacitor in conjunction with the diaphragm. The diaphragm is connected to an amplifier having a very high impedance. The high impedances mean that there is essentially a constant charge condition. Consequently when incident sound moves the diaphragm and the capacitance between it and the grid varies, the result will be a change of diaphragm voltage which can be amplified to produce an output.

As the condenser mechanism has a constant amplitude characteristic, a pressure or omni condenser microphone needs to use *stiffness control* to obtain a flat response. The resonant frequency is placed above the audio band. In a PG condenser microphone *resistance control* has to be used where a well-damped mid-band resonant frequency is used.

The condenser microphone requires active circuitry close to the capsule and this requires a source of DC power. This is often provided using the same wires as the audio output using the principle of *phantom powering* described in Section 5.7.

If the impedance seen by the condenser is not extremely high, charge can leak away when the diaphragm moves. This will result in poor output and phase shift at low frequencies. As the condenser microphone requires high impedances to work properly, it is at a disadvantage in damp conditions which means that in practice it has to be kept indoors in all but the most favourable weather. Some condenser microphones contain a heating element which is designed to drive out moisture. In older designs based on vacuum tubes, the heat from the tube would serve the same purpose. If a capacitor microphone has become damp, it may fail completely or create a great deal of output noise until it has dried out.

In the electret microphone a material is employed which can produce a constant electric field without power. It is the electrostatic equivalent of a permanent magnet. An electret is an extremely good insulator which has been heated in an intense electric field. A conductor moving in such a field will produce a high impedance output which will usually require to be locally amplified. The electret microphone can be made very light because no magnet is required. This is useful for hand-held and miniature designs.

In early designs the diaphragm itself was the polarized element, but this required an excessively heavy diaphragm. Later designs use a conductive diaphragm like that of a capacitor microphone and the polarized element is part of the backplate. These back-polarized designs usually offer a better frequency response. Whilst phantom power can be used, electret microphones are often powered by a small dry cell incorporated into the microphone body.

In variable directivity condenser microphones the double cardioid principle of Figure 5.6 is often used. However, the variable mixing of the two signals is achieved by changing the polarity and magnitude of the polarizing voltage on one diaphragm such that the diaphragm audio outputs can simply be added. Figure 5.13 shows that one diaphragm is permanently polarized with a positive voltage, whereas the other can be polarized with a range of voltages from positive, through zero, to negative.

When the polarizing voltages are the same, the microphone becomes omni, whereas if they are opposing it becomes an eight. Setting the variable voltage to zero allows the remaining diaphragm to function as a cardioid. A fixed set of patterns may be provided using a switch, or a continuously variable potentiometer may be supplied. Whilst this would appear to be more flexible it has

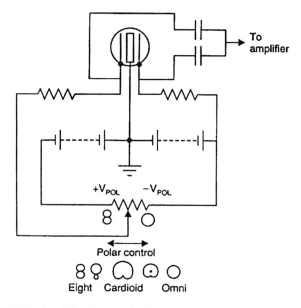

Figure 5.13 Variable directivity electrostatic microphone varies polarizing voltage and polarity on one diaphragm to change directivity.

the disadvantage of being less repeatable. It will be clear that this approach cannot be used with the electret microphone which usually has fixed directivity. The pressure zone microphone is an ideal application for the electret principle.

5.7 Phantom power

Balanced audio cabling and the star-quad technique were described in Section 2.8. Where active microphones (those containing powered circuitry) are used, it is common to provide the power using the balanced audio cable in reverse. Figure 5.14 shows how phantom powering works. As the audio signal is transmitted differentially, the DC power can be fed to the microphone without interfering with the returning audio signal. The cable screen forms the return path for the DC power. There is an alternative approach called A-B powering which does not use the screen as a return, but which requires DC-blocking capacitors.

The use of female connectors for audio inputs is because of phantom power. An audio input is a phantom power output and so requires the insulated contacts of the female connector. Many, but not all, passive microphones are wired such that their differential output is floating relative to the grounded case. These can be used with a cable carrying live phantom power. Provision is generally made to turn the phantom power off so that any type of passive or self-powered microphones can be used.

Some microphones contain valve amplifiers and these require heater and high tension (HT) supplies. This can only be provided by a special multi-core cable supplied with the microphone which has to be connected to a matching power supply unit. The audio output appears on sockets on the power unit.

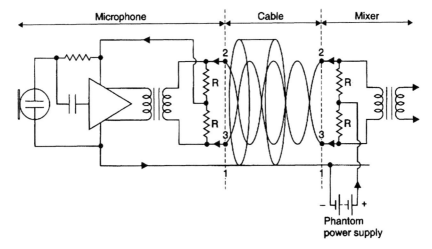

Figure 5.14 Phantom power system allows mixing console to power microphone down signal cable.

Variable directivity microphones may have the directivity control on the microphone body, or they may be provided with a control box which can be some distance away. This can offer practical advantages when a microphone is slung out of reach. A dedicated cable will be supplied which runs between the microphone and the control box. Some control boxes convert standard phantom power to suit the microphone and control circuitry, some have a compartment for a dry battery and some require an AC supply.

Microphones are frequently handled and the cable is pulled around by the microphone. The connector needs to be robust and provide suitable support and anchorage for the cable. Many microphone bodies have a standard XLR connector, but this is too large for some designs. Some manufacturers have unique connectors, some use a locking DIN connector made to a very high standard. Smaller microphones may have no connector at all but are provided with a captive cable having a line socket.

5.8 Stands and suspensions

Some microphones are physically large and heavy and hand use is not practicable. These types will incorporate some kind of mounting bracket or may have a very robust connector so that they can be supported by the plug. Most microphones are designed to be hand-held at least some of the time and this limits the body diameter. Generally no mounting brackets are fitted because these would be a nuisance when hand-holding and would detract from the appearance. Good microphones are expensive and manufacturers often give them a 'camera' finish to emphasize the quality. Mounting such microphones requires some kind of clamp fitted with a resilient lining to avoid damaging the finish. The clamp will usually have a female screw thread so that it can be installed on a stand.

Unfortunately there is no standard for the thread which is to be used and often a thread adaptor is needed. Thread adaptors usually come in two varieties: those which are always coming loose and those which are impossible to remove. Usually the type which is to hand is the opposite of what is needed.

Figure 5.15 shows a number of microphone stands. For desk mounting, a simple heavy base with a short upright and a spring-loaded rubber-jawed clamp is common. Alternatively a *gooseneck* can be used. This is a semi-flexible tube spiral wound from steel strip which can be formed to any convenient angle. Some goosenecks have internal wiring and connectors making for an elegant installation with a microphone designed to be supported by the connector.

The common microphone stand has a heavy base consisting of either a disk or several feet, with a vertical telescopically adjustable upright. These cannot be adjusted to suit certain musical instruments and people tend to strike the base with their feet. A non-vertical stand is useful for vocalists. More flexibility is available with a boom attachment which allows the microphone to be some distance horizontally from the base. The boom is more convenient for fine adjustments because it can be swung to and fro whereas the upright stand has to be moved bodily.

The cable should be firmly taped to the *base* of the stand. This will prevent the microphone being pulled from its clip should the cable be snagged. Fixing the cable at the base of the stand means the stand will be dragged along if someone trips over the cable. A higher fixing means the stand will tip over.

If any external vibrations reach a microphone they can cause relative movement between the diaphragm and the capsule body which will result in spurious output signals. Goosenecks do not absorb vibration and the rubber jaws of a normal clamp are only to protect the finish. Almost all microphones contain some flexible elements between the capsule and the body, but there is generally not much space for them and they are of limited effectiveness.

For high-quality work, many manufacturers supply an external flexible mount such as that shown in Figure 5.16. The microphone is clamped near its centre of mass to the inner part of the mount, and the outer part is fixed to whatever stand or support is being used. The two parts are joined by rubber cords

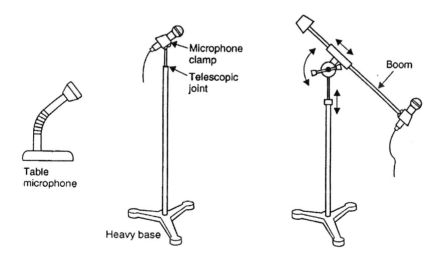

Figure 5.15 A range of microphone stands for various purposes.

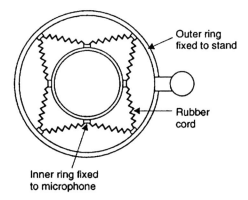

Figure 5.16 Rubber cord shock mount for microphones looks crude but is very effective.

which allow effective decoupling. It is important that the vibration isolation of the mount is not bypassed by a taut cable. The cable should be fixed to the stand and an isolation loop should be formed between the stand and the microphone.

For classical music recording it is often required to sling microphones high in the air. This is often achieved using thin high tensile wires to reduce the visual impact. With all installations of this kind it is essential that safety is considered at all times. A falling microphone could cause injury or damage to a valuable instrument. As an alternative to slinging, very large stands with booms are sometimes used. There is a danger of tipping with a heavy microphone on a long boom and the boom may be counterweighted. Sandbags or stage weights may be used on the feet of the stand to improve stability.

Sometimes resilient pads are fitted under the feet of the stand to prevent floor vibrations reaching the microphone. In this case it is important to adjust the counterweights so that the centre of mass of the boom/stand is in the centre so that an equal load is applied to each foot. If this is not done one pad will carry an excess load and will be flattened so that its isolation is impaired. There may also be a stability problem.

In film and television work the microphone will often have to be moved to follow the action. In a studio a dolly-mounted microphone crane or boom may be used. These devices vary in complexity, but in the most expensive kinds the dollies have rotating bodies so that they can travel in any direction and an arm which can slew, elevate and telescope without changing the orientation of the microphone. At the end of the arm the microphone can be panned and tilted. One operator will control the boom and pan/tilt and others will propel the dolly. There will be a comprehensive shock mount arrangement to prevent vibrations due to the operation of the boom from reaching the microphone.

Using a boom requires a good deal of skill as the operator has not only to estimate the field of view of the cameras to keep the microphone out of shot but also has to ensure that the boom does not cause shadows with the lighting being used.

As a alternative to a boom a hand-held *fishpole* can be used. This is a lightweight telescopic arm with a microphone mounted at a fixed angle. Only lightweight microphones can be used to reduce strain on the operator. Recently fishpoles made from aerospace composite materials such as carbon fibre have become available. These are much lighter and less tiring to use.

Miniature microphones are available which can be clipped onto clothing or hung around the neck on a light cord or *lavalier*.

5.9 Wind noise

The worst enemies of the microphone are wind and plosives from close speaking. Pressure gradient microphones are more sensitive than pressure microphones because pressure gradients can be caused when the body of the microphone obstructs the airflow. Wind generally causes turbulence as it flows around the microphone (see Section 4.17) and the amplitude can be extremely high. There is a distinct possibility of overloading the associated amplifiers. This is called *blasting* and the sound is actually interrupted as the amplifier recovers from driving its output voltage to the power rail. Turbulence is usually restricted to low frequencies and a high-pass filter is useful but such a filter will not prevent overloading of amplifier stages which precede it.

The only practical solution is to enclose the microphone in a specially made windshield. Turbulence will still take place, but it will be around the windshield instead. The windshield is generally cylindrical and consists of a metal or plastic grid supporting a porous material such as foam which damps the low-frequency noise due to turbulence but allows through the lower amplitudes in the audio spectrum. In extreme conditions the turbulence may be reduced by fitting the shield with an outside layer of long piled textile material. These windshields are often called 'Dougals' because of the resemblance to the puppet dog on the *Magic Roundabout* television programme.

The effect of plosives can be reduced with less drastic measures. Hand-held microphones can be fitted with a spherical foam ball or a light metal frame carrying a layer of silk or similar material can be placed in front of the microphone. In extreme cases the vocalist can be given a microphone to eat but the recording is actually made with a second microphone further away.

5.10 Radio microphones

In many applications the microphone cable is an unwanted constraint and microphones are used in conjunction with radio transmission. It is not easy to obtain high quality with radio microphones not least because of the potential for noise and interference in the radio link. In order to minimize such noise, techniques such as pre-emphasis and companding are used in the audio frequency domain and FM (see Section 1.1) is used in the radio frequency domain.

Designing a radio microphone is challenging because there are limits on size and weight yet complex processing at AF and RF is necessary. Power is limited because it has to be supplied from a battery. The electret principle is often used because it saves weight and takes up little space.

There are two approaches to radio microphone design. In the first, everything is packed into a single hand-held unit. This is an advantage where production constraints require the microphone to be passed from one user to the next. In the second approach the microphone itself is on a short lead and everything else is in a small box which can be pocketed or clipped to the wearer's belt. The second approach means that almost any microphone can be turned into a radio microphone and if the wiring to a miniature lapel microphone is routed inside the wearer's clothing a very discreet installation results.

Radio microphones have been built to work on a number of wavebands, including VHF and UHF. VHF units require a fairly long antenna which is usually a trailing wire. In UHF a short flexible rod is sufficient. The precise frequencies and licensing requirements vary from country to country and present a huge administrative problem for international artists. Clearly where several microphones are to be used in the same venue each must have to work on a different radio frequency.

There is very little control of the radio frequency environment and radio microphones are forced to use low power to allow reasonable battery life. The transmitter has to be omnidirectional because the wearer cannot be expected to aim a directional antenna at the receiver. Consequently multipath reception and shading are major problems and it is very difficult to avoid dead spots where there is so little signal that the reception is noisy. One solution is to use *diversity reception* in which two or more receiving antennae are used. When the signal picked up by one antenna fades the chances are that the other antenna is still receiving a strong signal.

References

1. *AES Anthology: Microphones*, New York: Audio Engineering Society
2. Bauer, B.B., A century of microphones, *J. Audio Eng. Soc.*, **35**, 246–258 (1967)

Chapter 6

Loudspeakers and headphones

Audio signals can only be produced to high standards if accurate monitoring is available, otherwise deficiencies go unnoticed and will not be rectified. The primary monitoring tool is the high-quality loudspeaker, but for practical reasons headphones may have to be used. This chapter considers the theory and practice of both.

6.1 Loudspeaker concepts

It is obvious that, like any piece of audio equipment, an ideal loudspeaker should produce a particle velocity proportional to the input waveform and so must have a wide, flat frequency response with no resonances and minimum linear and non-linear distortion. This, however, is only part of the design problem.

In all practical loudspeakers some form of diaphragm has to be vibrated which then vibrates the air.[1] There are contradictory requirements. As was seen in Section 3.17, the SPL which the loudspeaker can generate is determined by the volume velocity. If the frequency is halved, the displacement must be doubled either by doubling the area of the diaphragm or by doubling the travel or some combination of both. Figure 6.1 shows the requirements for various conditions. Clearly a powerful loudspeaker which is able to reproduce the lowest audio frequencies must have a large diaphragm capable of considerable travel. As human hearing is relatively insensitive at low-frequency a loudspeaker which can only reproduce the lowest audible frequencies at low power is a waste of time.

Unfortunately any diaphragm which is adequate for low-frequency use will be too heavy to move at the highest audio frequencies. Even if it could, it would have an extremely large value of ka at high frequencies and Section 3.19 showed that this would result in *beaming* or high directivity which is undesirable. One solution is to use a number of drive units which each handle only part of the frequency range. Those producing low frequencies arc called *woofers* and will have large diaphragms with considerable travel whereas those producing high frequencies are called *tweeters* and will have small diaphragms whose movement is seldom visible. In some systems, mid-range units or *squawkers* are also used, having characteristics mid-way between the other types. A frequency-dividing system or *crossover network* is required to limit the frequency range of signals to each drive unit.

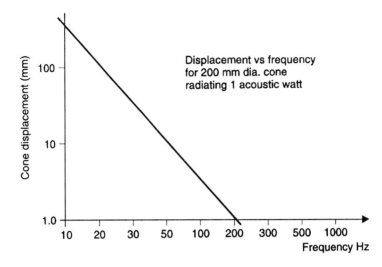

Figure 6.1 Volume velocity requirements dictate that low-frequency reproduction must involve large diaphragms or long travel.

A good microphone produces an accurate version of sounds *approaching* it from many directions. Even if a loudspeaker reproduced the microphone waveform exactly, the resulting sound is *leaving* in many directions. Spatially a single loudspeaker is producing sound travelling in exactly the opposite direction to the original. Consequently reproduction of the original soundfield is simply not possible.

Figure 6.2 shows the problem. Sound approaching a microphone (a) does so from a multiplicity of sources whereas sound leaving a single loudspeaker superimposes all of these sources into one. Consequently a monophonic or single loudspeaker is doomed to condense every sound source and its reverberation to a single point. When listening in anechoic conditions (b) this is exactly what happens. Whilst the waveform might be reproduced with great precision, the spatial characteristics of such a sound are quite wrong.

However, when listening in a room having a degree of reverberation a better result is achieved irrespective of the reverberation content of the signal. The reverberation in the mono signal has only time delay and no spatial characteristics whatsoever whereas the reverberation in the listening room has true spatial characteristics. The human listener is accustomed to ambient sound approaching from all directions in real life and when this does not happen in a reproduction system the result is unsatisfactory.

Thus in all real listening environments a considerable amount of reverberant sound is required in addition to the direct sound from the loudspeakers. Figure 6.2(c) shows that the reverberation of the listening room results in sound approaching the listener from all sides giving a closer approximation to the situation in (a). Clearly better reverberation will be obtained when the loudspeaker is out in clear space in the room. So-called bookcase loudspeakers mounted on walls or shelves can never give good results.

Better spatial accuracy requires more channels and more loudspeakers. Whilst the ideal requires an infinite number of loudspeakers, it will be seen

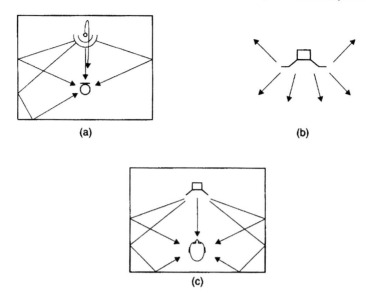

Figure 6.2 (a) Sound approaches microphone from many directions due to ambience and reverberation. (b) In anechoic conditions single loudspeaker produces exactly the opposite of (a). (c) Loudspeaker in reverberant conditions simulates situation of (a) at listener's ears.

that, with care, as few as two speakers can give a convincing spatial illusion. The improvement in spatial performance using two speakers is enormous. Tests[2] have shown that most people prefer stereo with poor bandwidth and significant distortion to pristine mono.

Two speakers can only give spatial accuracy for sound sources located between them. Reverberation in the listening room then provides ambient sound from all remaining directions. Clearly the resultant reverberant sound field can never be a replica of that at the microphone, but a plausible substitute is essential for realism and its absence results in an unsatisfactory result. Clearly the traditional use of heavily damped rooms for monitoring is suspect.

If realism is to be achieved, the polar diagram of the loudspeaker and its stability with frequency is extremely important. A common shortcoming with most drive units is that output becomes more directional with increasing frequency. Figure 6.3(a) shows that although the frequency response on-axis may be ruler flat giving a good quality direct sound, the frequency response off-axis may be quite badly impaired as at Figure 6.3(b). In the case of a multiple drive unit speaker, if the crossover frequency is too high, the low-frequency unit will have started beaming before it crosses over to the tweeter which widens the directivity again. Figure 6.3(c) shows that the off-axis response is then highly irregular. As the off-axis output excites the essential reverberant field the tonal balance of the reverberation will not match that of the direct sound.[3] The skilled listener can determine the crossover frequency, which by definition ought not to be possible in a good loudspeaker.

The resultant conflict between on- and off-axis tonality may only be perceived subconsciously and cause *listening fatigue* where the initial

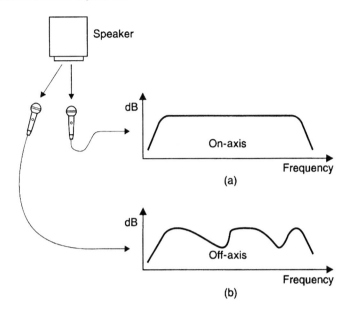

Figure 6.3 (a) Ideal on-axis response is achieved by many loudspeakers. (b) Off-axis response of most loudspeakers is irregular causing colouration of reverberant field.

impression of the loudspeaker is quite good but after a while one starts looking for excuses to stop listening. The hallmark of a good loudspeaker installation is that one can listen to it indefinitely and that of an excellent installation is where one does not want to stop.

Unfortunately such instances are rare. More often loudspeakers are used to having such poor off-axis frequency response that the only remedy is to make the room highly absorbent so that the off-axis sound never reaches the listener. This has led to the well-established myth that reflections are bad and that extensive treatment to make a room dead is necessary for good monitoring. This approach has no psychoacoustic basis and has simply evolved as a practical way of using loudspeakers having poor directivity. The problem is compounded by the fact that an absorbent room requires more sound power to obtain a given SPL. Consequently heavily treated rooms require high-power loudspeakers which have high distortion and often further sacrifice polar response in order to achieve that high power.

A conventional box-shaped loudspeaker with drive units in the front will suffer extensive shading of the radiation to the rear and thus will create a coloured reverberant field. Clearly a much more effective way of exciting reverberation with an accurate tonal balance is for the loudspeaker to emit sound to the rear as well as to the front. This is the advantage of the dipole loudspeaker which has a figure-of-eight polar diagram. Loudspeakers have also been seen with additional drive units facing upwards in order to improve the balance between direct and reverberant sound. These techniques work well but obviously in a dead room are a waste of time as the additional radiation will never reach the listener. The fault is in the room, not the speaker.

6.2 Loudspeaker mechanisms

The two transduction mechanisms used in microphones are both reversible and so can also be applied to loudspeakers. The electrodynamic loudspeaker produces a force by passing current through a magnetic field whereas the electrostatic loudspeaker produces force due to the action of an electric field upon a charge.

As with the microphone, the electrodynamic loudspeaker can be made using a ribbon or a moving coil. The ribbon loudspeaker is identical in construction to the ribbon microphone shown in Figure 5.12. Like the ribbon microphone, the ribbon loudspeaker has the advantage that the drive force is delivered direct to the diaphragm. The ribbon is not capable of much volume velocity and so is usually restricted to working at high frequencies, often with horn loading to improve efficiency. Large ribbon speakers with multiple magnet arrays have been seen, but these generally have directivity problems.

The moving coil loudspeaker[4] is by far the most common device. Figure 6.4 shows the structure of a typical low-cost unit containing an annular ferrite magnet. The magnet produces a radial field in which the coil operates. The coil drives the centre of the diaphragm which is supported by a *spider* allowing axial but not radial movement. The perimeter of the cone is supported by a flexible *surround*. The end of the coil is blanked off by a domed *dust cap* which is acoustically part of the cone. When the cone moves towards the magnet, air under the dust cap and the spider will be compressed and suitable vents must be provided to allow it to escape. If this is not done the air will force its way out at high speed causing turbulence and resulting in noise which is known as *chuffing*.

A major drawback of the moving coil loudspeaker is that the drive force is concentrated in one place on the diaphragm whereas the air load is distributed over the surface. This can cause the diaphragm to *break up*, which is undesirable in a woofer which should act as a rigid piston. Figure 6.5 shows that a more even application of coil thrust can be obtained by using a large coil so that the distance from the point where the drive force is applied is reduced. This distortion can be minimized in a low-frequency unit because extra mass is tolerable and the cone can be stiffer.

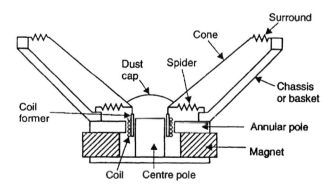

Figure 6.4 The components of a moving coil loudspeaker.

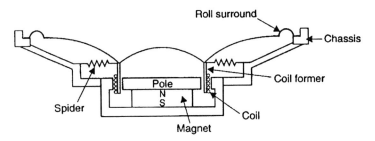

Figure 6.5 Woofer using large coil to distribute force more evenly.

The coil in a loudspeaker has appreciable resistance and this results in heating. At high power, the heat can only be dissipated by a large temperature rise. This has the effect of raising the coil resistance, reducing the sensitivity of the speaker. The result is known as thermal compression.

As the density of air is so low, the mass of air a loudspeaker actually moves is a few percent of the mass of the diaphragm. Consequently most of the drive power supplied to any kind of loudspeaker is wasted accelerating the diaphragm back and forth and the efficiency is very poor. There is no prospect of this situation ever being resolved in electromagnetic speakers. Where high SPL is required, a horn may be used to couple the diaphragm more efficiently to the air load by acting as an acoustic transformer. Figure 6.6 shows that a horn may be used with a ribbon or a moving coil driver.

Unfortunately the horn has a number of drawbacks. Unless large compared to the wavelength, the horn has no effect. Consequently the horn is restricted to mid-range frequencies and above. A further problem is that the mouth of the horn forms an impedance mismatch and standing waves occur down the horn, causing the frequency response to be irregular. Perhaps the greatest drawback of the horn is that the SPL at the diaphragm is considerably higher than at the mouth. Where sound pressures become an appreciable proportion of atmospheric pressure, the effect noted in Section 3.11 occurs in which the air becomes non-linear and distortion occurs. Consequently for high-quality applications the horn may have to be ruled out.

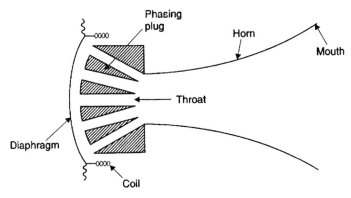

Figure 6.6 Horn loading may be used to improve the efficiency of moving coil or ribbon drivers. Horns can suffer from standing waves and distortion at high throat SPL.

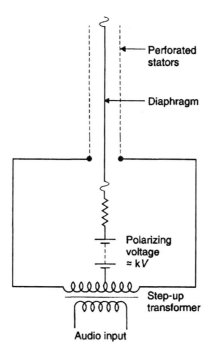

Figure 6.7 The electrostatic loudspeaker uses the force experienced by a charge in an electric field. The charge is obtained by polarizing the diaphragm.

The electrostatic loudspeaker is shown in Figure 6.7. A slightly conductive diaphragm is connected to a high-voltage DC supply so that it becomes charged. The high resistivity of the diaphragm prevents the charge moving around so that at all audio frequencies it can be considered fixed. Any charge placed in an electric field will experience a force. The electric field is provided by electrodes either side of the diaphragm which are driven in anti-phase, often by a centre-tapped transformer.

The advantage of the electrostatic loudspeaker is that the driving mechanism is fundamentally linear and the mechanical drive is applied uniformly all over the diaphragm. Consequently there is no reason why the diaphragm should break up. There is no heat dissipation mechanism in the electrostatic speaker and thermal compression is completely absent. The electrostatic loudspeaker has the further advantage that it is also inherently a dipole.

6.3 Directivity

One of the greatest challenges in a loudspeaker is to make the polar characteristics change smoothly with frequency in order to give an uncoloured reverberant field. Unfortunately crossing over between a number of drive units often does not achieve this. Figure 6.8 shows that at the crossover frequency both drive units separated by a are contributing equally to the radiation. If ka is small the system acts as a single driver and the polar diagram will be undisturbed. However, with typical values of a this will only be true below a few

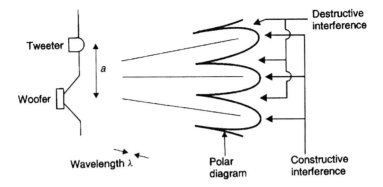

Figure 6.8 At the crossover frequency both drivers are operating and if *ka* is not small the polar diagram becomes highly irregular because of path-length differences.

hundred Hertz. If a crossover is attempted above that frequency a diffraction pattern will be created where the radiation will sum or cancel according to the path-length differences between the two drivers. This results in an irregular polar diagram and some quite undesirable off-axis frequency responses.[5]

Clearly the traditional loudspeaker with many different drive units is flawed. Certain moving coil and electrostatic transducers can approach the ideal with sufficient care. Figure 6.9(a) shows that if the flare angle of a cone-type moving coil unit is correct for the material, the forward component of the

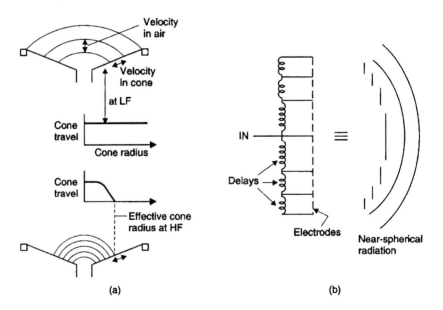

Figure 6.9 (a) A cone built as a lossy transmission line can reduce its diameter as a function of frequency, giving smooth directivity characteristics. (b) Using delay lines a flat electrostatic panel can be made to behave like a pulsating sphere.

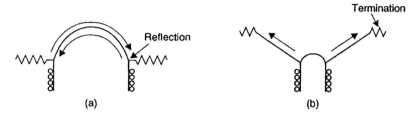

Figure 6.10 (a) Dome radiator has no way of terminating vibrations which have travelled across the dome. (b) Cone-type radiator uses surround as terminator.

speed of sound in the cone can be made slightly less than the speed of sound in the air, so that nearly spherical wavefronts can be launched. The cone is acting as a mechanical transmission line for vibrations which start at the coil former and work outwards. This should not be confused with cone breakup which is uncontrolled and detrimental. If frequency dependent loss is introduced into the transmission line, the higher the frequency the smaller is the area of the cone which radiates. Done correctly the result is a constant dispersion drive unit. There are vibrations travelling out across the cone surface and the cone surround must act as a matched terminator so that there can be no reflections.

The dome looks as if it ought to have a broad radiation pattern. Unfortunately this simplistic view is quite wrong.[1,6] The dome driver exhibits exactly the opposite of what is wanted. As frequency rises, the centre of the dome decouples and only the perimeter is fully driven, producing an annular radiator of maximum size. Consequently the polar characteristics of domes are poor and the off-axis response is likely to be impaired. Figure 6.10(a) shows that a further problem with domes is that vibrations launched from the coil travel right across the dome where they can reflect from the coil junction again. In cone-type radiators (b) the surround acts as a mechanical terminator and this problem is controlled. The only advantage of the dome driver is that a very large coil can be fitted which will allow high-power dissipation. This allows high SPL to be generated in dead rooms where the poor directivity will be concealed.

The elegant directivity solution in an electrostatic speaker first proposed by Walker of Quad[7] is to make the mechanically flat diaphragm behave like a sphere by splitting the electrode structure into concentric rings fed by lossy delay lines as shown in Figure 6.9(b). This produces what is known as a *phased array*. The outward propagation of vibrations across the diaphragm again simulates quite closely a sector of a pulsating sphere. Matched termination at the perimeter is required to prevent reflections. There is no reason why such an approach should not work with a segmented ribbon speaker.

Using either of these techniques allows the construction of a single drive unit which will work over the entire mid and treble range and display smooth directivity changes. A low *ka* crossover to a woofer completes the design. Such two-way speakers can display extremely good performance especially if implemented with active techniques.

Interestingly enough if strict polar response and distortion criteria are applied, the phased array electrostatic loudspeaker turns out to be capable of higher SPL than the moving coil unit. This is because the phased array approach allows the electrostatic loudspeaker to have a very large area diaphragm without beaming taking place. Consequently at mid and high frequencies it can achieve very large volume velocities.

6.4 The moving coil speaker

Unlike the electrostatic speaker, the moving coil speaker is not fundamentally linear and a number of mechanisms are responsible for less than ideal behaviour. There are two basic criteria for linearity. First, the drive force of the coil should depend only on the current and not on the coil position. Secondly, the restoring force of the suspension should be exactly proportional to the displacement. Both of these effects are worse at low frequencies where the cone travel is greatest.

Rice and Kellog discovered around 1925 that the displacement of a moving coil loudspeaker cone reaches a peak at the resonance, and falls at 6 dB/octave either side of that resonance as shown in Figure 6.11. This was explained in Section 2.10. Radiation is proportional to cone *velocity* which is obtained by differentiating the displacement. Differentiation tilts the response by 6 dB/octave. Consequently as (b) shows, the radiation is independent of frequency above resonance but falls at 12 dB/octave below.

Below resonance the motion is stiffness controlled. The displacement is proportional to coil current and velocity leads the current. Above resonance the system is mass controlled. The acceleration is proportional to coil current and velocity lags the current. Figure 6.12 shows the phase response through resonance. Note that at the resonant frequency the velocity is exactly in phase with the coil current. Because of this phase characteristic the polarity of a loudspeaker is a matter of opinion.

Manufacturers mark one terminal with a red spot or a + sign as an aid to wiring in the correct polarity. However, some use the convention that a positive DC voltage (e.g. from a battery) will cause forward motion of the cone, whereas others use the convention that the positive half-cycle of an AC voltage at a frequency above resonance will cause forward motion. It will be seen from Figure 6.12 that these two conventions are, of course, in phase opposition. The AC definition makes more sense as that is how the speaker is used; however, most manufacturers use the DC definition.

The phase reversal of a moving coil driver as it passes through resonance means that it is fundamentally incapable of reproducing the input waveform at frequencies near resonance. Clearly if it is intended to reproduce the input waveform accurately the fundamental resonance must be placed below the audio band at around 20 Hz or signal processing must be used to artificially lower the resonance.

Figure 6.13 shows that with an amplifier of low output resistance, the mechanical resonance is damped by the coil resistance. The moving coil motor acts as a transformer coupling the resonant system to the damping resistance. Increasing the flux density or the length of coil in the gap increases the effective ratio of the transformer and makes the coil resistance appear lower, increasing the damping. The peakiness of the resonance is adjusted in the design process by balancing the coil resistance, the strength of the magnet and the length of coil wire in the magnetic field. If the drive unit is not driven

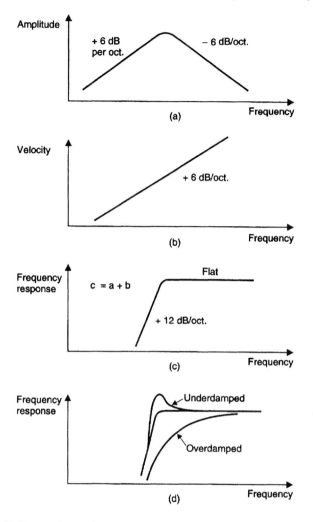

Figure 6.11 (a) The amplitude of a resonant system falls at 6 dB/octave away from the peak. (b) The velocity of the system in (a) is obtained by differentiating the displacement, resulting in a 6 dB/octave tilt. This gives a flat response region with a 12 dB/octave roll-off below resonance. (c) Moving coil motor acts as a transformer coupling resonant system to the coil resistance which acts as a damper. Motor design affects peakiness of resonance (d).

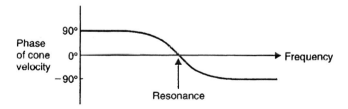

Figure 6.12 Phase response of loudspeaker passes through zero at resonance. The sharper the resonance the more rapid is the phase change.

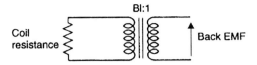

Figure 6.13 The Bl product of the motor coil determines the ratio of the coupling between the resonant system and the damping of the coil resistance.

by a low output impedance via low-resistance cables the resonance may be underdamped and a pronounced peak or *honk* may occur.

6.5 Magnets

The moving coil speaker obtains a drive force due to the coil current reacting against the magnetic field. Figure 6.14 shows that the radial magnetic field is provided by a permanent magnet, although early speakers were energized by a solenoid coil. If the moving coil motor is to be linear, the force produced must depend only on the current and not on the position. In the case of a tweeter, shown in (a), the travel is usually so small that the coil can be made the same length as the flux gap as the leakage flux effectively extends the gap.

In woofers the travel is significant and in order to obtain linearity the coil can be made shorter (b) or longer (c) than the flux gap. Clearly (c) is less efficient as power is wasted driving current through those parts of the coil which are not in the gap. However, Figure 6.14(b) requires a massive magnet

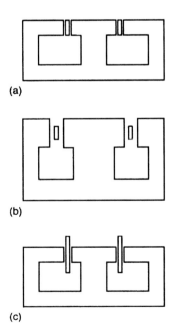

(a)

(b)

(c)

Figure 6.14 (a) Coil same length as gap. (b) Coil shorter than gap. (c) Coil longer than gap.

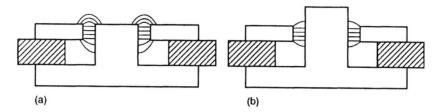

Figure 6.15 (a) Poor magnet design with asymmetrical flux. (b) Better design achieves flux symmetry.

structure and is little used because of the weight and cost. It is important that the flux distribution is symmetrical. Figure 6.15(a) shows a typical low-cost speaker with casual pole design. Figure 6.15(b) shows a better approach.

In permanent magnets the classical *B-H* curve is replaced with a *demagnetization curve* which is effectively the top-left quadrant of the *B-H* curve. The vertical axis is the remanence *B* in tesla and the value on that axis is the flux density which would be obtained if the magnet were immersed in a hypothetical infinitely permeable material needing no magnetomotive force to pass flux (analogous to a superconductor). The horizontal axis is the coercive force in kilAmps per metre and the value on that axis is the coercive force which would be observed if the magnet were immersed in a hypothetical totally impermeable material (analogous to an insulator). Figure 6.16 shows a set of demagnetization curves for a range of magnetic materials. There is a usefully linear region in each curve.

The maximum magnetomotive force (in kA) is analogous to the voltage in electricity, and is given by multiplying the coercive force *H* by the length of

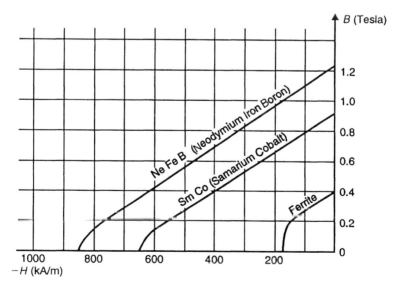

Figure 6.16 Demagnetization curves for various magnetic materials.

the magnet. The maximum flux density B has to be multiplied by the area of the magnet to obtain the maximum flux, ϕ. Figure 6.16 shows the relationship between magnetomotive force and ϕ. The magnet has an internal reluctance, corresponding to the source impedance in electricity given by the slope of the graph. Consequently the magnet will be most efficiently used when the load reluctance is equal to the source reluctance. Under these conditions the BH product (in joules per cubic metre) reaches a maximum value. Lines of BH product are shown on Figure 6.16.

It will be seen that for a given magnetic material the volume of the magnet is roughly proportional to the volume of the gap. The constant of proportionality is a function of the efficiency of the magnetic circuit and the energy of the magnet.

As has been seen, magnetic materials are measured by their B and H parameters which vary in magnitude and relative proportion. The first modern powerful magnetic material was Alnico, a cobalt-iron-nickel-aluminium-copper mix. Ticonal adds titanium to the mix, hence the name. These materials have high B in comparison to H, leading to a long columnar-shaped magnet as shown in Figure 6.17(a). Cobalt became expensive in the 1970s for political reasons and an alternative to Alnico had to be found. Ferrite is a low-cost material, but has a very low B compared to H. The only way a sufficiently strong magnet could be made was to put the magnet outside the coil in a ring (b) so that a large cross-sectional area could be used.

The volume of the air gap in which the coil operates is a critical factor. This is given by the gap width multiplied by the gap area. The wider the gap, the greater the magnetomotive force needed to drive flux across the gap and the longer the magnet needs to be. The gap width will be set by the radial thickness of the coil plus operating clearances. The gap depth and the pole radius set the pole area and the greater the area over which a given flux density is required, the greater the total flux and the greater the cross-sectional area of the magnet will need to be.

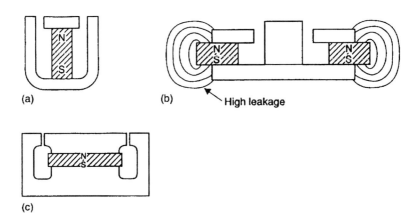

Figure 6.17 (a) Alnico requires columnar magnet. (b) Ferrite requires generous cross-sectional area and leaks badly. (c) Rare-earth magnet has very small volume and mass. Structures (a) and (c) have no external leakage.

The low-cost ferrite magnet is a source of considerable problems in the television environment because it produces so much stray flux. This can cause colour purity errors and distortion of the picture on a video monitor because it disturbs the magnetic deflection system. Whilst screening can be installed, a better solution for speakers to be used in television is to employ a different magnetic circuit design as shown in Figure 6.17(c). Such a design completely contains its own flux because the magnet is inside the magnetic circuit. The magnet has to be smaller, but sufficient flux is easily available using rare-earth magnets. As Figure 6.17(c) shows, the rare-earth magnet has a very high-energy product and only a small volume of material is required. This can easily be located inside the coil, leading to a very compact motor.

As there is no such thing as a magnetic insulator, a good deal of flux is lost to leakage where flux simply jumps from one pole to the next. Undercutting the pole pieces is of some help because it helps to keep as much air as possible in the shunt leakage path. However, the ferrite magnet fares poorly for leakage because its surface area is so large that the shunt reluctance is small. In ferrite designs often only a third of the magnet flux passes through the gap.

The rare-earth magnet can be so small that its shunt reluctance is significantly higher and there is much less leakage. Thus the magnet can be even smaller because the reduced leakage makes it more efficient. Whilst samarium-cobalt rare-earth magnets are expensive, the cost of neodymium-iron-boron magnets is falling and they are starting to become popular in loudspeakers because of their efficiency and low leakage. Rare-earth magnets may have 10 times the energy of ferrite, but because the leakage is much reduced it is often possible to use a magnet only one-twentieth the volume of a ferrite equivalent.

The lower weight allowed by neodymium means that the entire drive unit weighs about one-half of a ferrite equivalent. This means that portable equipment with a genuine bass response is a possibility. Neodymium magnets have a further advantage that they are highly conductive and thus resist flux modulation. This occurs when the flux due to current in the moving coil distorts the magnet flux. Flux modulation is a problem in ferrite because it is an insulator. Sometimes the pole pieces are copper-plated to help reduce the effect.

6.6 Coils

The coil must have good electrical conductivity and low mass. Most coils are copper, but aluminium and various techniques such as plating have been seen. An improvement in efficiency can be had using rectangular section wire which more effectively fills the gap. Hexagonal wire is also used. Rectangular wire has a further advantage in tweeters that the individual turns are more rigidly connected making for a stiffer coil.

In most cases the coil is made by winding onto a cylindrical former which delivers the coil thrust to the neck of the cone. In high-powered units the major difficulty is resistive heating which causes elevated temperatures in the coil and former. High temperatures can damage the coil insulation and cause decomposition of the coil former. Outgassing may cause the former to swell, making the coil assembly jam in the gap. Modern high-power speakers use composite materials for the coil former which can resist high temperatures.

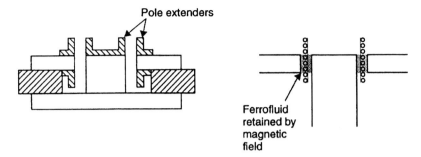

Figure 6.18 (a) Non-ferrous metal pole extenders help coil cooling. (b) Ferrofluid is retained by the flux and conducts coil heat to poles.

As the main cooling mechanism is by radiation to the magnet assembly, any measure which will improve that will be considered. Figure 6.18(a) shows the use of non-ferrous pole extenders which increase the area of the pole pieces rendering heat radiation more effective. Figure 6.18(b) shows the use of ferrofluid which is a colloidal suspension of ferrous particles in a liquid. The ferrofluid is retained in the gap by the magnetic field and reduces the thermal resistance from the coil to the magnet. The drive unit has to be appropriately designed to use ferrofluid.[8] If the interior of the magnet is sealed, the coil will raise the air pressure within as it enters the magnet and the air will be forced out through the ferrofluid with accompanying noises. This will be prevented by appropriate venting.

6.7 Cones

The job of the cone is to couple the motion of the coil to a large volume of air. In a woofer ka is small and the cone should be rigid so that it acts as a piston. As excessive mass reduces efficiency, the best materials for woofer cones are those which offer the best ratio of modulus of elasticity to density. Other factors include the ability to resist high temperatures reached by the coil former where it joins the neck of the cone, and the inherent damping of the material which prevents resonances.

Early cones were made of paper but these are easily damaged and are susceptible to moisture absorption which changes the speaker characteristics as a function of humidity. The conical shape was adopted because it allows a great increase in stiffness over a flat sheet of material. Even greater stiffness is possible if the cone is flared as shown in Figure 6.19.

True cone   Flared cone

Figure 6.19 A flared 'cone' is even stiffer than a true cone.

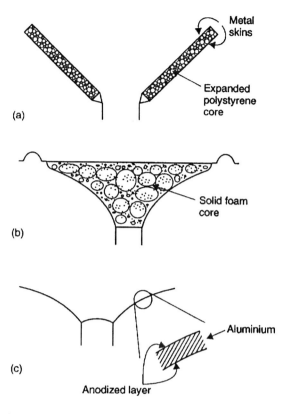

Figure 6.20 (a) Sandwich woofer cone. (b) Solid foam woofer diaphragm. (c) Anodized aluminium forms a sandwich.

Many different approaches have been tried for cone construction and in woofers at least, many of these have been successful. Figure 6.20 shows some possibilities for woofers. The sandwich construction uses two skins, typically of aluminium, with a low density core, typically of expanded polystyrene. This is sometimes taken to the extreme case where the cone is solid, having a flat front. Aluminium cones can be turned into sandwich structures by anodizing. The hard oxide layer due to the anodizing step forms a rigid skin and the aluminium forms the core. Thermoplastics have been used for cones and have the advantage that they can be moulded at low cost, but are unsuitable for high-power applications. Recently composite materials have become economic and their high stiffness to density ratio which makes them attractive in aerospace is equally useful in loudspeakers.

In tweeters the cone design is more difficult because of the need to allow the cone to act as a transmission line without allowing uncontrolled breakup. This makes the stiffness and damping critical. In many cases one single material is unsuitable and the cone is made from a suitably stiff material which has an appropriate internal speed of sound and a separate damping layer may then be provided.

6.8 Low-frequency reproduction

An unenclosed diaphragm acts as a dipole and these become extremely inefficient at low frequencies because air simply moves from the front to the back,
short circuiting the radiation. In practice the radiation from the two sides
of the diaphragm must be kept separate to create an efficient low-frequency
radiator. The concept of the *infinite baffle* is one in which the drive unit is
mounted in an aperture in an endless plate. Such a device cannot be made
and in practice the infinite baffle is folded round to make an *enclosure*. Sealed
enclosures are often and erroneously called infinite baffles, but they do not
have the same result. Unfortunately the enclosed air acts as a spring because
inward movement of the diaphragm reduces the volume, raising the pressure.

The stiffness of this air spring acts in parallel with the stiffness of the
diaphragm supports. The mass of the diaphragm and the total stiffness determines the frequency of *fundamental resonance* of the loudspeaker. To obtain
reproduction of lowest frequencies, the resonance must be kept low and this
implies a large box to reduce the stiffness of the air spring, and a high compliance drive unit. When the air stiffness dominates the drive unit stiffness, the
configuration is called *acoustic suspension*.

Acoustic suspension speakers were claimed to offer a more accurate compliance than the speaker's mechanical flexure, but this is not true. The air spring
in a sealed box is fundamentally non-linear. This is easy to see by considering
Figure 6.21. Here a diaphragm will displace a volume equal to the volume
of the box. If the diaphragm moves inwards, the pressure becomes infinite. If
the diaphragm moves outwards, the pressure is merely halved. It is simple to
calculate the distortion given the diaphragm displacement and the box volume.

In early attempts to give satisfactory low-frequency performance from
smaller boxes, a number of passive schemes have been tried. These include
the reflex cabinet shown in Figure 6.22(a) which has a port containing an air
mass. This is designed to resonate with the air spring at a frequency below
that of the fundamental resonance of the driver so that as the driver response
falls off the port output takes over. In some designs the air mass is replaced
by a compliantly mounted diaphragm having no coil, known as an ABR or
auxiliary bass radiator, Figure 6.22(b).

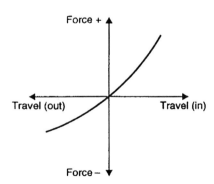

Figure 6.21 An air spring is non-linear producing more restoring force on compression than on
rarefaction.

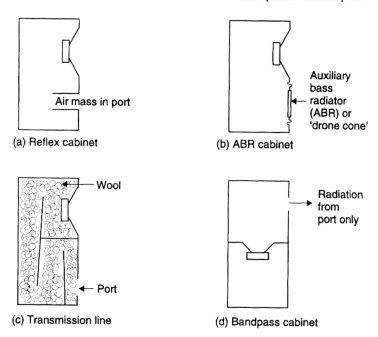

Figure 6.22 Various attempts to reproduce low frequencies. (a) Mass of air in reflex duct resonates with air spring in box. (b) Air mass replaced by undriven diaphragm or ABR. (c) Rear wave is phase shifted 180° in transmission line to augment front radiation. (d) Bandpass enclosure puts drive unit between two resonating chambers. None of these techniques can properly reproduce transients and active techniques have rendered them obsolete.

Another alternative is the transmission line speaker shown in Figure 6.22(c) in which the rear wave from the driver is passed down a long damped labyrinth which emerges at a port. The length is designed to introduce a 180° phase shift at the frequency where the port output is meant to augment the driver output. A true transmission line loudspeaker is quite large in order to make the labyrinth long enough. Some smaller models are available which claim to work on the transmission line principle but in fact the labyrinth is far too short and there is a chamber behind the drive unit which makes these heavily damped reflex cabinets.

More recently the *bandpass* enclosure (Figure 6.22(d)) has become popular, probably because suitable computer programs are now available to assist the otherwise difficult design calculations. The bandpass enclosure has two chambers with the drive unit between them. All radiation is via the port.

The reflex, ABR, bandpass and transmission line principles have numerous drawbacks the most serious of which are that the principle only works on continuous tone. Low-frequency transients suffer badly from linear distortion because the leading edge of the transients are removed and reproduced after the signal has finished to give the phenomenon of *hangover*. The low-frequency content of the sound lags behind the high frequencies in an unnatural way. In other words the input waveform is simply not reproduced by these tricks as is easily revealed by comparison with the original sound.

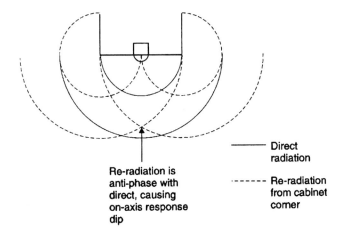

Re-radiation is
anti-phase with
direct, causing
on-axis response
dip

——— Direct
radiation

------ Re-radiation
from cabinet
corner

Figure 6.23 Attempting to radiate high frequency from a tweeter mounted in a large rectangular box produces frequency response irregularity due to diffraction from the box corners.

Different order filtering and different damping change the amount by which this lag takes place, but can never eliminate the problem which is most noticeable on transient musical information such as percussion and on effects such as door slams. It is quite impossible to use such a low-frequency technique in conjunction with a linear phase electrostatic high-frequency unit or a constant directivity mid-top moving coil unit because the quality mismatch is too obvious.

The only low-frequency structure which can approach the reproduction of the input waveform is the sealed box. In order to reproduce the lowest frequencies the box will need to be large to prevent the stiffness of the air spring raising the fundamental resonance. Unfortunately a large box forms a very poor place from which to radiate high frequencies. Figure 6.23 shows that when a high-frequency unit is fitted in a large box, diffraction results in re-radiation at the corners.[9,10]

When the direct and re-radiated sounds combine the result is constructive or destructive interference depending on the frequency. This causes ripples in the on-axis frequency response. The larger the box the further down the spectrum these ripples go. The effect can be reduced by making the cabinet smaller and making it curved with no sharp corners.

One approach to reducing the low-frequency enclosure size without resorting to resonant tricks is to use compound drive units. Figure 6.24 shows the isobaric loudspeaker in which two drivers are placed in tandem. The doubled cone mass allows a low resonant frequency with a smaller cabinet. The compliance of the air between the two drive units reduces the distortion due to the non-linearity of the air spring in the main enclosure. Although more complex to build, the isobaric speaker easily outperforms the approaches of Figure 6.22(a–d).

Further advances in performance can only be had by adopting active techniques which will be seen in Section 6.14.

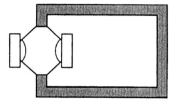

Figure 6.24 The isobaric or compound loudspeaker outperforms ported speakers because it is more faithful to the applied waveform.

6.9 Crossover networks

As the full-range drive unit is virtually impossible, practical loudspeakers need to employ several drive units, each optimized for a particular frequency range. A crossover network is needed so that signals of the right frequency are sent to the appropriate drive unit. Perhaps it is more accurate to say that the wrong frequencies are prevented from reaching drive units which cannot handle them.

Loudspeakers are highly traditional devices which continue to be built in the same way even though the restrictions which originally led to the approach have long since disappeared. At one time audio amplifiers were expensive and the multi-drive unit loudspeaker had to use a single amplifier to control cost. This meant that the crossover had to be performed at power level. Figure 6.25(a) shows a typical simple power level crossover. An inductor in series with the woofer reduces the input at high frequencies as its impedance rises whereas a capacitor in series with the tweeter increases the input at high frequencies as its impedance falls. The crossover slope is only 6 dB/octave which means that the two drive units will still receive significant signal levels outside their operating bandwidth. A steeper crossover can be obtained by using a second-order filter (Figure 6.25(b)) which achieves 12 dB/octave.

Unfortunately none of these simple approaches works properly. It is self-evident that the sum of the two crossover outputs ought to be the original

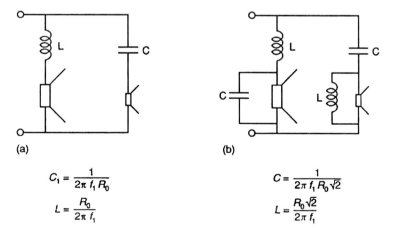

(a)

$$C_1 = \frac{1}{2\pi f_1 R_0}$$

$$L = \frac{R_0}{2\pi f_1}$$

(b)

$$C = \frac{1}{2\pi f_1 R_0 \sqrt{2}}$$

$$L = \frac{R_0 \sqrt{2}}{2\pi f_1}$$

Figure 6.25 (a) Simple 6 dB/octave crossover. (b) 12 dB/octave requires second-order filter.

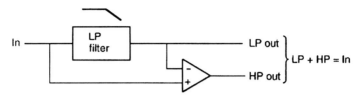

Figure 6.26 Active crossover with constant voltage characteristic gives precision that is impossible with passive crossovers.

signal. This is known as a *constant voltage* crossover.[11] However, passive crossover techniques simply cannot reach this goal and must be considered obsolete for high-quality applications.

Using analog computing techniques it is easy to create constant voltage crossovers. Figure 6.26 shows that a filter is used to prevent unwanted frequencies reaching a certain drive unit. If the output of that filter is subtracted from the input, the result is the frequencies which the drive unit can handle. By definition the sum of the two signals is the original. The crossover can be made symmetrical or asymmetrical and have slopes to suit the characteristics if the drive units. When properly engineered the crossover frequency simply cannot be determined by listening.

6.10 Enclosures

The loudspeaker enclosure has to fulfil two functions. At low frequencies it must contain the back radiation from the woofer and at high frequencies its shape determines the diffraction characteristics which will affect the on-axis frequency response and the directivity.

At low frequencies an enclosure must be rigid so that internal pressures do not cause flexing. As the area of the enclosure walls is an order of magnitude greater than the woofer cone area, the slightest flexing can cause appreciable colouration of the sound. In other industries, vessels to contain or resist pressure are invariably cylindrical or spherical. The aerosol can, the airliner and the submarine are good examples. The dogged adherence to the square box made from six flat panels illustrates that even today most loudspeakers are designed by tradition rather than physics.

At high frequencies the sharp corners on a square loudspeaker cause impedance changes and act as secondary radiators. This is also true of the woofer if this is exposed.[10] Many years ago Olsen[9] produced his definitive series of tests on speaker enclosure shapes, but this has been largely ignored.

Fortunately a small number of manufacturers are making loudspeakers in enclosures which rely on curvature for both stiffness and diffraction control. These are not necessarily expensive to make using moulding techniques, but the result is much less colouration of the sound and an opportunity for a more interesting visual appearance.

6.11 Electrostatic loudspeakers

The electrostatic loudspeaker is less common than the moving coil speaker and perhaps because of that is widely misunderstood, although the excellent treatment by the late Peter Baxandall[12] leaves little to doubt. If cost or compactness is the only criterion, the electrostatic speaker is ruled out, but if quality is the criterion the electrostatic demands attention because it can produce a combination of low distortion and directivity which moving coil designers only dream about.

Because the electrostatic is a rarity, most audio engineers are accustomed to the inevitable distortion of moving coil designs which makes the sound artificially bright. When faced with an electrostatic for the first time the reaction is universally that there is 'no treble', whereas in fact the truth is that there is 'no distortion'. Truly low distortion speakers sound initially unexciting because they add nothing to the sound, but a good measure of their quality is that it is possible to listen indefinitely without fatigue.

A further difficulty is that the electrostatic speaker is a dipole and needs a relatively undamped room in which it can excite the reverberation. In a simple comparison with a traditional moving coil box monitor in a dead control room the electrostatic will be unsatisfactory, although the fault is in the room. When the control room is designed correctly the electrostatic will deliver surprisingly high SPL.

Electrostatics drive the diaphragm by the action of an electric field upon a charge. Figure 6.27 shows that the diaphragm is supported between conducting plates or *stators* which are perforated to allow the sound out. Clearly the most uniform electric field is obtained with solid electrodes, whereas the best escape of sound is obtained with the stators absent. Some compromise is needed with a solidity of around 50%.

Figure 6.27 shows that stators can be made from parallel metal rods on a frame or by using a punched printed circuit board. The printed circuit arrangement makes it easy to subdivide the stators electrically to fabricate a phased array. As there is actual air movement through the holes in the stator, the shape of the holes is important, with the round shape of the rod construction avoiding turbulence which can occur if the corners of punched holes are not rounded. This is known from cinema screen design. Sharp corners are also best avoided because they encourage electrical breakdown. As high voltages are inherent in the process, dust attraction is a problem which is usually overcome by the provision of dust shields made from extremely thin plastic sheet. The assembly is then encased in a protective framework which prevents damage and allows the ignorant to live longer.

The diaphragm is generally made from Mylar which is tensioned on a frame. The diaphragm is rendered very slightly conductive (about 10^{10} ohms per square) so that it can be connected to a polarizing supply to charge it. This may be done using a proprietary coating but home constructors may use graphite applied by hand. The high resistivity is necessary to prevent charge moving around the diaphragm so that *constant charge* operation results. Figure 6.28(a)

Figure 6.27 Electrostatic loudspeaker stators can be punched or fabricated from rods.

(a) Low frequency   **(b) High frequency**

Figure 6.28 (a) At low frequency, diaphragm moves in a curved shape. (b) At high frequency, diaphragm is mass controlled and moves parallel.

shows that at low frequencies the diaphragm moves in a curved shape because it is constrained by the frame. Were it not for the high resistivity, charge would rush to the apex of the curve.

Note that at high frequencies the diaphragm is mass controlled and remains substantially flat as it moves as in Figure 6.28(b). The force on the diaphragm is a function of the stator voltage only and so the speaker is perfectly linear. The uniform application of drive force and air load on the diaphragm makes for an extremely low distortion speaker having inherent phase linearity which puts most moving coil transducers to shame.

When the diaphragm is polarized it will attempt to attract the two stators. When it is central the attraction is symmetrical and cancels out. However, when the diaphragm is off-centre the attraction from the nearer electrode will be greater. The result is that the polarized diaphragm acts as if it had negative compliance. In all practical speakers the positive compliance due to the tension in the diaphragm must exceed the negative compliance otherwise the diaphragm will simply stick to one of the stators.

In practice the diaphragm tension and its mass determine the fundamental resonance. Upon applying the polarizing voltage the resonant frequency moves down. This is a useful result and the resonance shift can even be used to measure the diaphragm tension. The negative compliance effect limits the span of a diaphragm between supports so in practice large diaphragms are made from smaller units. Figure 6.29(a) shows that these may be quite independent whereas (b) shows that a single diaphragm may be fitted with intermediate supports.

The polarizing voltage is generally of the order of 5 kV on a full-range speaker, but this is quite safe because the supply is arranged to have a very high source impedance and inadvertent bodily contact is only irritating. The same is not true of the stator drive signals which are also in the kiloVolts region but with low source impedance. Touching these when high SPL is being produced may result in death or at least a change of underwear.

The drive force on the diaphragm is the product of the charge and the electric field. The electric field is given by the inter-stator drive voltage divided by the spacing. Again a compromise is needed because the highest field is obtained with a close spacing which precludes long diaphragm travel. If the spacing

(a)   **(b)**

Figure 6.29 (a) Large diaphragm made from small units. (b) Single-piece diaphragm with regular supports.

is increased it will be found that the drive voltage becomes so high that breakdown occurs.

The stator drive signals are generally provided using step-up transformers although a few home-built designs have appeared for direct-drive high-voltage amplifiers based on radio transmitter vacuum tubes. These work in Class A and usually dissipate several kiloWatts restricting them to enthusiast applications. Commercially available designs use transformers which must be of high quality. Recent developments in magnetic materials and transformer design have led to significant improvements in transformer performance. In an active design the transformer can be included in the amplifier feedback loop which yields further performance improvement.

A conventional electrostatic loudspeaker appears to the amplifier as a capacitive load. The loudspeaker itself is extremely efficient as there is no resistive heat dissipation, but the down side is that the amplifier has to drive current between the stators simply to charge and discharge the capacitance. This current is in quadrature to the voltage and so the amplifier sees a complex load and dissipates a lot of power because of the current flow. Amplifiers which are not designed for this duty may be unstable or may even fail. In contrast the delay lines used in the Quad ESL-63 are transmission lines and make the speaker appear resistive at most audio frequencies. An amplifier which cannot drive a Quad is grim indeed.

The light diaphragm of the electrostatic loudspeaker results in excessively high fundamental resonance if placed in a conventional enclosure. Consequently most electrostatic speakers are employed as dipoles, which has been seen to be advantageous, except at low frequencies where acoustic short circuiting and the short diaphragm travel allowed by the fixed electrodes bring things to a halt before significant SPL can be achieved. Consequently the electrostatic principle is best employed as a mid and high-frequency dipole in conjunction with a low distortion moving coil low-frequency unit.

Although limited in travel, the electrostatic speaker can be made in very large area panels using phased array structures to avoid beaming. When used as a mid-range and treble unit a large electrostatic panel can easily exceed 100 dB SPL with linear and non-linear distortion lower than any other technique. Until recently this approach was impracticable because it was impossible to provide a matching woofer with low enough distortion. However, recently developed active crossover and woofer techniques allow low distortion linear phase performance which can match the accuracy of an electrostatic panel. The Celtic Legend precision monitor speaker shown in Figure 6.30 is an example of this new breed of powerful hybrid electrostatics.

Electrostatic loudspeakers need protection against overload, otherwise electrical breakdown can occur which can damage the diaphragm and stators. Mechanical overload is not hazardous. If the diaphragm moves so far that it touches the stator, the high resistivity of the diaphragm means that negligible current can flow. Some manufacturers take the further precaution of insulating the stators. In this case the only detrimental effect will be audible distortion as the diaphragm is arrested.

However, if the voltage between the stators is allowed to exceed safe limits a flashover can occur which will punch a hole in the diaphragm. If the air then ionizes an arc can form which will burn the stators. In practical speakers this is avoided by installing limiters or soft clippers which restrict the voltage

Figure 6.30 The Celtic Legend active hybrid speaker uses a phased array mid/top electrostatic panel and a precision moving coil woofer to produce high SPL with low distortion.

at the primary of the step-up transformers, the secondary or both. In the Quad ELS-63 there is a further safety measure in the form of an antenna which detects RF radiation from a corona discharge and clamps the audio input. There is also need for protection in the case of passive speakers where drive from the amplifier might be applied when the polarizing supply has not been turned on.

6.12 Power amplifiers

Audio power amplifiers have a relatively simple task in that they have to make an audio signal bigger and its source impedance lower. Unfortunately they also have to be reasonably efficient. Figure 6.31 shows some approaches to power amplifiers. The Class A amplifier has potentially the lowest distortion because the entire waveform is handled by the same devices. The high standing current required results in a serious heat dissipation problem which deters all but enthusiasts. The Class B amplifier has different devices to provide the positive and negative halves of the cycle. This reduces dissipation but it is extremely difficult to prevent crossover distortion taking place at low signal levels where the handover from one polarity to the other takes place. The Class D or switching amplifier is extremely efficient because its output devices are either on or off. A variable output is obtained by controlling the duty cycle

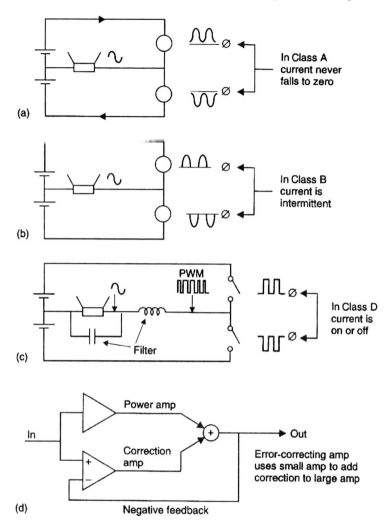

Figure 6.31 (a) Class A amplifier has low distortion but heavy standing current produces heat problem. (b) Class B amplifier is more efficient but introduces crossover distortion. (c) Class D amplifier is very efficient and ideal for driving active woofers. (d) Error-correcting amplifier uses small, wide bandwidth Class A amplifier to compensate for distortions of Class B stage.

of the switching and filtering the output. Whilst switching amplifiers cannot produce low enough distortion for high-quality mid and treble drive, they offer an ideal solution to efficient generation of low frequencies. Such amplifiers save on cost and weight because they can use a smaller power supply and need no massive heatsinks.

In most amplifiers linearity is achieved using negative feedback which compares the output with the input to create a correction signal. In most semiconductor amplifiers the open loop characteristics are not very good and a lot of feedback is necessary, requiring high open loop gain which is not always available at all frequencies. It is not possible to make a racehorse by putting feedback around a donkey, consequently when a fast slewing input arrives, the amplifier may go open loop and produce transient intermodulation distortion (TIM). This has resulted in negative feedback being unpopular in high-end hi-fi circles when it is merely poor design that is responsible.

In the *error-correcting* amplifier topology the output of a small Class A high-performance amplifier is added to that from a powerful but average Class B amplifier. The drive to the high-power amplifier is arranged so that it always delivers enough current to the load to prevent the small amplifier from clipping. Careful feedback arrangements allow the small amplifier to cancel the distortions of the large one. The low distortion of a Class A amplifier can then be combined with the reduced heat dissipation of a Class B amplifier.

In the current dumping topology due to Walker[13] the error-correcting process was used to allow the amplifier to be built without adjustments and to remain drift free for its working life. Essentially current dumping works using a reverse crossover network as Figure 6.32 shows. The large inaccurate amplifier drives the load via an inductor whereas the small accurate amplifier drives the load via a resistor. The inductor prevents the low output impedance of the large amplifier shorting the output of the small amplifier and avoids the dissipation which would result if a resistor were to be used. The large amplifier has a dead band so that at low signal levels only the small amplifier operates. As the level rises the large amplifier comes to the aid of the small amplifier by dumping extra current into the load. The negative feedback is arranged so that when the large amplifier operates the gain is reduced so that the transfer function of the amplifier combination remains the same.

In the error-cancelling approach due to Sandman[14] the low-powered amplifier shown in Figure 6.33 drives the load through one side of a bridge. The large amplifier attempts to balance the bridge so that the small amplifier sees an infinite load impedance. Only when the large amplifier makes an error does the small amplifier have to deliver significant power to correct it. The later Class AA design by Technics is surprisingly similar.

With very few exceptions, audio amplifiers are voltage sources with a very low output impedance. No attempt is made to impedance match the loudspeaker to the amplifier in the conventional sense as there is no transmission

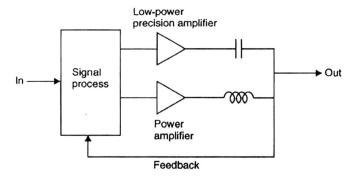

Figure 6.32 The current dumping amplifier combines a small precision amplifier with a large amplifier to obtain a unit which needs no adjustment.

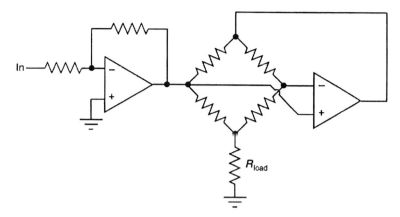

Figure 6.33 Bridged amplifier due to Sandman attempts to make load seen by small amplifier infinite.

line involved. A lower impedance loudspeaker simply takes more power for a given amplifier output voltage. When driven to the extreme an amplifier will run out of supply voltage if driving a high impedance, whereas it will current limit if driving a low impedance. There is an optimum load impedance into which the amplifier can deliver the most power when current and voltage limiting coincide. In transistor amplifiers this will be fixed, whereas valve amplifiers have output transformers and can deliver full power into a variety of load impedances simply by changing the transformer ratio.

Clearly, however good an amplifier it cannot function properly if the power supply voltage sags when transient loads are applied. Consequently good quality audio amplifiers invariably have a well-specified power supply which can deliver all conceivable requirements. The traditional transformer/rectifier/capacitor type supply is slowly giving way to the switched mode supply which has superior voltage regulation and in high-powered sizes also costs less.

6.13 Speaker cables

The subject of loudspeaker cables is almost guaranteed to spark a heated debate in which the laws of physics are temporarily set aside to improve the spectacle. This author frankly has no time for such theatricalities and prefers to cling to reality.

The job of a loudspeaker cable is simply to deliver electricity from an amplifier to a loudspeaker. It is undeniable that the cable is in series with the signal path, and could in principle affect the sound quality, but in practice the losses are dwarfed by shortcomings elsewhere.

Clearly a real cable will have resistance, inductance and capacitance and so it is hardly surprising that the waveform at the amplifier end will be slightly different to that at the speaker end. Obviously this difference can be minimized by the use of a sufficiently heavy conductor cross-section to make the series resistance low.

Much of the myth of the audibility of cables stems from the fact that a conventional loudspeaker forms a complex, frequency dependent and non-linear load. Complex is used in the mathematical sense to suggest that the current is not in phase with the voltage. This is obvious from the fact that each drive unit in a speaker has a fundamental resonance and contains moving masses which can store energy and feed it back to the amplifier at a later time. The impedance and phase angle of a typical low-cost speaker varies enormously with frequency. Where the impedance is high, little current flows and the cable causes a small loss. Where the impedance is low, more current flows and a greater loss is experienced.

Consequently if one were to measure the frequency response at the speaker end of a cable, it would be slightly irregular even if the amplifier is beyond reproach. However, it does not follow that this is the frequency response of the cable. The remedy is to equip the loudspeaker with a complex conjugate compensation circuit[15] which is a network of passive components which makes the impedance of the loudspeaker constant at all frequencies. The same cable will then appear to have a flat response. Unfortunately few loudspeakers use such compensation because it costs money.

In the same way that the speaker impedance characteristic can give the cable an apparent frequency response, apparent distortion is also possible. The cone of a non-linear speaker drive unit moves at a velocity which is not proportional to the input waveform and so the back EMF is not proportional to the applied EMF. Consequently the current which flows contains distortion components. These do not affect the terminal voltage of the amplifier because it is a voltage source with a low output impedance. However, if this distortion current flows in the finite resistance of a cable, the effect is that the cable appears to be causing distortion! Again the solution is to improve the loudspeaker.

Vendors of specialist speaker cables have developed an extensive range of claims to explain how their products improve quality. As the quality loss is caused by the loudspeaker it is difficult to see how these can be true. Many of the explanations rely on the assumption that a speaker cable is a transmission line and all sorts of nonsense about matching and impedance is trotted out which is just plausible to the novice. Unfortunately at all practicable lengths, a speaker cable simply is not a transmission line.[16]

A true transmission line causes significant delay to the signal so that several cycles of the signal may be present between the input and output. The source and load impedance must both be matched to the characteristic impedance of the line at all signal frequencies. Unfortunately speaker cables aren't used in this way. The electrical wavelength at 20 kHz is 15 km or 9 miles and so getting a phase shift out of a typical length of speaker cable is difficult. The output impedance of power amplifiers tends to zero and the input impedance of speakers is variable.

Another phenomenon which is invoked is the *skin effect* which is the tendency of high-frequency signals to flow in the surface of a conductor rather than throughout the section. Unfortunately the magnitude of skin effect at the highest audio frequency is so small that it can be totally neglected.

In practice the distortion currents in cables tend to be due to the woofer and the practice of *bi-wiring* is useful to combat it. As Figure 6.34 shows, the loudspeaker crossover network is split so that separate input terminals are available for woofer and tweeter. The amplifier is connected to the two sets

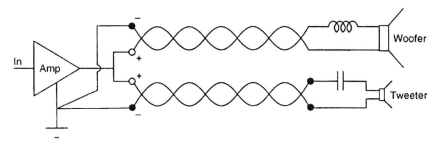

Figure 6.34 Bi-wiring removes heavy woofer currents from tweeter wiring and is a step towards the fully active speaker.

of terminals by different sets of cable. Complex load currents flowing in the woofer circuit cannot then cause voltage drops in the tweeter wiring. Exotic cables are quite unnecessary and heavy gauge AC power cord works well. For those who must have designer cables the author is prepared to autograph lengths of mains cable for a small fee.

Whatever the reader may believe about cables, there can be no doubt that elimination of the cable must eliminate the problem. This is exactly what an active loudspeaker achieves. As the drive units and amplifiers are in the same place, the dilemma of which cable to use is gone which is a great relief. As active loudspeakers are superior in every way to the passive speaker with a remote amplifier, devoting energies to developing speaker cables is doubly futile.

6.14 Active loudspeakers

With modern electronic techniques all passive schemes with their inevitable compromises and shortcomings must be considered obsolete. The traditional wideband power amplifier having low distortion and a flat frequency response which expects to see a loudspeaker with the same characteristics is both overly optimistic and an unnecessary restriction. What is required is a system whereby the radiated sound resembles the original line level signal. What goes on inside the box to achieve this is the designer's problem.

An active loudspeaker containing its own amplifiers can easily introduce equalization and signal processing which can artificially move the fundamental resonance down in frequency and achieve any desired damping factor. The crossover can be performed precisely and at low cost at line level instead of imprecisely at power level using large expensive components. Inverse distortion can be applied to compensate for drive unit deficiencies. The advantage of this approach is that the speaker can be made relatively phase linear and will not suffer from hangover. The smaller cabinet also allows the radiation characteristics of the mid and high frequencies to be improved. The sensitivities of the drive units do not need to be matched so that there is more freedom in their design. In the absence of separate amplifier cabinets less space is needed overall and the dubious merits of exotic speaker cables become irrelevant.

Figure 6.35 shows the block diagram of a modern active loudspeaker. The line-level input passes to a crossover filter which routes low and high

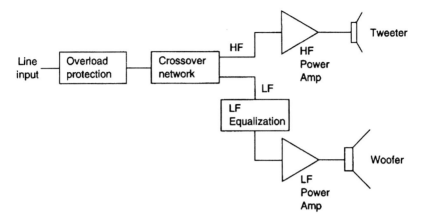

Figure 6.35 In an active loudspeaker the crossover function is performed at line level and separate amplifiers are used for each drive unit. The low-frequency response is controlled by equalization and electronic damping allowing accurate transient reproduction.

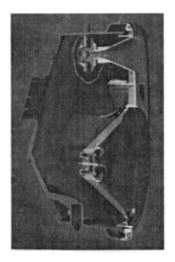

Figure 6.36 Modern active speaker design. (a) JBL PA speaker uses die cast baffle including driver chassis, port and cooling for amplifier. (b) Celtic Legend uses massive heatsink which is decorative, structural, thermal and screens active circuitry.

frequencies to each driver. Each drive unit has its own power amplifier. The low-frequency power amplifier is preceded by a compensation network which can electronically lower the resonance of the low-frequency driver and determine the damping factor. In some units the low-frequency diaphragm is fitted with a feedback transducer so that distortion can be further reduced.

There is more to successful active loudspeaker design than just putting the amplifiers in the cabinet as this is missing a number of opportunities. Figure 6.36(a) shows an active PA speaker by JBL. The baffle is a single piece die casting which incorporates both drive unit chassis and reflex port and also acts as a heatsink. The amplifier is mounted adjacent to the reflex port which is internally finned so that cooling automatically increases with low-frequency level.

In the Celtic Legend shown in Figure 6.36(b) the entire back of the woofer enclosure consists of a massive curved casting which dissipates heat as well as acting as an enclosure stiffening member and an electronics housing and EMC screen. The heatsink assembly and the enclosed power transformer also acts as a reactive mass reducing the amplitude of vibrations at the woofer chassis.

6.15 Headphones

Whilst headphones cannot deliver the spatial realism that stereophonic loud-speakers allow they are nevertheless a practical necessity in many applications. In professional use headphones are the only way of monitoring a signal in the presence of other sounds or of communicating with, for example, a cameraman in the presence of an open microphone. Such headphones must shut out ambient sound and prevent their own internally produced sound from escaping. This requires a design which entirely surrounds the pinna with a rigid cup and a flexible seal to the side of the head as Figure 6.37(a) shows. This type of headphone is known as *circum-aural*.

The seal is usually obtained using a partially filled vinyl bag containing silicone oil which will accommodate the irregularities of the wearer's head. The seal is slightly compliant, but even if it were rigid the wearer's flesh is also compliant. Consequently it is impossible to stop the circum-aural cup vibration to external sound by rigid coupling and the only solution is to make

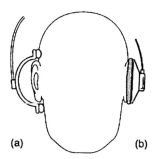

Figure 6.37 (a) Circum-aural headphone seals to the side of the head for maximum isolation. (b) Supra-aural headphone is more comfortable but gives poor isolation.

it heavy. The weight of the cups is taken by a soft band which rests across the top of the wearer's head.

Circum-aural headphones are very effective but they are heavy and perspiration under the seals makes them uncomfortable after long periods. A further problem is that the effective seal of the cup around the ear raises the acoustic impedance and this means that sounds caused by muscle action and blood flow which are normally inaudible become obvious and some people find this irritating. This is known as the *occlusion effect* and is the same phenomenon which produces sound when a sea-shell is held to the ear.

Where effective sealing is not essential or even undesirable, the *supra-aural* headphone is popular. Figure 6.37(b) shows that with this construction the body of the headphone sits lightly upon the pinna with a layer of acoustically open foam between. This much lighter construction is more comfortable to wear and allows external sounds to be heard and eliminates the occlusion effect. However, it will also allow the sound produced by the headphones to leak out. A further problem is that the imperfect seal of the supra-aural headphone makes reproduction of the lowest frequencies difficult. Low-frequency level is also dependent on how the headphones are worn as the degree of compression of the foam will affect the frequency response.

Headphone-drive mechanisms are much the same as in loudspeakers and both moving coil and electrostatic designs are available. Electrostatic headphones work very well because of their minimal phase distortion and are quite light, but the cables needed for safe high voltage insulation tend to be bulky and a separate power unit is required. Consequently the majority of headphones rely on the moving coil principle. As weight is an issue headphones embraced rare-earth magnets as soon as they were economic.

Whilst the mechanics of a headphone may be superficially similar to those of a loudspeaker, the acoustics are quite different. In the case of the loudspeaker the listener is in the far field, whereas with the headphone the listener is in the near field. The presence of the listener's head and pinnae are essential to the functioning of the headphone which sounds tinny and feeble if it is taken off. Headphone designers have to use dummy heads which simulate the acoustic conditions when the headphones are being worn in order to make meaningful tests.

In particular with a circum-aural headphone the transducer diaphragm is working into a small sealed pressure chamber in which ka remains small up to several kHz. Under these conditions the pressure is proportional to diaphragm displacement and independent of frequency whereas in a loudspeaker the pressure is proportional to the diaphragm acceleration. Consequently the frequency response of a headphone is tilted by 12 dB/octave with respect to that of a loudspeaker. Figure 6.38(a) shows the mechanical response of a resonant diaphragm reaching a peak. When installed in a headphone the response is flat below resonance but falls at 12 dB/octave above as shown in (b). Fortunately headphones do not need to be very efficient as they are radiating into such a small volume and heavy damping can be used to flatten the response.

The presence of air leaks in a circum-aural headphone or the deliberate introduction of leaks in supra-aural headphones results in a low-frequency roll-off as shown in (c). This is often compensated by the introduction of passive membranes which resonate to boost output.

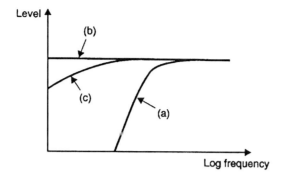

Figure 6.38 (a) Diaphragm response in free air. (b) Response in headphone being worn. (c) Air leaks cause low-frequency loss.

References

1. Kelly, S., Transducer drive mechanisms, in Borwick, J. (ed.), *Loudspeaker and Headphone Handbook*, Chapter 2, Focal Press: Oxford (1994)
2. Olson, H.F., Stereophonic sound reproduction in the home. *J. Audio Eng. Soc.*, **6**, No. 2, 80–90 (1958)
3. Moir, J., Speaker directivity and sound quality. *Wireless World*, **85**, No. 1526 (Oct 1979)
4. Rice, C.W. and Kellog, E.W., Notes on the development of a new type of hornless loudspeaker. *JAIEE*, **12**, 461–480 (1925)
5. Shorter, D.E.L., A survey of performance criteria and design considerations for high-quality monitoring loudspeakers. *Proc. IEE*, **105**, Part B, No. 24, 607–623 (Nov 1958)
6. Kates, J.M., *Radiation from a dome, in AES Anthology: Loudspeakers, Vol. 1–Vol. 25*, New York: Audio Engineering Society (1978)
7. Walker, P.J., New developments in electrostatic loudspeakers. *J. Audio Eng. Soc.*, **28**, No. 11, 795–799 (Nov 1980)
8. Bottenberg, W., Melillo, L. and Raj, K., The dependence of loudspeaker design parameters on the properties of magnetic fluids, in *AES Anthology: Loudspeakers, Vol. 26–Vol. 31*, New York: Audio Engineering Society (1984)
9. Olson, H.F., *Acoustical Engineering*, Philadelphia: Professional Audio Journals Inc. (1991)
10. Kates, J.M., Loudspeaker cabinet reflection effects. *J. Audio Eng. Soc.*, **27**, No. 5 (1979)
11. Small, R.H., Constant voltage crossover design, in *AES Anthology: Loudspeakers, Vol. 1–Vol. 25*, New York: Audio Engineering Society (1978)
12. Baxandall, P.J., Electrostatic loudspeakers, in Borwick, J. (ed.), *Loudspeaker and Headphone Handbook*, Chapter 3, Oxford: Focal Press (1994)
13. Walker, P.J., Current dumping. *Wireless World*, 560–562 (Dec 1975)
14. Sandman, A.M., *Wireless World*, 38–39 (Sep. 1982)
15. Thiele, A.N., Optimum passive loudspeaker dividing networks. *Proc. IREE* (July 1975)
16. Greiner, R.A., Amplifier–Loudspeaker interfacing, in *AES Anthology: Loudspeakers, Vol. 26–Vol. 31*, New York: Audio Engineering Society (1984)

Chapter 7

Stereophony

Much greater spatial realism is obtained by using more than one loudspeaker. Generally a reproduction system having some spatial realism will be preferred to a technically superior monophonic system. This chapter treats the directional ability of human hearing as the basis on which spatial illusions can be created.

7.1 History of stereophony

Stereophonic sound is actually older than most people think. The first documented public demonstration was in 1881 when Ader used pairs of telephone lines to convey stereo from the Paris Opera.[1] A multi-channel cylinder recorder was marketed in 1898.

Probably the largest single contributor to stereophony was Alan Dower Blumlein whose 1931 patent application[2] contains the essential elements of stereo systems used today. It included the use of velocity microphones, panning, M-S and width control as well as describing the 45/45 stereo disk-cutting technique.

The outbreak of the Second World War delayed the introduction of stereo for consumer purposes and it was not until the mid-1950s that prerecorded reel-to-reel stereo tapes became available. These were closely followed by the stereophonic vinyl disk and the FM multiplex stereo broadcasting system. From that point on, consumer audio was essentially assumed to be stereo.

For a comprehensive anthology of important stereophonic papers the reader is recommended to read 'Stereophonic techniques'.[3]

7.2 Directionality in hearing

The human listener can determine reasonably well where a sound is coming from. An understanding of the mechanisms of direction sensing is important for the successful implementation of spatial illusions such as stereophonic sound. As Figure 7.1 shows, having a pair of spaced ears allows a number of mechanisms. At (a) a phase shift will be apparent between the two versions of a tone picked up at the two ears unless the source of the tone is dead ahead (or behind). At (b) the distant ear is shaded by the head resulting in reduced response compared to the nearer ear. At (c) a transient sound arrives later at the more distant ear.

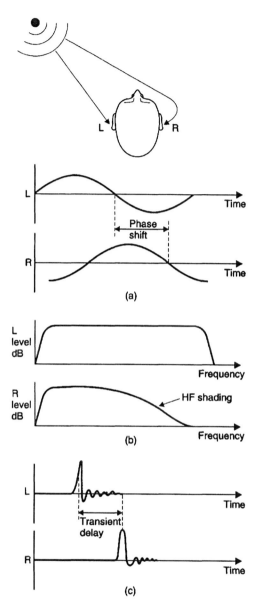

Figure 7.1 Having two spaced ears is cool. (a) Off-centre sounds result in phase difference. (b) Distant ear is shaded by head producing loss of high frequencies. (c) Distant ear detects transient later.

If the phase-shift mechanism (a) is considered, then it will be clear that there will be considerable variation in this effect with frequency. At a low frequency such as 30 Hz, the wavelength is around 11.5 metres. Even if heard from the side, the ear spacing of about 0.2 metres will result in a phase shift of only 6° and so this mechanism must be quite weak at low frequencies.

At high frequencies such as 10 kHz, the ear spacing is many wavelengths and variations in the path-length difference will produce a confusing and complex phase relationship. The problem with tones or single frequencies is that they produce a sinusoidal waveform, one cycle of which looks much like another leading to ambiguities in the time between two versions. This is shown in Figure 7.2(a).

Pure tones are extremely difficult to localize, especially as they will often excite room-standing waves which give a completely misleading impression of the location of the sound source. Consequently the phase-comparison mechanism must be restricted to frequencies where the wavelength is short enough to give a reasonable phase shift, but not so short that complete cycles of shift are introduced. This suggests a frequency limit of around 1500 Hz which has been confirmed by experiment.

The shading mechanism of Figure 7.1(b) will be a function of the omnipresent *ka* suggesting that at low and middle frequencies sound will diffract round the head sufficiently well that there will be no significant difference between the level at the two ears. Only at high frequencies does sound become directional enough for the head to shade the distant ear causing what is called an inter-aural intensity difference (IID). At very high frequencies, the shape of the pinnae must have some effect on the sound which is a function of direction. It is thought that the pinnae allow some discrimination in all axes.

Phase differences are only useful at low frequencies and shading only works at high frequencies. Fortunately real-world sounds are timbral or broadband and often contain transients, especially those sounds which indicate a potential hazard. Timbral, broadband and transient sounds differ from tones in that

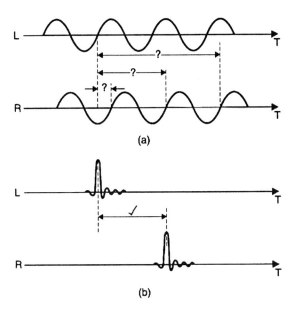

(a)

(b)

Figure 7.2 (a) Pure tones cause ambiguity in timing differences. (b) Transients have no ambiguity and are easier to localize.

they contain many different frequencies. A transient has a unique aperiodic waveform which Figure 7.2(b) shows has the advantage that there can be no ambiguity in the inter-aural delay (IAD) between two versions. Figure 7.3 shows the time difference for different angles of incidence for a typical head. Note that a 1° change in sound location causes an IAD of around 10 μs. The smallest detectable IAD is a remarkable 6 μs.

Chapter 3 showed that the basilar membrane is a frequency analysis device which produces nerve impulses from different physical locations according to which frequencies are present in the incident sound. Clearly when a timbral or transient sound arrives from one side, many frequencies will be excited simultaneously in the nearer ear, resulting in a pattern of nerve firings. This will closely be followed by a similar excitation pattern in the further ear. Shading may change the relative amplitudes of the higher frequencies, but it will not change the pattern of frequency components present.

A timbral waveform is periodic at the fundamental frequency but the presence of harmonics means that a greater number of nerve firings can be compared between the two ears. As the statistical deviation of nerve firings with respect to the incoming waveform is about 100 μs the only way an IAD of 6 μs can be perceived is if the timing of many nerve firings is correlated in some way.

The broader the range of frequencies in the sound source, the more effective this process will be. Analysis of the arrival time of transients is a most effective lateral direction-sensing mechanism. This is easily demonstrated by wearing a blindfold and having a helper move around the room making a variety of noises. The helper will be easier to localize when making clicking noises than when humming. It is easy to localize a double bass despite the low fundamental as it is a harmonically rich instrument.

It must be appreciated that human hearing can locate a number of different sound sources simultaneously. The hearing mechanism must be constantly comparing excitation patterns from the two ears with different delays. Strong correlation will be found where the delay corresponds to the inter-aural delay for a given source. This is apparent in the binaural threshold of hearing which

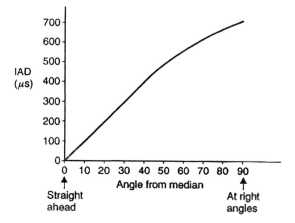

Figure 7.3 The inter-aural delay (IAD) for various arrival directions.

is 3–6 dB better than monaural at around 4 kHz. This delay-varying mechanism will take time and it is to be expected that the ear would then be slow to react to changes in source direction. This is indeed the case and experiments have shown that oscillating sources can only be tracked up to 2–3 Hz.[4] The ability to locate bursts of noise improves with burst duration up to about 700 ms.

The inter-aural phase, delay and level mechanisms vary in their effectiveness depending on the nature of the sound to be located. A fixed response to each mechanism would be ineffective. For example, on a low-frequency tone, the time-of-arrival mechanism is useless whereas on a transient it excels. The different mechanisms are quite separate on one level, but at some point in the brain's perception a fuzzy logic or adaptive decision has to be made as to how the outcome of these mechanisms will be weighted to make the final judgement of direction.

7.3 Hearing in reverberant conditions

Chapter 3 introduced the finite time resolution of the ear due to the resonant behaviour of the basilar membrane. This means that the perception we have of when a sound stops and starts is not very precise. This is just as well because we live in a reverberant world which is filled with sound reflections. If we could separately distinguish every diffcrent reflection in a reverberant room we would hear a confusing cacophony. In practice we hear very well in reverberant surroundings, far better than microphones can, because of the transform nature of the ear and the way in which the brain processes nerve signals.

When two or more versions of a sound arrive at the ear, provided they fall within a time span of about 30 ms, they will not be treated as separate sounds, but will be fused into one sound. Only when the time separation reaches 50–60 ms do the delayed sounds appear as echoes from different directions. As we have evolved to function in reverberant surroundings, reflections do not impair our ability to locate the source of a sound. The fusion will be impaired if the spectra of the two sounds are too dissimilar.

A moment's thought will confirm that the *first* version of a transient sound to reach the ears must bo the one which has travelled by the shortest path. Clearly this must be the direct sound rather than a reflection. Consequently the ear has evolved to attribute source direction from the time of arrival difference at the two ears of the first version of a transient. Later versions which may arrive from elsewhere simply add to the perceived loudness but do not change the perceived location of the source.

This phenomenon is known as the precedence or Haas effect after the Dutch researcher who investigated it. Haas found that the precedence effect is so powerful that even when later arriving sounds are artificially amplified (a situation which does not occur in nature) the location still appears to be that from which the first version arrives. Figure 7.4 shows how much extra level is needed to overcome the precedence effect as a function of arrival delay.

Experiments have been conducted in which the delay and intensity clues are contradictory to investigate the way the weighting process works. The same sound is produced in two locations but with varying relative delay and shading

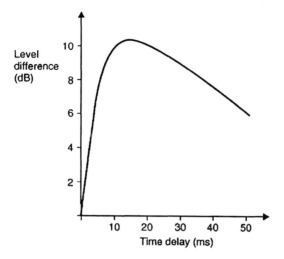

Figure 7.4 The precedence effect is powerful. This curve shows the extra level which is needed in a later sound to overcome the precedence effect.

dependent level. The way in which the listener perceives an apparent sound direction reveals how the directional clues are weighted.

Within the maximum inter-aural delay of about 700 µs the precedence effect does not function and the perceived direction can be pulled away from that of the first arriving source by an increase in level. Figure 7.5 shows that this area is known as the time-intensity trading region. Once the maximum inter-aural delay is exceeded, the hearing mechanism knows that the time difference must be due to reverberation and the trading ceases to change with level.

It is important to realize that in real life the hearing mechanism expects a familiar sound to have a familiar weighting of phase, time of arrival and

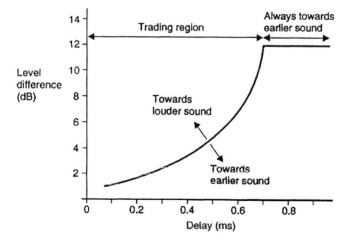

Figure 7.5 Time-intensity trading occurs within the inter-aural delay period.

shading clues. A high-quality sound reproduction system must do the same if a convincing spatial illusion is to be had. Consequently a stereo system which attempts to rely on just one of these effects will not sound realistic. Worse still is a system which relies on one effect to be dominant but where another is contradictory. Time-intensity trading is an interesting insight into the operation of the hearing mechanism, but it cannot be used in quality sound reproduction because although the ear is fooled, it *knows* it is being fooled because it is attempting to resolve conflicting stimuli and the result is inevitably listening fatigue.

In the presence of an array of simultaneous sound sources the hearing mechanism has an ability to concentrate on one of them based on its direction. The brain appears to be able to insert a controllable time delay in the nerve signals from one ear with respect to the other so that when sound arrives from a given direction the nerve signals from both ears are coherent. Sounds arriving from other directions are incoherent and are heard less well. This is known as *attentional selectivity*[5] but is more usually referred to as the *cocktail-party effect*.

Monophonic systems prevent the use of this effect completely because the first version of all sounds reaching the listener come from the same loudspeaker. Stereophonic systems allow the cocktail-party effect to function in that the listener can concentrate on specific sound sources in a reproduced stereophonic image with the same facility as in the original sound.

One of the most compelling demonstrations of stereo is to make a stereophonic recording of a crowded function in a reverberant room. On replaying several times it will be possible to employ attentional selectivity to listen to a different conversation each time. Upon switching to mono it will be found that the result is completely unintelligible. A corollary of this experiment is that if a striking difference between stereo and mono is not heard, there is a defect in the equipment or the listener.

One of the most frustrating aspects of hearing impairment is that hearing loss in one ear destroys the ability to use the cocktail-party effect. In quiet surroundings many people with hearing loss can follow what is said in normal tones. In a crowded room they are at a serious disadvantage because they cannot select a preferred sound source.

Laterally separated ears are ideal for determining the location of sound sources in the plane of the earth's surface, which is after all where most sound sources emanate. Our ability to determine height in sound is very poor. As the ears are almost exactly half-way back on each side of the head it is quite possible for sound sources ahead or behind, above or below to produce almost the same relative delay, phase shift and shading resulting in an ambiguity. This leads to the concept of the *cone of confusion* where all sources on a cone with the listener at the vertex will result in the same IAD.

There are two main ways in which the ambiguity can be resolved. If a plausible source of sound can be seen, then the visual clue will dominate. Experience will also be used. People who look up when they hear birdsong may not be able to determine the height of the source at all, they may simply know, as we all do, that birds sing in trees.

A second way of resolving front/back ambiguity is to turn the head slightly. This is often done involuntarily and most people are not aware they are using the technique. In fact when people deliberately try harder to locate a sound they often keep their head quite still making the ambiguity worse. Section 1.2

showed that intensity stereo recordings are fundamentally incompatible with headphone reproduction. A further problem with headphones is that they turn with the wearer's head, disabling the ability to resolve direction by that means.

The convolutions of the pinna also have some effect at high frequencies where the wavelength of sound is sufficiently short. The pinna produces a comb-filtering spectral modification which is direction dependent.

Figure 7.6 shows that when standing, sounds from above reach the ear directly and via a ground reflection which has come via a longer path. (There is also a smaller effect due to reflection from the shoulders.) At certain frequencies the extra path length will correspond to a 180° phase shift, causing cancellation at the ear. The result is a frequency response consisting of evenly spaced nulls which is called *comb filtering*. A moving object such as a plane flying over will suffer changing geometry which will cause the frequency of the nulls to fall towards the point where the overhead position is reached.

The result depends on the kind of aircraft. In a piston-engined aircraft, the sound is dominated by discrete frequencies such as the exhaust note. When such a discrete spectrum is passed through a swept comb filter the result is simply that the levels of the various components rise and fall. The Doppler effect has the major effect on the pitch which appears to drop suddenly as the plane passes. In the case of a jet plane, the noise emitted on approach is broadband white noise from the compressor turbines. As stated above, the Doppler effect has no effect on aperiodic noise. When passed through a swept comb filter, the pitch of the output appears to fall giving the characteristic descending whistle of a passing jet.

The direction-sensing ability has been examined by making binaural recordings using miniature microphones actually placed down the ear canals of a volunteer. When these are played back on headphones to the person whose ears were used for the recording, full localization of direction including front/rear and height discrimination is obtained. However, the differences between people's ears are such that the results of playing the recording to someone else are much worse. The same result is obtained if a dummy head is used.

Whilst binaural recordings give very realistic spatial effects, these effects are only obtained on headphones and consequently the technique is unsatisfactory for signals intended for loudspeaker reproduction and cannot used in prerecorded music, radio or television.

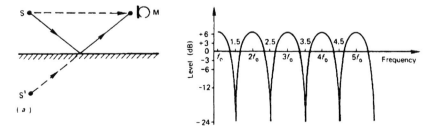

Figure 7.6 Comb-filtering effect produced by interference between a sound wave and a delayed version of itself. (a) Microphone M receives direct wave from source S and a delayed wave from image sound source S^1. (b) Frequency response shows alternating peaks and troughs resembling the teeth of a comb.

When considering the localization ability of sound, it should be appreciated that vision can produce a very strong clue. If only one person can be seen speaking in a crowd, then any speech heard must be coming from that person. The same is true when watching films or television. This is a bonus because it means that the utmost precision is not required in the spatial accuracy of stereo or surround sound accompanying television pictures. However, if the inaccuracy is too great fatigue may result and the viewer may have a displacement of his localization for some time afterwards.[6]

7.4 The stereophonic illusion

The most popular technique for giving an element of spatial realism in sound is *stereophony*, nowadays abbreviated to stereo, based on two simultaneous audio channels feeding two spaced loudspeakers. Figure 7.7 shows that the optimum listening arrangement for stereo is where the speakers and the listener are at different points of a triangle which is almost equilateral.

Stereophony works by creating differences of phase and time of arrival of sound *at the listener's ears*. It was shown above that these are the most powerful hearing mechanisms for determining direction. Figure 7.8(a) shows that this time of arrival difference is achieved by producing the same waveform at each speaker simultaneously, but with a difference in the relative level, rather than phase. Each ear picks up sound from both loudspeakers and sums the waveforms.

The sound picked up by the ear on the same side as the speaker is in advance of the same sound picked up by the opposite ear. When the level emitted by the left loudspeaker is greater than that emitted by the right, it will be seen from Figure 7.8(b) that the sum of the signals received at the left ear is a waveform which is phase advanced with respect to the sum of the waveforms received at the right ear. If the waveforms concerned are transient the result will be a time-of-arrival difference. These differences are interpreted as being due to a sound source left of centre.

The stereophonic illusion only works properly if the two loudspeakers are producing in-phase signals. In the case of an accidental phase reversal, the

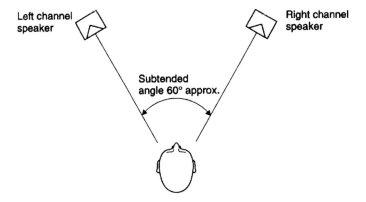

Figure 7.7 Configuration used for stereo listening.

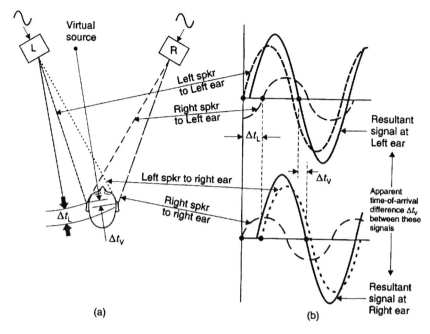

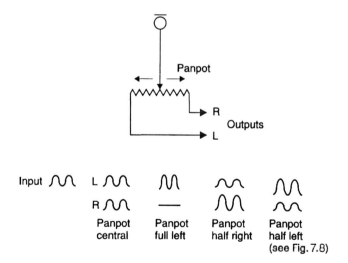

Figure 7.8 (a) Stereo illusion is obtained by producing the same waveform at both speakers, but with different amplitudes. (b) As both ears hear both speakers but at different times, relative level causes apparent time shift at the listener. Δt_L = inter-aural delay due to loudspeaker; Δt_V = inter-aural delay due to virtual source.

Figure 7.9 The panpot distributes a monophonic microphone signal into two stereo channels allowing the sound source to be positioned anywhere in the image.

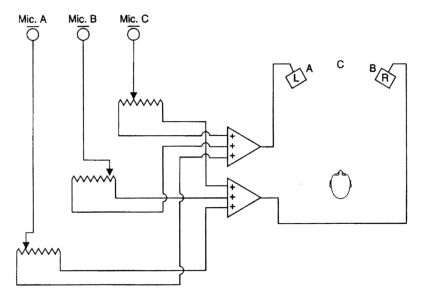

Figure 7.10 Multi-channel mixing technique pans multiple sound sources into one stereo image.

spatial characteristic will be ill-defined and lack images. At low frequencies the two loudspeakers are in one another's near field and so anti-phase connection results in bass cancellation.

As the apparent position of a sound source between the two speakers can be controlled solely by the relative level of the sound emitted by each one, stereo signals in this format are called *intensity stereo*. In intensity stereo it is possible to 'steer' a monophonic signal from a single microphone into a particular position in a stereo image using a form of differential gain control. Figure 7.9 shows that this device, known as a *panoramic potentiometer* or panpot for short, will produce equal outputs when the control is set to the centre. If the panpot is moved left or right, one output will increase and the other will reduce, moving or *panning* the stereo image to one side.

If the system is perfectly linear, more than one sound source can be panned into a stereo image, with each source heard in a different location. This is done using a stereo mixer, shown in Figure 7.10, in which monophonic inputs pass via panpots to a stereo mix bus. The majority of pop records are made in this way, usually with the help of a multitrack tape recorder with one track per microphone so that mixing and panning can be done at leisure.

7.5 Stereo microphones

Panpotted audio can never be as realistic as the results of using a stereo microphone because the panpot causes all of the sound to appear at one place in the stereo image. In the real world the direct sound should come from that location but reflections and reverberation should come from elsewhere. Artificial reverberation has to be used on panpotted mixes.

The job of a stereo microphone is to produce two audio signals which have no phase or time differences but whose relative levels are a function of the direction from which sound arrives. The most spatially accurate technique involves the use of directional microphones which are coincidentally mounted but with their polar diagrams pointing in different directions. This configuration is known variously as a *crossed pair* or a *coincident pair*. Figure 7.11 shows a stereo microphone constructed by crossing a pair of figure-of-eight microphones at 90°.

The output from the two microphones will be equal for a sound source straight ahead, but as the source moves left, the output from the left-facing microphone will increase and the output from the right-facing microphone will reduce. When a sound source has moved 45° off-axis, it will be in the response null of one of the microphones and so only one loudspeaker will emit sound. Thus the fully left or fully right reproduction condition is reached at ±45°. The angle between nulls in L and R is called the acceptance angle which has some parallels with the field of view of a camera.

Sounds between 45 and 135° will be emitted out of phase and will not form an identifiable image. Important sound sources should not be placed in this region. Sounds between 135 and 225° are in-phase and are mapped onto the frontal stereo image. The all-round pickup of the crossed eight makes it particularly useful for classical music recording where it will capture the ambience of the hall.

Other polar diagrams can be used, for example the crossed cardioid, shown in Figure 7.12, is popular. There is no obvious correct angle at which cardioids should be crossed, and the actual angle will depend on the application. Commercially available stereo microphones are generally built on the side-fire principle with one capsule vertically above the other. The two capsules

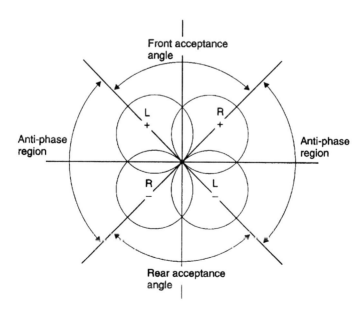

Figure 7.11 Crossed eight stereo microphone. Note acceptance angle between nulls.

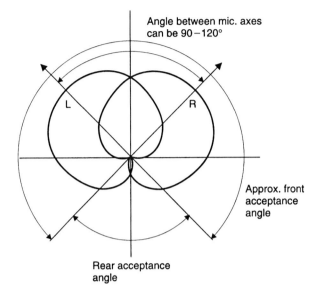

Figure 7.12 Crossed cardioid microphone.

can be independently rotated to any desired angle. Usually the polar diagrams of the two capsules can be changed.

In the soundfield microphone, four capsules are fitted in a tetrahedron. By adding and subtracting proportions of these four signals in various ways it is possible to synthesize a stereo microphone having any acceptance angle and to point it in any direction relative to the body of the microphone. This can be done using the control box supplied with the microphone. Although complex the soundfield microphone has the advantage that it can be electrically steered and so no physical access is needed after it is slung. If all four outputs are recorded, the steering process can be performed in post production by connecting the control box to the recorder output.

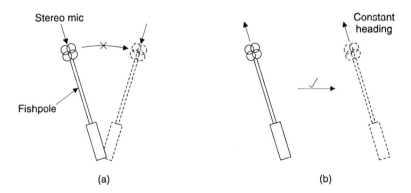

Figure 7.13 (a) Swinging a fishpole causes sound image to rotate. Tracking has to be used as in (b).

Clearly the use of stereo will make it obvious if a microphone has been turned. In applications such as TV sound the microphone is turned as a side-effect of swinging a boom or fishpole, as in Figure 7.13(a). This is undesirable in stereo, and more precise handling techniques are necessary to keep the microphone heading constant as in (b).

7.6 Headphone stereo

It should be clear that the result of Figure 7.8 cannot be obtained with headphones because these prevent both ears receiving both channels. As a result there is no stereophonic image and the sound appears, quite unrealistically, to be inside the listener's head. Consequently headphones are quite useless for monitoring the stereo image in signals designed for loudspeaker reproduction.

Highly realistic results can be obtained on headphones using the so-called *dummy head* microphone which Figure 7.14 shows is a more or less accurate replica of the human head with a microphone at each side. Clearly the two audio channels simply move the listener's ears to the location of the dummy. Unfortunately dummy head signals are incompatible with loudspeaker reproduction and are not widely used.

Headphones can be made compatible with signals intended for loudspeakers using a shuffler.[7] This device, shown in Figure 7.15, simulates the cross-coupling of loudspeaker listening by feeding each channel to the other ear via a delay and a filter which attenuates high frequencies to simulate the effect of head shading. The result is a sound image which appears in front of the listener so that decisions regarding the spatial position of sources can be made. Although the advantages of the shuffler have been known for decades, the information appears to have eluded most equipment manufacturers.

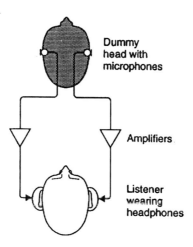

Dummy
head with
microphones

Amplifiers

Listener
wearing
headphones

Figure 7.14 In a dummy head stereo system the microphones and headphones literally bring the listener's ears to the original soundfield. Unfortunately the signal format is incompatible with loudspeaker reproduction.

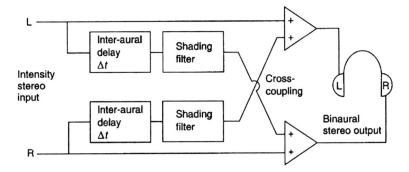

Figure 7.15 The shuffler converts L, R loudspeaker stereo into signals suitable for headphone reproduction by simulating the cross-coupling, delay and shading of loudspeaker reproduction.

7.7 Alternative microphone techniques

Other microphone techniques will be found in stereo, such as the use of spaced omnidirectional microphones. Spaced microphones result in time-of-arrival differences and frequency-dependent phase shifts in the output signals. When reproduced by loudspeakers and both signals are heard by both ears the mechanism of Figure 7.8 simply cannot operate. Instead the result shown in Figure 7.16 is obtained. Consider a continuous off-axis source well to the left. This will result in two microphone signals having a small amplitude difference but a large time difference. If initially it is assumed that the source frequency is such that the path-length difference is an integer number of wavelengths

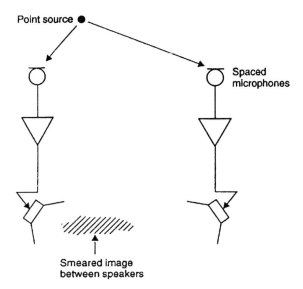

Figure 7.16 Spaced microphones cannot form images in loudspeaker reproduction because the phase differences between them are frequency dependent. Consequently the result is image smear which gives a spacious effect but without identifiable images.

there will be no phase difference. This will be reproduced by two speakers as a source slightly left of centre because of the small amplitude difference.

If the frequency is reduced slightly, the wavelength becomes longer and the right microphone signal will begin to lead in phase, moving the phantom image right. If the frequency is increased slightly, the wavelength becomes shorter and the right microphone signal lags, moving the image left. Consequently the apparent location of a source is highly frequency dependent and the result is that all sources having a harmonic content appear to be very wide.

Where the distance between microphones corresponds to an odd multiple of a half wavelength, the two microphone signals are out of phase. As both ears hear both loudspeakers the result is a dip in frequency response and no image at all. This comb-filtering effect is a serious drawback of spaced microphone techniques and makes monophonic compatibility questionable.

A central source will give identical signals which create a central phantom image. However, the slightest movement of the source off-axis results in time differences which pull the image well to the side. The resulting hole-in-the-middle effect is often counteracted by a third central microphone which is equally fed into the two channels.

This technique cannot be described as stereophonic *reproduction* because it is impossible for the listener to locate a particular source. Instead a spacious *effect* is the result. As there is no scientific basis for the technique it is hardly surprising that there is no agreement on the disposition of the microphones and a great many spacings from a few centimetres to a few metres are used. One rule of thumb which has emerged is that the spacing should be no more than one-third of the source width.

7.8 M-S stereo

In audio production the apparent width of the stereo image may need to be adjusted, especially in television to obtain a good audio transition where there has been a change of shot or to match the sound stage to the picture. This can be done using *M-S stereo* and manipulating the difference between the two channels. Figure 7.17(a) shows that the two signals from the microphone, L and R, are passed through a *sum and difference* unit which produces two signals, M and S. The M or Mid signal is the sum of L and R whereas the S or Side signal is the difference between L and R. The sums and differences are divided by two to keep the levels correct.

The result of this sum-difference process can be followed in Figure 7.17(b). A new polar diagram is drawn which represents the sum of L and R for all angles of approach. It will be seen that this results in a forward-facing eight, as if a monophonic microphone had been used, hence the term M or Mid. If the same process is performed using L-R, the result is a sideways facing eight, hence the term S or Side. In L, R format the acceptance angle is between the nulls whereas in M-S format the acceptance angle is between the points where the M and S polar diagrams cross.

The S signal can now be subject to variable gain. Following this a second sum and difference unit is used to return to L, R format for monitoring. The S gain control effectively changes the size of the S polar diagram without affecting the M polar diagram. Figure 7.17(c) shows that reducing the S gain makes the acceptance angle wider whereas increasing the S gain makes it

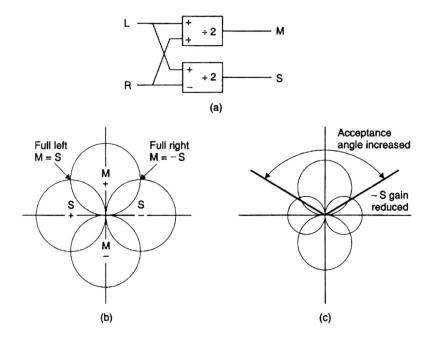

Figure 7.17 (a) M-S working adds and subtracts the L, R stereo signals to produce Mid and Side signals. In the case of a crossed eight, the M-S format is the equivalent of forward and sideways facing eights. (c) Changing the S gain alters the acceptance angle of the stereo microphone.

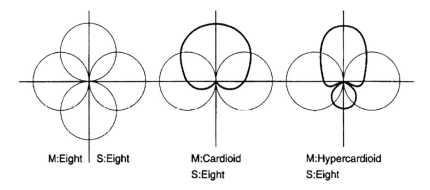

Figure 7.18 Various M-S microphone configurations. Note that the S microphone must always be an eight.

smaller. Clearly if the S gain control is set to unity, there will be no change to the signals.

Whilst M-S stereo can be obtained by using a conventional L, R microphone and a sum and difference network, it is clear from Figure 7.17(b) that M-S signals can be obtained directly using a suitable microphone. Figure 7.18 shows a number of M-S microphones in which the S capsule is always an eight. A variety of responses (other than omni) can be used for the M capsule.

The M-S microphone technique has a number of advantages. The narrowing polar diagram at high frequencies due to diffraction is less of a problem because the most prominent sound source will generally be in the centre of the stereo image and this is directly on the axis of the M capsule. An image width control can easily be built into an M-S microphone. A favourite mono microphone can be turned into an M-S microphone simply by mounting a side-facing eight above it.

On a practical note it is often necessary to use a stereo microphone with a fishpole. In some shots the microphone will be above the action, but in a close-up it may also be used below shot. Inverting the microphone in this way will interchange Left and Right channels. If an M-S microphone is used, the fishpole operator can operate a simple phase reverse switch in the S channel which will reverse the channels. Using balanced signals a passive phase reverse switch is possible and can easily be built into an XLR barrel connector.

7.9 Mono compatibility

Whilst almost all fixed audio equipment is now stereo, the portable radio or television set may well remain monophonic for some time to come and it will be necessary to consider the mono listener when making stereo material. There is a certain amount of compatibility between intensity stereo and mono systems. If the S gain of a stereo signal is set to zero, only the M signal will pass. This is the component of the stereo image due to sounds from straight ahead and is the signal used when monophonic audio has to be produced from stereo.

Sources positioned on the extreme edges of the sound stage will not appear as loud in mono as those in the centre and any anti-phase ambience will cancel out, but in most cases the result is adequate. Clearly an accidental situation in which one channel is phase reversed is catastrophic in mono as the centre of the image will be cancelled out. Stereo signals from spaced microphones generally have poor mono compatibility because of comb filtering.

One characteristic of stereo is that the viewer is able to concentrate on a sound coming from a particular direction using the cocktail-party effect. Thus it will be possible to understand dialogue which is quite low in level even in the presence of other sounds in a stereo mix. In mono the listener will not be able to use spatial discrimination and the result may be reduced intelligibility which is particularly difficult for those with hearing impairments.[8] Consequently it is good practice to monitor stereo material in mono to check for acceptable dialogue.

A mono signal can also be reproduced on a stereo system by creating identical L and R signals, producing a central image only. Whilst there can be no real spatial information most people prefer mono on two speakers to mono on a single speaker.

7.10 Stereo metering

In stereo systems it is important that the left and right channels display the same gain after line up. It is also important that the left and right channels are not inadvertently exchanged, and that both channels have the same polarity.

Often an indication of the width of the stereo image is useful. In some stereo equipment a *twin PPM* is fitted, having two needles which operate coaxially. One is painted red (L) and the other green (R). In stereo line-up tone, the left channel may be interrupted briefly so that it can be distinguished from the right channel. The interruptions are so brief that the PPM reading is unaffected.

Unfortunately the twin PPM gives no indication that the unacceptable out-of-phase condition exists. A better solution is the *twin-twin PPM* which is two coaxial PPMs, one showing L, R and one showing M, S (see Section 7.8). When lining up for identical channel gain, obtaining an S null is more accurate. Some meters incorporate an S gain boost switch so that a deeper null can be displayed.

When there is little stereo width, the M reading will exceed the S reading. Equal M and S readings indicate a strong source at one side of the sound stage. When an anti-phase condition is met, the S level will exceed the M level. The M needle is usually white, and the S needle is yellow. This is not very helpful under dim incandescent lighting which makes both appear yellow. Exasperated users sometimes dismantle the meter and put black stripes on the S needle. In modern equipment the moving coil meter is giving way to the bargraph meter which is easier to read.

The audio vectorscope is a useful tool which gives a lot of spatial information although it is less useful for level measurements. If an oscilloscope is connected in X, Y mode so that the M signal causes vertical beam deflection

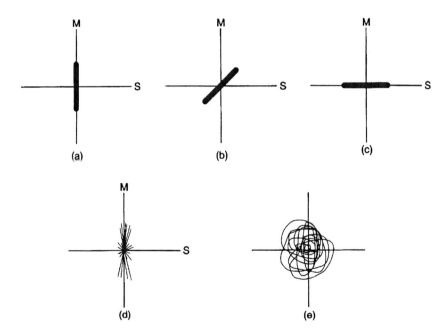

Figure 7.19 An audio vectorscope deflects the beam vertically for M inputs and horizontally for S inputs. (a) Mono signal L = R. (b) Fully right. (c) Anti-phase condition. (d) Coincident microphones with dominant central image. (e) Spaced omni microphones.

and the S signal causes lateral deflection, Figure 7.19 shows that the result will be a trace which literally points to the dominant sound sources in the stereo image. Visual estimation of the width of the stereo image is possible. An out-of-phase condition causes the trace to become horizontal. Non-imaging stereo from, for example, spaced microphones causes the trace to miss the origin because of phase differences between the channels.

Whilst self-contained audio vectorscopes are available, it is possible to employ a unit which synthesizes a video signal containing the vectorscope picture. This can then be keyed into the video signal of a convenient picture monitor. Some units also provide L, R, M, S bargraphs or virtual meters in the video.

7.11 Surround sound

Whilst the stereo illusion is very rewarding when well executed, it can only be enjoyed by a small audience and often a television set will be watched by more people than can fit into the acceptable stereo reproduction area. Surround sound is designed to improve the spatial representation of audio in film and TV.

Figure 7.20 shows the 5.1 channel system proposed for advanced television sound applications. In addition to the conventional L and R stereo speakers at the front, a centre speaker is used. When normal L, R stereo is heard from off-centre, the image will be pulled towards the nearer speaker. The centre speaker is primarily to pull central images back for the off-centre listener. In most television applications it is only the dialogue which needs to be centralized and consequently the centre speaker need not be capable of the full frequency range.

Rear L and R speakers are provided to allow reasonably accurate reproduction of sound sources from any direction, making a total of five channels.

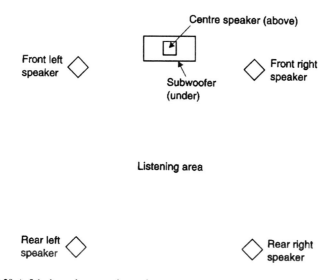

Figure 7.20 A 5.1 channel surround sound system.

A narrow bandwidth subwoofer channel is also provided to produce low frequencies for the inevitable earthquakes and explosions. The restricted bandwidth means that six full channels are not required, hence the term 5.1.

5.1 channel systems require the separate channels to be carried individually to the viewer. This is easy in digital compressed systems such as MPEG, but not in most consumer equipment such as VCRs which only have two audio

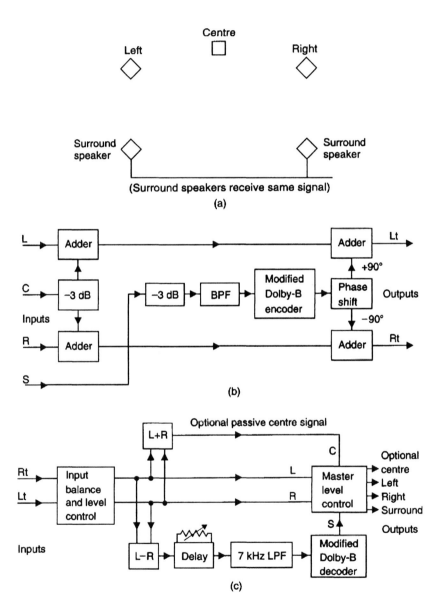

Figure 7.21 (a) Dolby surround sound speaker layout. (b) Dolby surround encoder; see text. (c) Simple passive decoder.

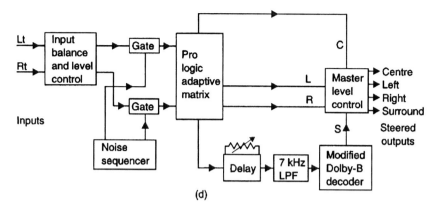

Figure 7.21 (*continued*) (d) Pro-logic decoder uses variable matrices driven by analysis of dominant sound source location.

channels. The Dolby Surround Sound system is designed to give some of the advantages of discrete multi-channel surround sound whilst only requiring two audio channels.

Figure 7.21(a) shows that Dolby Surround derives four channels, L, Centre, R and a surround channel which can be reproduced from two or more rear speakers. In fact a similar speaker arrangement to five channel discrete can be used. Figure 7.21(b) shows a Dolby Surround encoder. The centre channel is attenuated 3 dB and added equally to L and R. The surround signal is attenuated 3 dB and band limited prior to being encoded with a modified Dolby-B process. The resultant signal is then phase shifted so that a 180° phase relationship is formed between the components which are added to the Lt and Rt signals.

In a simple passive decoder (c) Lt and Rt are used to drive L and R speakers directly. In the absence of a C speaker, the L and R speakers will reproduce the C signal centrally. The added anti-phase surround signal will fail to produce an image. If a C speaker is provided, adding Lt and Rt will produce a suitable signal to drive it, although this will result in narrowing of the frontal image. Subtracting Lt and Rt will result in sum of the surround signal and any difference between the original L, R signals. This is band limited then passed through the Dolby-B decoder to produce a surround output.

The Dolby-B-like processing is designed to reduce audibility of L minus R breakthrough on the S signal, particularly on sibilants. The degree of compression is less than that in true Dolby-B to prevent difficulties when Lt and Rt are used as direct speaker drive signals.

In the Pro-logic decoder, shown in (d), the sum and difference stages are replaced with variable matrices which can act like a two-dimensional panpot or steering system. A simple decoder performs poorly when a single point source of sound is present in quiet surroundings whereas a steered decoder will reduce the crosstalk output from unwanted channels. The steering is done by analysing the input signals to identify dominant sound sources. Figure 7.21(d) shows that comparison of the Lt and Rt signals will extract left/right dominance, whereas comparison of sum and difference signals will extract front/rear dominance.

7.12 Microphone criteria for stereo

The additional requirement for spatial accuracy places more constraints on the performance of microphones used in stereo. Pairs of microphones must be identical in output level, frequency and phase response otherwise imaging will be impaired. In crossed coincident working the central sound source will be off-axis to both microphones and these must have good directivity characteristics. If the microphone diaphragm is too large the high-frequency response will not be maintained off-axis and the centre of the stereo image will appear dull and treble components will tend to smear towards the loudspeakers. M-S microphones avoid the central dullness problem but may substitute dullness at the edges of the image. Soundfield microphones are advantageous if difficulties of this kind are encountered. It should be stressed that many microphone users are quite unaware of the shortcomings of their microphones because their monitoring loudspeakers are simply incapable of revealing the imaging problems because of their own imaging difficulties.

7.13 Loudspeaker criteria for stereo

The accuracy required for stereo reproduction is much greater than for mono. If there is any non-linearity in the system, different sound sources will intermodulate and produce phantom sound sources which appear to come from elsewhere in the stereo image than either of the original sounds. As these phantom sources are spatially separate from the genuine sources, they are easier to detect.

Where non-ideal speakers are used, it is important that the two speakers are absolutely identical. If the frequency and phase responses are not identical, the location of the apparent sound source will move with frequency. Where a harmonically rich source is involved, it will appear to be wider than it really is. This is known as *smear*.

If the loudspeakers suffer from beaming at high frequency then a proper stereo image will only be obtained at a small 'sweet spot' and all high frequencies will appear to emanate from the speakers, not from a central image. Small movements by the listener may cause quite large image changes. Irregular polar diagrams will also destroy stereo imaging. Such irregularities often occur near the crossover frequencies. Placing the drive units in a vertical line will prevent the horizontal polar diagram becoming too irregular. However, this idea is completely destroyed if such a speaker is placed on its side; a technique which is seen with depressing regularity. This may be because many loudspeakers are so mediocre that turning them on their side does not make them any worse.

Loudspeakers and amplifiers used for stereo mixing must have very low distortion and a precisely tracking frequency and phase response. Loudspeakers must have smoothly changing directivity characteristics. In practice a great deal of equipment fails to meet these criteria.

In many applications in television the level of background noise is high and/or the room acoustic is deficient, making conventional monitoring difficult. In many cases there is little space left for loudspeakers once all of the video equipment is installed. One solution is the *close-field* monitor which is

Figure 7.22 In the Celtic Cabar loudspeaker, active technology is used to obtain full frequency range from a very compact enclosure. Rare-earth magnets need no screening but do not disturb picture monitors.

designed and equalized so that the listener can approach very close to it. The term near field is often and erroneously used to describe close-field monitors. Near field has a specific meaning which has been defined in Chapter 3. The essence of close-field monitoring is that direct sound reaches the listener so much earlier than the reverberant sound that the room acoustic becomes less important. In stereo close-field monitoring the loudspeakers are much closer together, even though the same angle is subtended to the listener.

Figure 7.22 shows the Cabar loudspeaker (courtesy of Celtic Audio) which is a precision stereo close-field monitor only 6 inches (15 cm) in diameter, this unit uses active drive and rare-earth magnets to obtain response down to 30 Hz without stray magnetic fields. A line level crossover drives one amplifier per drive unit and these are aligned so that a correct stereo image will be obtained only 1 metre away. In this case the direct sound also has significantly higher level than ambient sound so that monitoring can even take place in noisy surroundings.

References

1. Hertz, B., 100 years with stereo: the beginning. *J. Audio Eng. Soc.*, **29**, No. 5, 368–372 (1981)
2. British Patent Specification 394,325 (1931)
3. Stereophonic Techniques. *AES Anthology*, New York: AES (1986)
4. Moore, B.C., *An Introduction to the Psychology of Hearing*, Sec. 6.12, London: Academic Press (1989)
5. Cao, Y., Sridharan, S. and Moody, M., Co-talker separation using the cocktail-party effect. *J. Audio Eng. Soc.*, **44**, No. 12, 1084–1096 (1996)
6. Moore, B.C., *An Introduction to the Psychology of Hearing*, Sec. 6.13, London: Academic Press (1989)
7. Thomas, M.V., Improving the stereo headphone sound image. *J. Audio Eng. Soc.*, **25**, No. 7/8, 474–478 (Jul/Aug 1977)
8. Harvey, F.K. and Uecke, E.H., Compatibility problems in two channel stereophonic recordings. *J. Audio Eng. Soc.*, **10**, No. 1, 8–12 (Jan 1962)

6. Moore, B.C., *An Introduction to the Psychology of Hearing*, Sec. 6.13, London: Academic Press (1989)
7. Thomas, M.V., Improving the stereo headphone sound image. *J. Audio Eng. Soc.*, **25**, No. 7/8, 474–478 (Jul/Aug 1977)
8. Harvey, F.K. and Uecke, E.H., Compatibility problems in two channel stereophonic recordings. *J. Audio Eng. Soc.*, **10**, No. 1, 8–12 (Jan 1962)

Chapter 8

Digital audio signals

Digital techniques have been adopted very rapidly and have changed the nature of sound reproduction considerably, especially in recording. Unfortunately digital technology relies on an utterly different set of principles from analog. Many audio engineers who were highly experienced in analog audio failed to learn these new principles and found themselves unable to make considered judgements when digital audio was involved. The careful reader of this chapter will have no such difficulty.

8.1 Introduction to digital audio

In an analog system, information is conveyed by some infinite variation of a continuous parameter such as the voltage on a wire or the strength of flux on a tape. In a recorder, distance along the medium is a further, continuous, analog of time.

Those characteristics are the main weakness of analog signals. Within the allowable bandwidth, *any* waveform is valid. If the speed of the medium is not constant, one valid waveform is changed into another valid waveform; a timebase error cannot be detected in an analog system. In addition, a voltage error simply changes one valid voltage into another; noise cannot be detected in an analog system. It is a characteristic of analog systems that degradations cannot be separated from the original signal, so nothing can be done about them. At the end of a system a signal carries the sum of all degradations introduced at each stage through which it passed. This sets a limit to the number of stages through which a signal can be passed before it is useless.

An ideal digital audio channel has the same characteristics as an ideal analog channel: both of them are totally transparent and reproduce the original applied waveform without error. Needless to say in the real world ideal conditions seldom prevail, so analog and digital equipment both fall short of the ideal. Digital audio simply falls short of the ideal by a smaller distance than does analog and at lower cost, or, if the designer chooses, can have the same performance as analog at much lower cost.

There is one system, known as pulse code modulation (PCM), which is in virtually universal use. Figure 8.1 shows how PCM works. The time axis is represented in a discrete, or stepwise manner and the waveform is carried by measurement at regular intervals. This process is called sampling and the

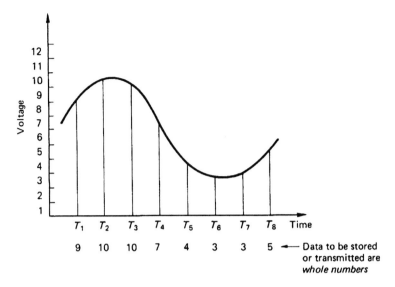

Figure 8.1 In pulse code modulation (PCM), the analog waveform is measured periodically at the sampling rate. The voltage (represented here by the height) of each sample is then described by a whole number. The whole numbers are stored or transmitted rather than the waveform itself.

frequency with which samples are taken is called the sampling rate or sampling frequency F_s. The sampling rate is generally fixed and every effort is made to rid the sampling clock of jitter so that every sample will be made at an exactly even time step. If there is any subsequent timebase error, the instants at which samples arrive will be changed but the effect can be eliminated by storing the samples temporarily in a memory and reading them out using a stable, locally generated clock. This process is called timebase correction and all properly engineered digital audio systems must use it. Clearly timebase error is not reduced; it is totally eliminated. As a result there is little point measuring the wow and flutter of a digital recorder; it doesn't have any.

Figure 8.1 also shows that each sample is also discrete, or represented in a stepwise manner. The length of the sample, which will be proportional to the voltage of the audio waveform, is represented by a whole number. This process is known as quantizing and results in an approximation, but the size of the error can be controlled by using more steps until it is negligible. The advantage of using whole numbers is that they are not prone to drift. If a whole number can be carried from one place to another without numerical error, it has not changed at all.

Essentially, digital audio carries the original waveform numerically. The number of the sample is an analog of time, and the magnitude of the sample is an analog of the pressure at the microphone. By describing audio waveforms numerically, the original information has been expressed in a way which is better able to resist unwanted changes.

As both axes of the digitally represented waveform are discrete, the waveform can be accurately restored from numbers as if it were being drawn on graph paper. If we require greater accuracy, we simply choose paper with smaller squares. Clearly more numbers are then required and each one could change over a larger range.

Digital audio has two main advantages, but it is not possible to say which is the most important.

1. The quality of reproduction of a well-engineered digital audio system is independent of the medium and depends only on the quality of the conversion processes and any compression technique.
2. The conversion of audio to the digital domain allows tremendous opportunities which were denied to analog signals.

Someone who is only interested in sound quality will judge the former the most relevant. If good quality converters can be obtained, all of the shortcomings of analog recording can be eliminated to great advantage. When a digital recording is copied, the same numbers appear on the copy: it is not a dub, it is a clone. If the copy is indistinguishable from the original, there has been no generation loss. Digital recordings can be copied indefinitely without loss of quality.

Once audio is in the digital domain, it becomes data, and as such is indistinguishable from any other type of data. Systems and techniques developed in other industries for other purposes can be used for audio. Computer equipment is available at low cost because the volume of production is far greater than that of professional audio equipment. Disk drives and memories developed for computers can be put to use in audio products. A word processor adapted to handle audio samples becomes a workstation. There seems to be little point in waiting for a tape to wind when a disk head can access data in milliseconds. The difficulty of locating the edit point and the irrevocable nature of tape-cut editing are hardly worth considering when the edit point can be located by viewing the audio waveform on a screen or by listening at any speed to audio from a memory. The edit can be simulated and trimmed before it is made permanent.

Communications networks developed to handle data can happily carry digital audio over indefinite distances without quality loss. Digital audio broadcasting (DAB) makes use of these techniques to eliminate the interference, fading and multipath reception problems of analog broadcasting. At the same time, more efficient use is made of available bandwidth.

8.2 Binary

Binary is the most minimal numbering system, which has only two digits, 0 and 1. Binary digits are universally contracted to bits. These are readily conveyed in switching circuits by an 'on' state and an 'off' state. With only two states, there is little chance of error.

Figure 8.2 shows that in binary, the bits represent one, two, four, eight, sixteen, etc. A multidigit binary number is commonly called a word, and the number of bits in the word is called the wordlength. The right-hand bit is called the Least Significant Bit (LSB) whereas the bit on the left-hand end of the word is called the Most Significant Bit (MSB). Clearly more digits are required in binary than in decimal, but they are more easily handled. A word of eight bits is called a byte, which is a contraction of 'by eight'.

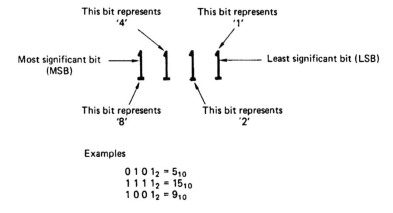

Examples

$$0\ 1\ 0\ 1_2 = 5_{10}$$
$$1\ 1\ 1\ 1_2 = 15_{10}$$
$$1\ 0\ 0\ 1_2 = 9_{10}$$

Figure 8.2 In a binary number, the digits represent increasing powers of two from the LSB. Also defined here are MSB and wordlength. When the wordlength is eight bits, the word is a byte. Binary numbers are used as memory addresses, and the range is defined by the address wordlength. Some examples are shown here.

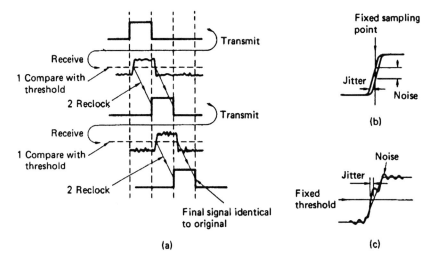

Figure 8.3 (a) A binary signal is compared with a threshold and reclocked on receipt, thus the meaning will be unchanged. (b) Jitter on a signal can appear as noise with respect to fixed timing. (c) Noise on a signal can appear as jitter when compared with a fixed threshold.

In a digital audio system, the length of the sample is expressed by a binary number of typically 16 bits. The signals sent have two states, and change at predetermined times according to some stable clock. Figure 8.3 shows the consequences of this form of transmission. If the binary signal is degraded by noise, this will be rejected by the receiver, which judges the signal solely by whether it is above or below the half-way threshold, a process known as slicing. The signal will be carried in a channel with finite bandwidth, and this limits the slew rate of the signal; an ideally upright edge is made to slope.

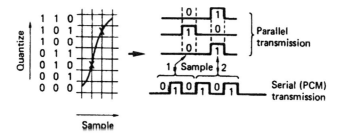

Figure 8.4 When a signal is carried in numerical form, either parallel or serial, the mechanisms of Figure 8.3 ensure that the only degradation is in the conversion processes.

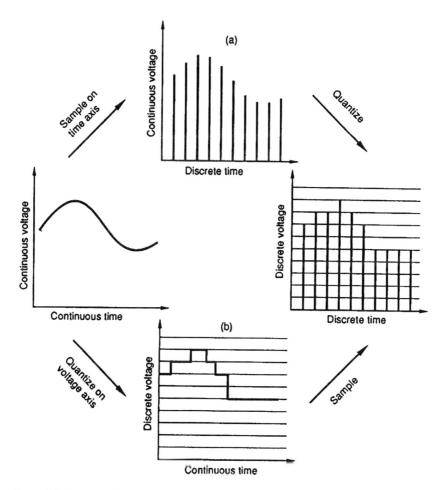

Figure 8.5 Since sampling and quantizing are orthogonal, the order in which they are performed is not important. In (a) sampling is performed first and the samples are quantized. This is common in audio converters. In (b) the analog input is quantized into an asynchronous binary code. Sampling takes place when this code is latched on sampling clock edges. This approach is universal in video converters.

Noise added to a sloping signal can change the time at which the slicer judges that the level passed through the threshold. This effect is also eliminated when the output of the slicer is reclocked. However many stages the binary signal passes through, it still comes out the same, only later.

Audio samples which are represented by whole numbers can be reliably carried from one place to another by such a scheme, and if the number is correctly received, there has been no loss of information en route.

There are two ways in which binary signals can be used to carry audio samples and these are shown in Figure 8.4. When each digit of the binary number is carried on a separate wire this is called parallel transmission. The state of the wires changes at the sampling rate. Using multiple wires is cumbersome, particularly where a long wordlength is in use, and a single wire can be used where successive digits from each sample are sent serially. This is the definition of pulse code modulation. Clearly the clock frequency must now be higher than the sampling rate. Whilst digital transmission of audio eliminates noise and timebase error, there is a penalty that a single high-quality audio channel requires around one million bits per second. Clearly digital audio could only come into use when such a data rate could be handled economically. Further applications become possible when means to reduce the data rate become economic.

8.3 Conversion

The input to a converter is a continuous-time, continuous-voltage waveform, and this is changed into a discrete-time, discrete-voltage format by a combination of sampling and quantizing. These two processes are totally independent and can be performed in either order and discussed quite separately in some detail. Figure 8.5(a) shows an analog sampler preceding a quantizer, whereas (b) shows an asynchronous quantizer preceding a digital sampler. Ideally, both will give the same results; in practice each has different advantages and suffers from different deficiencies. Both approaches will be found in real equipment.

8.4 Sampling and aliasing

The sampling process originates with a regular pulse train which is shown in Figure 8.6(a) to be of constant amplitude and period. The audio waveform amplitude-modulates the pulse train in much the same way as the carrier is modulated in an AM radio transmitter. One must be careful to avoid overmodulating the pulse train as shown in (b) and this is achieved by applying a DC offset to the analog waveform so that silence corresponds to a level half-way up the pulses as at (c). Clipping due to any excessive input level will then be symmetrical.

In the same way that AM radio produces sidebands or images above and below the carrier, sampling also produces sidebands although the carrier is now a pulse train and has an infinite series of harmonics as shown in Figure 8.7(a). The sidebands repeat above and below each harmonic of the sampling rate as shown in (b).

The sampled signal can be returned to the continuous-time domain simply by passing it into a low-pass filter. This filter has a frequency response which

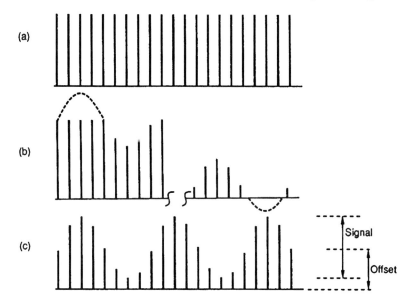

Figure 8.6 The sampling process requires a constant-amplitude pulse train as shown in (a). This is amplitude-modulated by the waveform to be sampled. If the input waveform has excessive amplitude or incorrect level, the pulse train clips as shown in (b). For an audio waveform, the greatest signal level is possible when on offset of half the pulse amplitude is used to centre the waveform as shown in (c).

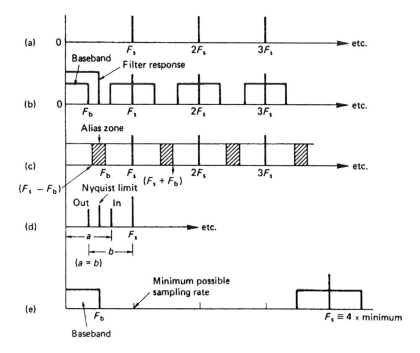

Figure 8.7 (a) Spectrum of sampling pulses. (b) Spectrum of samples. (c) Aliasing due to sideband overlap. (d) Beat-frequency production. (e) 4× oversampling.

prevents the images from passing, and only the baseband signal emerges, completely unchanged. If considered in the frequency domain, this filter is called an anti-image or reconstruction filter.

If an input is supplied having an excessive bandwidth for the sampling rate in use, the sidebands will overlap (c) and the result is aliasing, where certain output frequencies are not the same as their input frequencies but instead become difference frequencies (d). It will be seen from Figure 8.7 that aliasing does not occur when the input frequency is equal to or less than half the sampling rate, and this derives the most fundamental rule of sampling, which is that the sampling rate must be at least twice the highest input frequency.

Whilst aliasing has been described above in the frequency domain, it can be described equally well in the time domain. In Figure 8.8(a) the sampling rate is obviously adequate to describe the waveform, but at (b) it is inadequate and aliasing has occurred.

In practice it is necessary also to have a low-pass, or anti-aliasing filter at the input to prevent frequencies of more than half the sampling rate from reaching the sampling stage.

If ideal low-pass anti-aliasing and anti-image filters are assumed, an ideal spectrum shown at Figure 8.9(a) is obtained. The impulse response of a phase-linear ideal low-pass filter is a sin x/x waveform in the time domain, and this is shown in (b). Such a waveform passes through zero volts periodically. If the cut-off frequency of the filter is one-half of the sampling rate, the impulse passes through zero *at the sites of all other samples*. It can be seen from (c) that at the output of such a filter, the voltage at the centre of a sample is due to that sample alone, since the value of *all* other samples is zero at that instant. In other words the continuous-time output waveform must pass through the peaks of the input samples. In between the sample instants, the output of the filter is the sum of the contributions from many impulses, and the waveform smoothly joins the tops of the samples.

The ideal filter with a vertical 'brick-wall' cut-off slope is impossible to implement and in practice a filter with a finite slope has to be accepted as shown in Figure 8.10. The cut-off slope begins at the edge of the required band, and consequently the sampling rate has to be raised a little to drive aliasing products to an acceptably low level.

Every signal which has been through the digital domain has passed through both an anti-aliasing filter and a reconstruction filter. These filters must be

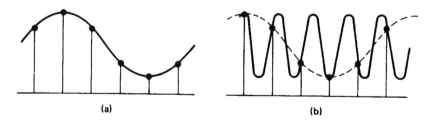

Figure 8.8 In (a) the sampling is adequate to reconstruct the original signal. In (b) the sampling rate is inadequate, and reconstruction produces the wrong waveform (dashed). Aliasing has taken place.

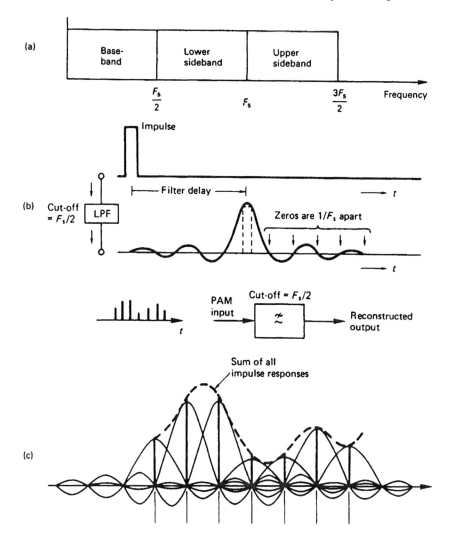

Figure 8.9 If ideal 'brick wall' filters are assumed, the efficient spectrum of (a) results. An ideal low-pass filter has an impulse response shown in (b). The impulse passes through zero at intervals equal to the sampling period. When convolved with a pulse train at the sampling rate, as shown in (c), the voltage at each sample instant is due to that sample alone as the impulses from all other samples pass through zero there.

carefully designed in order to prevent audible artifacts, particularly those due to lack of phase linearity as they may be audible.[1-3] The nature of the filters used has a great bearing on the subjective quality of the system.

Much effort can be saved in analog filter design by using oversampling which is considered later in this chapter. Strictly oversampling means no more than that a higher sampling rate is used than is required by sampling theory. In the loose sense an 'oversampling converter' generally implies that some combination of high sampling rate and various other techniques has been

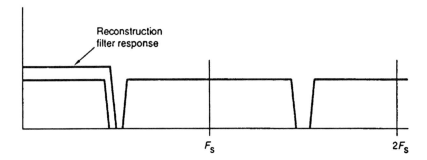

Figure 8.10 As filters with finite slope are needed in practical systems, the sampling rate is raised slightly beyond twice the highest frequency in the baseband.

applied. The audible superiority and economy of oversampling converters has led them to be almost universal.

8.5 Choice of sampling rate

At one time, video recorders were adapted to store audio samples by creating a pseudo-video waveform which could convey binary as black and white levels. The sampling rate of such a system is constrained to relate simply to the field rate and field structure of the television standard used, so that an integer number of samples can be stored on each usable TV line in the field. In 60 Hz video, there are 35 blanked lines, leaving 490 lines per frame, or 245 lines per field for samples. If three samples are stored per line, the sampling rate becomes $60 \times 245 \times 3 = 44.1$ kHz. This sampling rate was adopted for Compact Disc. Although CD has no video circuitry, the first equipment used to make CD masters was video based.

For professional products, there is a need to operate at variable speed for pitch correction. When the speed of a digital recorder is reduced, the offtape sampling rate falls, and with a minimal sampling rate the first image frequency can become low enough to pass the reconstruction filter. If the sampling frequency is raised to 48 kHz without changing the response of the filters, the speed can be reduced without this problem. The currently available DVTR formats offer only 48 kHz audio sampling and so this is the only sampling rate practicable for video installations.

For landlines to FM stereo broadcast transmitters having a 15 kHz audio bandwidth, the sampling rate of 32 kHz is more than adequate, and has been in use for some time in the United Kingdom and Japan. This frequency is also in use in the NICAM 728 stereo TV sound system and in DAB.

8.6 Sampling clock jitter

Figure 8.11 shows the effect of sampling clock jitter on a sloping waveform. Samples are taken at the wrong times. When these samples have passed through a system, the timebase correction stage prior to the DAC will remove the jitter, and the result is shown at (b). The magnitude of the unwanted signal is proportional to the slope of the audio waveform and so the amount of jitter

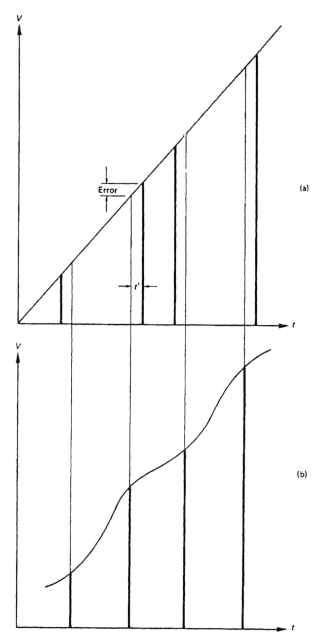

Figure 8.11 The effect of sampling timing jitter on noise, and calculation of the required accuracy for a 16-bit system. (a) Ramp sampled with jitter has error proportional to slope. (b) When jitter is removed by later circuits, error appears as noise added to samples. For a 16-bit system there are $2^{16}Q$, and the maximum slope at 20 kHz will be $20\,000\pi \times 2^{16}Q$ per second. If jitter is to be neglected, the noise must be less than $\frac{1}{2}Q$, thus timing accuracy t' multiplied by maximum slope $= \frac{1}{2}Q$ or $20\,000\pi \times 2^{16}Qt' = \frac{1}{2}Q$

$$\therefore t' = \frac{1}{2 \times 20\,000 \times \pi \times 2^{16}} = 121 \text{ ps}$$

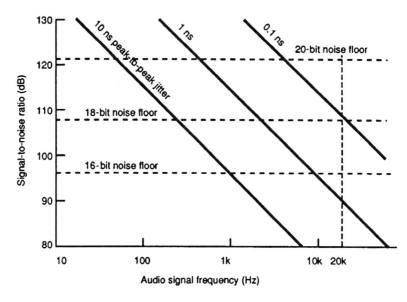

Figure 8.12 Effects of sample clock jitter on signal-to-noise ratio at different frequencies, compared with the theoretical noise floors of systems with different resolutions. (After W.T. Shelton, with permission).

which can be tolerated falls at 6 dB/octave. Figure 8.12 shows the effect of differing amounts of random jitter with respect to the noise floor of various wordlengths. Note that even small amounts of jitter can degrade a 20-bit converter to the performance of a good 16-bit unit.

The allowable jitter is measured in picoseconds, as shown in Figure 8.11, and clearly steps must be taken to eliminate it by design. Converter clocks must be generated from clean power supplies which are well decoupled from the power used by the logic because a converter clock must have a signal-to-noise ratio of the same order as that of the audio.

If an external clock source is used, it cannot be used directly, but must be fed through a well-designed, well-damped phase-locked loop which will filter out the jitter. The phase-locked loop must be built to a higher accuracy standard than in most applications.

Although it has been documented for many years, attention to control of clock jitter is not as great in actual hardware as it might be. It accounts for much of the slight audible differences between converters reproducing the same data. A well-engineered converter should substantially reject jitter on an external clock and should sound the same when reproducing the same data irrespective of the source of the data. A remote converter which sounds different after changing the type of digital cable feeding it is a dud. Unfortunately many external DACs fall into this category, as the steps outlined above have not been taken.

8.7 Aperture effect

The reconstruction process of Figure 8.9 only operates exactly as shown if the impulses are of negligible duration. In many DACs this is not the case,

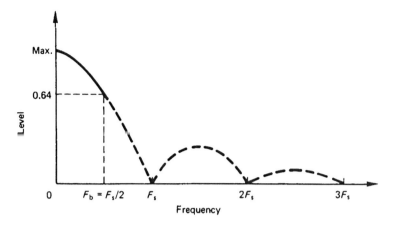

Figure 8.13 Frequency response with 100% aperture nulls at multiples of sampling rate. Area of interest is up to half sampling rate.

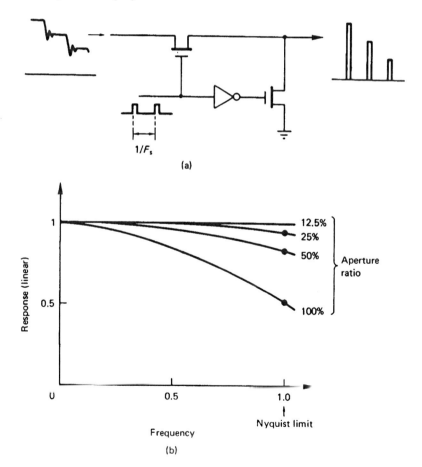

Figure 8.14 (a) Resampling circuit eliminates transients and reduces aperture ratio. (b) Response of various aperture ratios.

and many keep the analog output constant for a substantial part of the sample period or even until a different sample value is input. This produces a waveform which is more like a staircase than a pulse train. The case where the pulses have been extended in width to become equal to the sample period is known as a zero-order hold system and has a 100% aperture ratio.

Whereas pulses of negligible width have a uniform spectrum, which is flat within the audio band, pulses of 100% aperture ratio have a sin x/x spectrum which is shown in Figure 8.13. The frequency response falls to a null at the sampling rate, and as a result is about 4 dB down at the edge of the audio band. An appropriate equalization circuit can render the overall response flat once more. An alternative is to use resampling which is shown in Figure 8.14. Resampling passes the zero-order hold waveform through a further synchronous sampling stage which consists of an analog switch which closes briefly in the centre of each sample period. The output of the switch will be pulses which are narrower than the original. If, for example, the aperture ratio is reduced to 50% of the sample period, the first frequency response null is now at twice the sampling rate, and the loss at the edge of the audio band is reduced. As the figure shows, the frequency response becomes flatter as the aperture ratio falls. The process should not be carried too far, as with very small aperture ratios there is little energy in the pulses and noise can be a problem. A practical limit is around 12.5% where the frequency response is virtually ideal.

8.8 Quantizing

Quantizing is the process of expressing some infinitely variable quantity by discrete or stepped values. In audio the values to be quantized are infinitely variable voltages from an analog source. Strict quantizing is a process which operates in the voltage domain only.

Figure 8.15(a) shows that the process of quantizing divides the voltage range up into quantizing intervals Q, also referred to as steps S. In digital audio the quantizing intervals are made as identical as possible so that the binary numbers are truly proportional to the original analog voltage. Then the digital equivalents of mixing and gain changing can be performed by adding and multiplying sample values.

Whatever the exact voltage of the input signal, the quantizer will locate the quantizing interval in which it lies. In what may be considered a separate step, the quantizing interval is then allocated a code value which is typically some form of binary number. The information sent is the number of the quantizing interval in which the input voltage lay. Whereabouts that voltage lay within the interval is not conveyed, and this mechanism puts a limit on the accuracy of the quantizer. When the number of the quantizing interval is converted back to the analog domain, it will result in a voltage at the centre of the quantizing interval as this minimizes the magnitude of the error between input and output. The number range is limited by the wordlength of the binary numbers used. In a 16-bit system there are 65 536 different quantizing intervals.

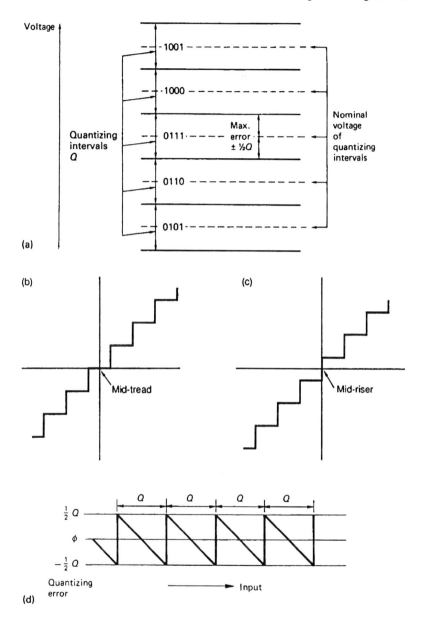

Figure 8.15 Quantizing assigns discrete numbers to variable voltages. All voltages within the same quantizing interval are assigned the same number which causes a DAC to produce the voltage at the centre of the intervals shown by the dashed lines in (a). This is the characteristic of the mid-tread quantizer shown in (b). An alternative system is the mid-riser system shown in (c). Here zero volts analog falls between two codes and there is no code for zero. Such quantizing cannot be used prior to signal processing because the number is no longer proportional to the voltage. Quantizing error cannot exceed $\pm\frac{1}{2}Q$ as shown in (d).

8.9 Quantizing error

It is possible to draw a transfer function for such an ideal quantizer followed by an ideal DAC, and this is also shown in Figure 8.15. A transfer function is simply a graph of the output with respect to the input. In audio, when the term linearity is used, this generally means the straightness of the transfer function. Linearity is a goal in audio, yet it will be seen that an ideal quantizer is anything but linear.

Figure 8.15(b) shows the transfer function is somewhat like a staircase, and zero volts analog, corresponding to all zeros digital or muting, is half-way up a quantizing interval, or on the centre of a tread. This is the so-called mid-tread quantizer which is universally used in audio. Figure 8.15(c) shows the alternative mid-riser transfer function which causes difficulty in audio because it does not have a code value at muting level and as a result the numerical code value is not proportional to the analog signal voltage.

Quantizing causes a voltage error in the audio sample which is given by the difference between the actual staircase transfer function and the ideal straight line. This is shown in Figure 8.15(d) to be a sawtooth-like function which is periodic in Q. The amplitude cannot exceed $\pm\frac{1}{2}Q$ peak-to-peak unless the input is so large that clipping occurs.

Quantizing error can also be studied in the time domain where it is better to avoid complicating matters with the aperture effect of the DAC. For this reason it is assumed here that output samples are of negligible duration. Then impulses from the DAC can be compared with the original analog waveform and the difference will be impulses representing the quantizing error waveform. This has been done in Figure 8.16. The horizontal lines in the drawing are the boundaries between the quantizing intervals, and the curve is the input waveform. The vertical bars are the quantized samples which reach to the centre of the quantizing interval. The quantizing error waveform shown at (b) can be thought of as an unwanted signal which the quantizing process adds to the perfect original. If a very small input signal remains within one quantizing interval, the quantizing error *is* the signal.

As the transfer function is non-linear, ideal quantizing can cause distortion. As a result practical digital audio devices deliberately use non-ideal quantizers to achieve linearity. The quantizing error of an ideal quantizer is a complex function, and it has been researched in great depth.[4] [6] It is not intended to go into such depth here. The characteristics of an ideal quantizer will only be pursued far enough to convince the reader that such a device cannot be used in quality audio applications.

As the magnitude of the quantizing error is limited, its effect can be minimized by making the signal larger. This will require more quantizing intervals and more bits to express them. The number of quantizing intervals multiplied by their size gives the quantizing range of the converter. A signal outside the range will be clipped. Provided that clipping is avoided, the larger the signal the less will be the effect of the quantizing error.

Where the input signal exercises the whole quantizing range and has a complex waveform (such as from orchestral music), successive samples will have widely varying numerical values and the quantizing error on a given sample will be independent of that on others. In this case the size of the quantizing error will be distributed with equal probability between the limits.

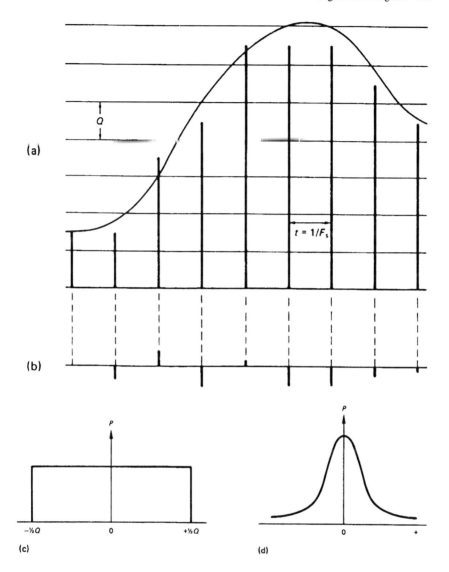

Figure 8.16 At (a) an arbitrary signal is represented to finite accuracy by PAM needles whose peaks are at the centre of the quantizing intervals. The errors caused can be thought of as an unwanted signal (b) added to the original. In (c) the amplitude of a quantizing error needle will be from $-\frac{1}{2}Q$ to $+\frac{1}{2}Q$ with equal probability. Note, however, that white noise in analog circuits generally has Gaussian amplitude distribution, shown in (d).

Figure 8.16(c) shows the resultant uniform probability density. In this case the unwanted signal added by quantizing is an additive broadband noise uncorrelated with the signal, and it is appropriate in this case to call it quantizing noise. This is not quite the same as thermal noise which has a Gaussian probability. The difference is of no consequence as in the large signal case the noise is masked by the signal. Under these conditions, a meaningful signal-to-noise ratio can be calculated as follows:

In a system using n-bit words, there will be 2^n quantizing intervals. The largest sinusoid which can fit without clipping will have this peak-to-peak amplitude. The peak amplitude will be half as great, i.e. $2^{n-1}Q$ and the rms amplitude will be this value divided by $\sqrt{2}$.

The quantizing error has an amplitude of $\frac{1}{2}Q$ peak which is the equivalent of $Q/\sqrt{12}$ rms. The signal-to-noise ratio for the large signal case is then given by:

$$20 \log_{10} \frac{\sqrt{12} \times 2^{n-1}}{\sqrt{2}} \, dB$$

$$= 20 \log_{10}(\sqrt{6} \times 2^{n-1}) \, dB$$

$$= 20 \log_{10} \left(2^n \times \frac{\sqrt{6}}{2} \right) \, dB$$

$$= 20n \log 2 + 20 \log \frac{\sqrt{6}}{2} \, dB$$

$$= 6.02n + 1.76 \, dB$$

By way of example, a 16-bit system will offer around 98.1 dB SNR.

Whilst the above result is true for a large complex input waveform, treatments which then assume that quantizing error is *always* noise give results which are at variance with reality. The expression above is only valid if the probability density of the quantizing error is uniform. Unfortunately at low levels, and particularly with pure or simple waveforms, this is simply not the case.

At low audio levels, quantizing error ceases to be random, and becomes a function of the input waveform and the quantizing structure as Figure 8.16 showed. Once an unwanted signal becomes a deterministic function of the wanted signal, it has to be classed as a distortion rather than a noise. Distortion can also be predicted from the non-linearity, or staircase nature, of the transfer function. With a large signal, there are so many steps involved that we must stand well back, and a staircase with 65 000 steps appears to be a slope. With a small signal there are few steps and they can no longer be ignored.

The non-linearity of the transfer function results in distortion, which produces harmonics. Unfortunately these harmonics are generated *after* the anti-aliasing filter, and so any which exceed half the sampling rate will alias. Figure 8.17 shows how this results in anharmonic distortion within the audio band. These anharmonics result in spurious tones known as birdsinging. When the sampling rate is a multiple of the input frequency the result is harmonic distortion. This is shown in Figure 8.18. Where more than one frequency is present in the input, intermodulation distortion occurs, which is known as granulation.

As the input signal is further reduced in level, it may remain within one quantizing interval. The output will be silent because the signal is

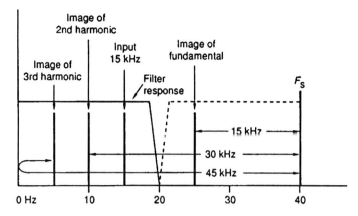

Figure 8.17 Quantizing produces distortion *after* the anti-aliasing filter; thus the distortion products will fold back to produce anharmonics in the audio band. Here the fundamental of 15 kHz produces second and third harmonic distortion at 30 and 45 kHz. This results in aliased products at $40 - 30 = 10$ kHz and at $40-50 = (-)5$ kHz.

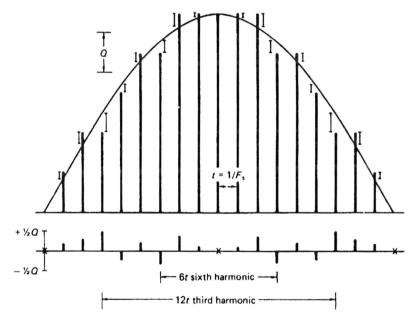

Figure 8.18 Mathematically derived quantizing error waveform for sine wave sampled at a multiple of itself. The numerous autocorrelations between quantizing errors show that there are harmonics of the signal in the error, and that the error is not random, but deterministic.

now the quantizing error. In this condition, low-frequency signals such as air-conditioning rumble can shift the input in and out of a quantizing interval so that the quantizing distortion comes and goes, resulting in noise modulation.

Needless to say any one of the above effects would preclude the use of an ideal quantizer for high-quality work. There is little point in studying the adverse effects further as they should be and can be eliminated completely in practical equipment by the use of dither. The importance of correctly dithering a quantizer cannot be emphasized enough, since failure to dither irrevocably distorts the converted signal: there can be no process which will subsequently remove that distortion.

The signal-to-noise ratio derived above has no relevance to practical audio applications as it will be modified by the dither and by any noise shaping used.

8.10 Introduction to dither

At high signal levels, quantizing error is effectively noise. As the audio level falls, the quantizing error of an ideal quantizer becomes more strongly correlated with the signal and the result is distortion. If the quantizing error can be decorrelated from the input in some way, the system can remain linear but noisy. Dither performs the job of decorrelation by making the action of the quantizer unpredictable and gives the system a noise floor like an analog system.

The first documented use of dither was by Roberts[7] in picture coding. In this system, pseudo-random noise with rectangular probability and a peak-to-peak amplitude of Q was added to the input signal prior to quantizing, but was subtracted after reconversion to analog. This is known as subtractive dither and was investigated by Schuchman[8] and much later by Sherwood.[9] Subtractive dither has the advantages that the dither amplitude is non-critical, the noise has full statistical independence from the signal[5] and has the same level as the quantizing error in the large signal undithered case.[10] Unfortunately, it suffers from practical drawbacks, since the original noise waveform must accompany the samples or must be synchronously recreated at the DAC. This is virtually impossible in a system where the audio may have been edited or where its level has been changed by processing, as the noise needs to remain synchronous and be processed in the same way. All practical digital audio systems use non-subtractive dither where the dither signal is added prior to quantization and no attempt is made to remove it at the DAC.[11] The introduction of dither prior to a conventional quantizer inevitably causes a slight reduction in the signal-to-noise ratio attainable, but this reduction is a small price to pay for the elimination of non-linearities. The technique of noise shaping in conjunction with dither will be seen to overcome this restriction and produce performance in excess of the subtractive dither example above.

The ideal (noiseless) quantizer of Figure 8.15 has fixed quantizing intervals and must always produce the same quantizing error from the same signal. In Figure 8.19 it can be seen that an ideal quantizer can be dithered by linearly adding a controlled level of noise either to the input signal or to the reference voltage which is used to derive the quantizing intervals. There are several ways of considering how dither works, all of which are equally valid.

The addition of dither means that successive samples effectively find the quantizing intervals in different places on the voltage scale. The quantizing error becomes a function of the dither, rather than a predictable function of the input signal. The quantizing error is not eliminated, but the subjectively unacceptable distortion is converted into a broadband noise which is more benign to the ear.

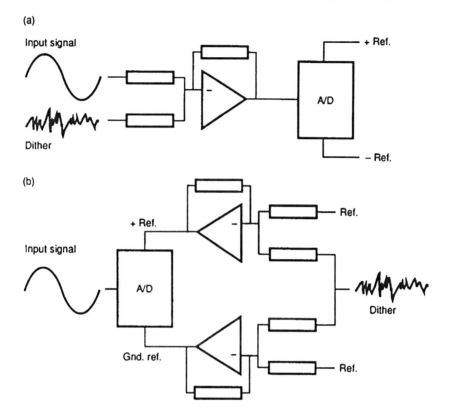

Figure 8.19 Dither can be applied to a quantizer in one of two ways. In (a) the dither is linearly added to the analog input signal, whereas in (b) it is added to the reference voltages of the quantizer.

Some alternative ways of looking at dither are shown in Figure 8.20. Consider the situation where a low level input signal is changing slowly within a quantizing interval. Without dither, the same numerical code is output for a number of sample periods, and the variations within the interval are lost. Dither has the effect of forcing the quantizer to switch between two or more states. The higher the voltage of the input signal within a given interval, the more probable it becomes that the output code will take on the next higher value. The lower the input voltage within the interval, the more probable it is that the output code will take the next lower value. The dither has resulted in a form of duty cycle modulation, and the resolution of the system has been extended indefinitely instead of being limited by the size of the steps.

Dither can also be understood by considering what it does to the transfer function of the quantizer. This is normally a perfect staircase, but in the presence of dither it is smeared horizontally until with a certain amplitude the average transfer function becomes straight.

In an extension of the application of dither, Blesser[12] has suggested digitally generated dither which is converted to the analog domain and added to the input signal prior to quantizing. That same digital dither is then subtracted

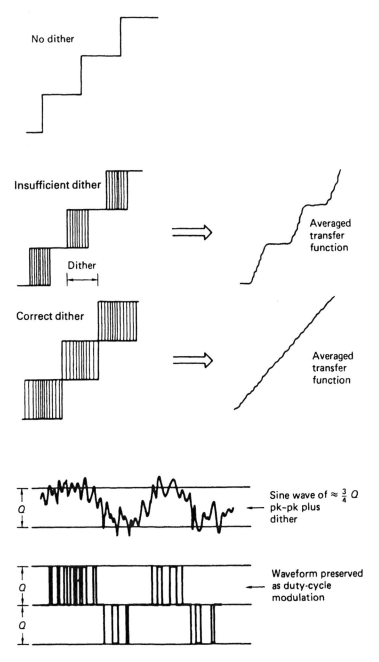

Figure 8.20 Wideband dither of the appropriate level linearizes the transfer function to produce noise instead of distortion. This can be confirmed by spectral analysis. In the voltage domain, dither causes frequent switching between codes and preserves resolution in the duty cycle of the switching.

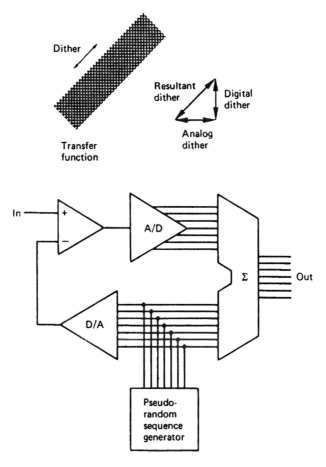

Figure 8.21 In this dither system, the dither added in the analog domain shifts the transfer function horizontally, but the same dither is subtracted in the digital domain, which shifts the transfer function vertically. The result is that the quantizer staircase is smeared diagonally as shown top left. There is thus no limit to dither amplitude, and excess dither can be used to improve differential linearity of the converter.

from the digital quantizer output. The effect is that the transfer function of the quantizer is smeared diagonally (Figure 8.21). The significance of this diagonal smearing is that the amplitude of the dither is not critical. However much dither is employed, the noise amplitude will remain the same. If dither of several quantizing intervals is used, it has the effect of making all the quantizing intervals in an imperfect converter appear to have the same size.

8.11 Requantizing and digital dither

The advanced ADC technology which is detailed later in this chapter allows 18- and 20-bit resolution to be obtained, with perhaps more in the future. The situation then arises that an existing 16-bit device such as a digital recorder needs to be connected to the output of an ADC with greater wordlength. The words need to be shortened in some way.

When a sample value is attenuated, the extra low-order bits which come into existence below the radix point preserve the resolution of the signal and the dither in the least significant bit(s) which linearizes the system. The same word extension will occur in any process involving multiplication, such as digital filtering. It will subsequently be necessary to shorten the wordlength. Clearly the high order bits cannot be discarded in two's complement as this would cause clipping of positive half cycles and a level shift on negative half cycles due to the loss of the sign bit. Low-order bits must be removed instead. Even if the original conversion was correctly dithered, the random element in the low-order bits will now be some way below the end of the intended word. If the word is simply truncated by discarding the unwanted low-order bits or rounded to the nearest integer the linearizing effect of the original dither will be lost.

Shortening the wordlength of a sample reduces the number of quantizing intervals available without changing the signal amplitude. As Figure 8.22 shows, the quantizing intervals become larger and the original signal is *requantized* with the new interval structure. This will introduce requantizing distortion having the same characteristics as quantizing distortion in an ADC. It then is obvious that when shortening the wordlength of a 20-bit converter to 16 bits, the four low-order bits must be removed in a way that displays the same overall quantizing structure as if the original converter had been only of 16-bit wordlength. It will be seen from Figure 8.22 that truncation cannot be used

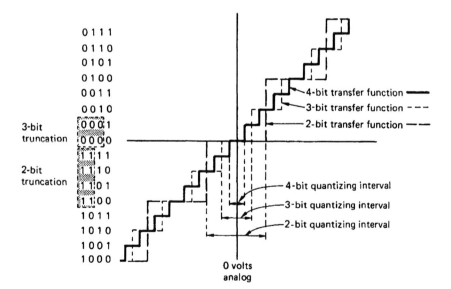

Figure 8.22 Shortening the wordlength of a sample reduces the number of codes which can describe the voltage of the waveform. This makes the quantizing steps bigger, hence the term requantizing. It can be seen that simple truncation or omission of the bits does not give analogous behaviour. Rounding is necessary to give the same result as if the larger steps had been used in the original conversion.

because it does not meet the above requirement but results in signal dependent offsets because it always rounds in the same direction. Proper two's complement numerical rounding is essential in audio applications.

Requantizing by numerical rounding accurately simulates analog quantizing to the new interval size. Unfortunately the 20-bit converter will have a dither amplitude appropriate to quantizing intervals one-sixteenth the size of a 16-bit unit and the result will be highly non-linear.

In practice, the wordlength of samples must be shortened in such a way that the requantizing error is converted to noise rather than distortion. One technique which meets this requirement is to use digital dithering[13] prior to rounding. This is directly equivalent to the analog dithering in an ADC. It will be shown later in this chapter that in more complex systems noise shaping can be used in requantizing just as well as it can in quantizing.

Digital dither is a pseudo-random sequence of numbers. If it is required to simulate the analog dither signal of Figures 8.19 and 8.20, then it is obvious that the noise must be bipolar so that it can have an average voltage of zero. Two's complement coding must be used for the dither values as it is for the audio samples.

Figure 8.23 shows a simple digital dithering system (i.e. one without noise shaping) for shortening sample wordlength. The output of a two's complement pseudo-random sequence generator of appropriate wordlength is added to input samples prior to rounding. The most significant of the bits to be discarded is examined in order to determine whether the bits to be removed sum to more or less than half a quantizing interval. The dithered sample is either rounded down, i.e. the unwanted bits are simply discarded, or rounded up, i.e. the unwanted bits are discarded but one is added to the value of the new short word. The rounding process is no longer deterministic because of the added dither which provides a linearizing random component.

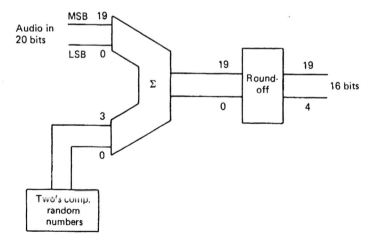

Figure 8.23 In a simple digital dithering system, two's complement values from a random number generator are added to low-order bits of the input. The dithered values are then rounded up or down according to the value of the bits to be removed. The dither linearizes the requantizing.

If this process is compared with that of Figure 8.19 it will be seen that the principles of analog and digital dither are identical; the processes simply take place in different domains using two's complement numbers which are rounded or voltages which are quantized as appropriate. In fact quantization of an analog dithered waveform is identical to the hypothetical case of rounding after bipolar digital dither where the number of bits to be removed is infinite, and remains identical for practical purposes when as few as eight bits are to be removed. Analog dither may actually be generated from bipolar digital dither (which is no more than random numbers with certain properties) using a DAC.

8.12 Dither techniques

The intention here is to treat the processes of analog and digital dither as identical except where differences need to be noted. The characteristics of the noise used are rather important for optimal performance, although many sub-optimal but nevertheless effective systems are in use. The main parameters of interest are the peak-to-peak amplitude, the amplitude probability distribution function (pdf) and the spectral content.

The most comprehensive ongoing study of non-subtractive dither has been that of Vanderkooy and Lipshitz[11,13,14] and the treatment here is based largely upon their work.

8.12.1 Rectangular pdf dither

The simplest form of dither (and therefore easiest to generate) is a single sequence of random numbers which have uniform or rectangular probability. The amplitude of the dither is critical. Figure 8.24(a) shows the time-averaged transfer function of one quantizing interval in the presence of various amplitudes of rectangular dither. The linearity is perfect at an amplitude of $1Q$ peak-to-peak and then deteriorates for larger or smaller amplitudes. The same will be true of all levels which are an integer multiple of Q. Thus there is no freedom in the choice of amplitude.

With the use of such dither, the quantizing noise is not constant. Figure 8.24(b) shows that when the analog input is exactly centred in a quantizing interval (such that there is no quantizing error) the dither has no effect and the output code is steady. There is no switching between codes and thus no noise. On the other hand when the analog input is exactly at a riser or boundary between intervals, there is the greatest switching between codes and the greatest noise is produced. Mathematically speaking, the first moment, or mean error is zero but the second moment, which in this case is equal to the variance, is not constant. From an engineering standpoint, the system is linear but suffers noise modulation: the noise floor rises and falls with the signal content and this is audible in the presence of low-frequency signals.

The dither adds an average noise amplitude of $Q/\sqrt{12}$ rms to the quantizing noise of the same level. In order to find the resultant noise level it is necessary to add the powers as the signals are uncorrelated. The total power is given by:

$$2 \times \frac{Q^2}{12} = \frac{Q^2}{6} \tag{8.1}$$

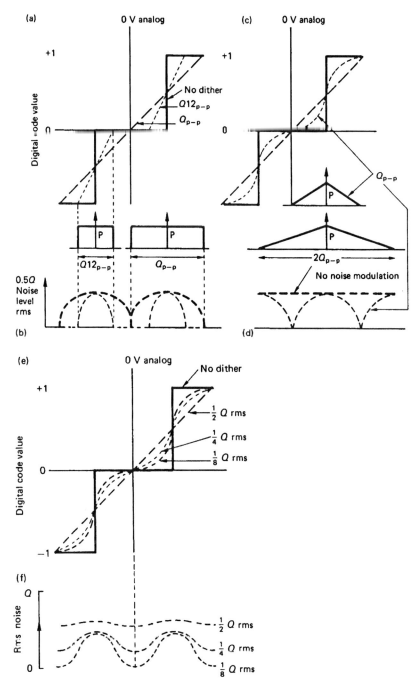

Figure 8.24 Use of rectangular probability dither can linearize, but noise modulation (b) results. Triangular pdf dither (c) linearizes, but noise modulation is eliminated as at (d). Gaussian dither (e) can also be used, almost eliminating noise modulation at (f).

and the rms voltage is $Q/\sqrt{6}$. Another way of looking at the situation is to consider that the noise power doubles and so the rms noise voltage has increased by 3 dB in comparison with the undithered case. Thus for an n-bit wordlength, using the same derivation as expression (8.1) above, the signal-to-noise ratio for Q peak-to-peak rectangular dither will be given by:

$$6.02n - 1.24\,\text{dB} \tag{8.2}$$

Unlike the undithered case, this is a true signal-to-noise ratio and linearity is maintained at all signal levels. By way of example, for a 16-bit system 95.1 dB SNR is achieved. The 3 dB loss compared to the undithered case is a small price to pay for linearity.

8.12.2 Triangular pdf dither

The noise modulation due to the use of rectangular-probability dither is undesirable. It comes about because the process is too simple. The undithered quantizing error is signal dependent and the dither represents a single uniform-probability random process. This is only capable of decorrelating the quantizing error to the extent that its mean value is zero, rendering the system linear. The signal dependence is not eliminated, but is displaced to the next statistical moment. This is the variance and the result is noise modulation. If a further uniform-probability random process is introduced into the system, the signal dependence is displaced to the next moment and the second moment or variance becomes constant.

Adding together two statistically independent rectangular probability functions produces a triangular probability function. A signal having this characteristic can be used as the dither source.

Figure 8.24(c) shows the averaged transfer function for a number of dither amplitudes. Linearity is reached with a peak-to-peak amplitude of $2Q$ and at this level there is no noise modulation. The lack of noise modulation is another way of stating that the noise is constant. The triangular pdf of the dither matches the triangular shape of the quantizing error function.

The dither adds two noise signals with an amplitude of $Q/\sqrt{12}$ rms to the quantizing noise of the same level. In order to find the resultant noise level it is necessary to add the powers as the signals are uncorrelated. The total power is given by:

$$3 \times \frac{Q^2}{12} = \frac{Q^2}{4}$$

and the rms voltage is $Q/\sqrt{4}$. Another way of looking at the situation is to consider that the noise power is increased by 50% in comparison to the rectangular dithered case and so the rms noise voltage has increased by 1.76 dB. Thus for an n-bit wordlength, using the same derivation as expressions (8.1) and (8.2) above, the signal-to-noise ratio for Q peak-to-peak rectangular dither will be given by:

$$6.02n - 3\,\text{dB} \tag{8.3}$$

Continuing the use of a 16-bit example, a SNR of 93.3 dB is available which is 4.8 dB worse than the SNR of an undithered quantizer in the large signal case. It is a small price to pay for perfect linearity and an unchanging noise floor.

8.12.3 Gaussian pdf dither

Adding more uniform probability sources to the dither makes the overall probability function progressively more like the Gaussian distribution of analog noise. Figure 8.24(d) shows the averaged transfer function of a quantizer with various levels of Gaussian dither applied. Linearity is reached with $\frac{1}{2}Q$ rms and at this level noise modulation is negligible. The total noise power is given by:

$$\frac{Q^2}{4} + \frac{Q^2}{12} = \frac{3 \times Q^2}{12} + \frac{Q^2}{12} = \frac{Q^2}{3}$$

and so the noise level will be $Q/\sqrt{3}$ rms. The noise level of an undithered quantizer in the large signal case is $Q/\sqrt{12}$ and so the noise is higher by a factor of:

$$\frac{Q}{\sqrt{3}} \times \frac{\sqrt{12}}{Q} = \frac{Q}{\sqrt{3}} \times \frac{2\sqrt{3}}{Q} = 2 = 6.02\,\text{dB}$$

Thus the SNR is given by:

$$6.02(n-1) + 1.76\,\text{dB} \tag{8.4}$$

A 16-bit system with correct Gaussian dither has an SNR of 92.1 dB.

This is inferior to the figure in expression (8.3) by 1.1 dB. In digital dither applications, triangular probability dither of $2Q$ peak-to-peak is optimum because it gives the best possible combination of nil distortion, freedom from noise modulation and SNR. Using dither with more than two rectangular processes added is detrimental. Whilst this result is also true for analog dither, it is not practicable to apply it to a real ADC as all real analog signals contain thermal noise which is Gaussian. If triangular dither is used on a signal containing Gaussian noise, the results derived above are not obtained. ADCs should therefore use Gaussian dither of $Q/2$ rms and performance will be given by expression (8.4).

It should be stressed that all of the results in this section are for conventional quantizing and requantizing. The use of techniques such as oversampling and/or noise shaping require an elaboration of the theory in order to give meaningful SNR figures.

8.13 Basic digital-to-analog conversion

This direction of conversion will be discussed first, since ADCs often use embedded DACs in feedback loops.

The purpose of a digital-to-analog converter is to take numerical values and reproduce the continuous waveform that they represent. Figure 8.25 shows the major elements of a conventional conversion subsystem, i.e. one in which oversampling is not employed. The jitter in the clock needs to be removed with a VCO or VCXO. Sample values are buffered in a latch and fed to the

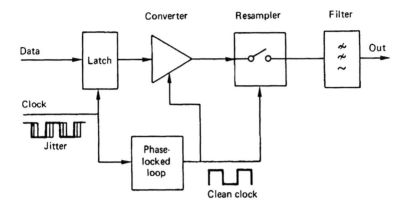

Figure 8.25 The components of a conventional converter. A jitter-free clock drives the voltage conversion, whose output may be resampled prior to reconstruction.

converter element which operates on each cycle of the clean clock. The output is then a voltage proportional to the number for at least a part of the sample period. A resampling stage may be found next, in order to remove switching transients, reduce the aperture ratio or allow the use of a converter which takes a substantial part of the sample period to operate. The resampled waveform is then presented to a reconstruction filter which rejects frequencies above the audio band.

This section is primarily concerned with the implementation of the converter element. There are two main ways of obtaining an analog signal from PCM data. One is to control binary-weighted currents and sum them; the other is to control the length of time a fixed current flows into an integrator. The two methods are contrasted in Figure 8.26. They appear simple, but are of no use for audio in these forms because of practical limitations. In Figure 8.26(c), the binary code is about to have a major overflow, and all the low-order currents are flowing. In Figure 8.26(d), the binary input has increased by one, and only the most significant current flows. This current must equal the sum of all the others plus one. The accuracy must be such that the step size is within the required limits. In this simple four-bit example, if the step size needs to be a rather casual 10% accurate, the necessary accuracy is only one part in 160, but for a 16-bit system it would become one part in 655 360, or about 2 ppm. This degree of accuracy is almost impossible to achieve, let alone maintain in the presence of ageing and temperature change.

The integrator-type converter in this four-bit example is shown in Figure 8.26(e); it requires a clock for the counter which allows it to count up to the maximum in less than one sample period. This will be more than 16 times the sampling rate. However, in a 16-bit system, the clock rate would need to be 65 536 times the sampling rate, or about 3 GHz. Whilst there may be a market for a CD player which can defrost a chicken, clearly some refinements are necessary to allow either of these converter types to be used in audio applications.

One method of producing currents of high relative accuracy is *dynamic element matching*.[15,16] Figure 8.27 shows a current source feeding a pair of

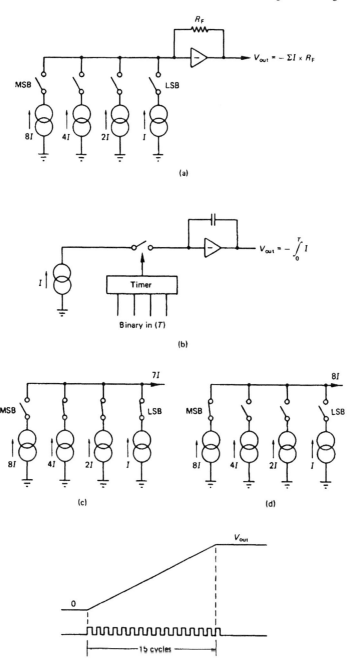

Figure 8.26 Elementary conversion: (a) weighted current DAC; (b) timed integrator DAC; (c) current flow with 0111 input; (d) current flow with 1000 input; (e) integrator ramps up for 15 cycles of clock for input 1111.

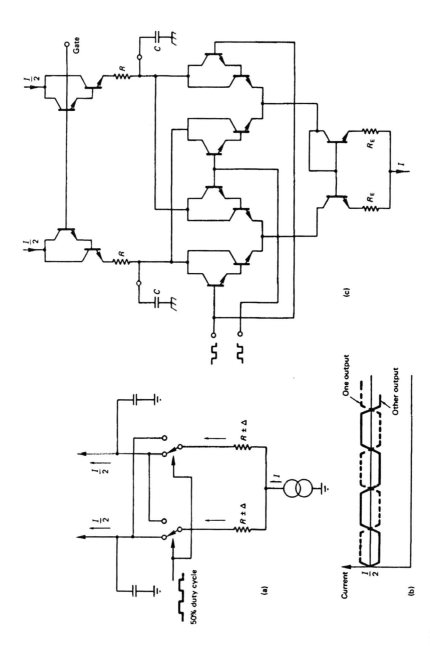

Figure 8.27 Dynamic element matching. (a) Each resistor spends half its time in each current path. (b) Average current of both paths will be identical if duty cycle is accurately 50%. (c) Typical monolithic implementation. Note clock frequency is arbitrary.

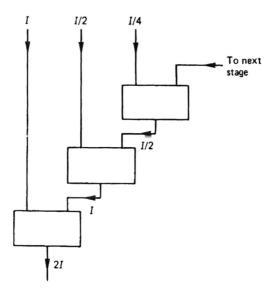

Figure 8.28 Cascading the current dividers of Figure 8.27 produces a binary-weighted series of currents.

nominally equal resistors. The two will not be the same owing to manufacturing tolerances and drift, and thus the current is only approximately divided between them. A pair of change-over switches places each resistor in series with each output. The average current in each output will then be identical, provided that the duty cycle of the switches is exactly 50%. This is readily achieved in a divide-by-two circuit. The accuracy criterion has been transferred from the resistors to the time domain in which accuracy is more readily achieved. Current averaging is performed by a pair of capacitors which do not need to be of any special quality. By cascading these divide-by-two stages, a binary-weighted series of currents can be obtained, as in Figure 8.28. In practice, a reduction in the number of stages can be obtained by using a more complex switching arrangement. This generates currents of ratio 1:1:2 by dividing the current into four paths and feeding two of them to one output, as shown in Figure 8.29. A major advantage of this approach is that no trimming is needed in manufacture, making it attractive for mass production. Freedom from drift is a further advantage.

To prevent interaction between the stages in weighted-current converters, the currents must be switched to ground or into the virtual earth by change-over switches. The on-resistance of these switches is a source of error, particularly the MSB, which passes most current. A solution in monolithic converters is to fabricate switches whose area is proportional to the weighted current, so that the voltage drops of all the switches arc the same. The error can then be removed with a suitable offset. The layout of such a device is dominated by the MSB switch since, by definition, it is as big as all the others put together.

The practical approach to the integrator converter is shown in Figure 8.30 and Figure 8.31 where two current sources whose ratio is 256:1 are used;

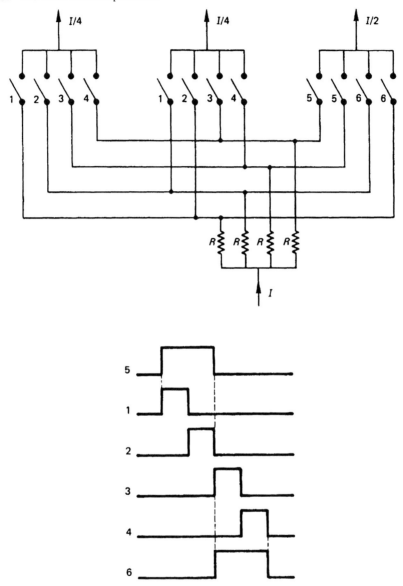

Figure 8.29 More complex dynamic element-matching system. Four drive signals (1, 2, 3, 4) of 25% duty cycle close switches of corresponding number. Two signals (5, 6) have 50% duty cycle, resulting in two current shares going to right-hand output. Division is thus into 1:1:2.

the larger is timed by the high byte of the sample and the smaller is timed by the low byte. The necessary clock frequency is reduced by a factor of 256. Any inaccuracy in the current ratio will cause one quantizing step in every 256 to be of the wrong size as shown in Figure 8.32, but current tracking is easier to achieve in a monolithic device. The integrator

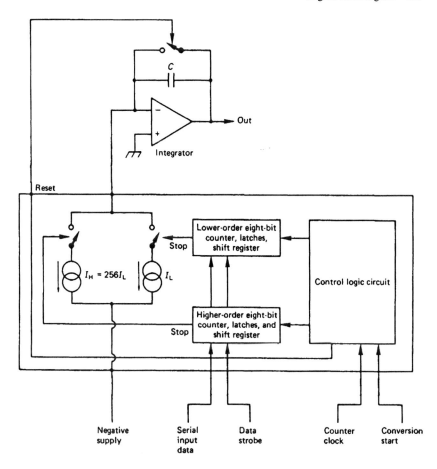

Figure 8.30 Simplified diagram of Sony CX-20017. The high-order and low-order current sources (I_H and I_L) and associated timing circuits can be seen. The necessary integrator is external.

capacitor must have low dielectric leakage and relaxation, and the operational amplifier must have low bias current as this will have the same effect as leakage.

The output of the integrator will remain constant once the current sources are turned off, and the resampling switch will be closed during the voltage plateau to produce the pulse amplitude modulated output. Clearly this device cannot produce a zero-order hold output without an additional sample-hold stage, so it is naturally complemented by resampling. Once the output pulse has been gated to the reconstruction filter, the capacitor is discharged with a further switch in preparation for the next conversion. The conversion count must take place in rather less than one sample period to permit the resampling and discharge phases. A clock frequency of about 20 MHz is adequate for a 16-bit 48 kHz unit, which permits the ramp to complete in 12.8 μs, leaving 8 μs for resampling and reset.

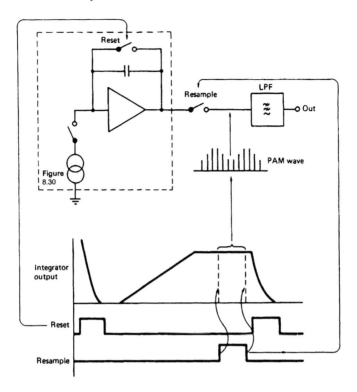

Figure 8.31 In an integrator converter, the output level is only stable when the ramp finishes. An analog switch is necessary to isolate the ramp from subsequent circuits. The switch can also be used to produce a PAM (pulse amplitude modulated) signal which has a flatter frequency response than a zero-order hold (staircase) signal.

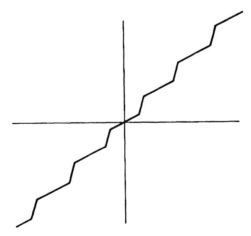

Figure 8.32 Imprecise tracking in a dual-slope converter results in the transfer function shown here.

8.14 Basic analog-to-digital conversion

A conventional analog-to-digital subsystem is shown in Figure 8.33. Following the anti-aliasing filter there will be a sampling process. Many of the ADCs described here will need a finite time to operate, whereas an instantaneous sample must be taken from the input. The solution is to use a track/hold circuit. Following sampling the sample voltage is quantized. The number of the quantized level is then converted to a binary code, typically two's complement. This section is concerned primarily with the implementation of the quantizing step.

The general principle of a quantizer is that different quantized voltages are compared with the unknown analog input until the closest quantized voltage is found. The code corresponding to this becomes the output. The comparisons can be made in turn with the minimal amount of hardware, or simultaneously.

The flash converter is probably the simplest technique available for PCM and DPCM conversion. The principle is shown in Figure 8.34. The threshold voltage of every quantizing interval is provided by a resistor chain which is fed by a reference voltage. This reference voltage can be varied to determine the sensitivity of the input. There is one voltage comparator connected to every reference voltage, and the other input of all of these is connected to the analog input. A comparator can be considered to be a one-bit ADC. The input voltage determines how many of the comparators will have a true output. As one comparator is necessary for each quantizing interval, then, for example, in an eight-bit system there will be 255 binary comparator outputs, and it is necessary to use a priority encoder to convert these to a binary code. Note that the quantizing stage is asynchronous; comparators change state as and when the variations in the input waveform result in a reference voltage being crossed. Sampling takes place when the comparator outputs are clocked into a subsequent latch. This is an example of quantizing before sampling as was illustrated in Figure 8.5. Although the device is simple in principle, it contains a lot of circuitry and can only be practicably implemented on a chip. A 16-bit

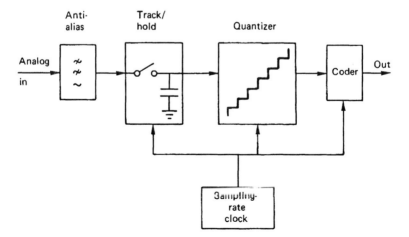

Figure 8.33 A conventional analog-to-digital subsystem. Following the anti-aliasing filter there will be a sampling process, which may include a track/hold circuit. Following quantizing, the number of the quantized level is then converted to a binary code, typically two's complement.

device would need a ridiculous 65 535 comparators, and thus these converters are not practicable for direct audio conversion, although they will be used to advantage in the DPCM and oversampling converters described later in this chapter. The analog signal has to drive a lot of inputs which results in a significant parallel capacitance, and a low-impedance driver is essential to avoid restricting the slewing rate of the input. The extreme speed of a flash converter is a distinct advantage in oversampling. Because computation of all bits is performed simultaneously, no track/hold circuit is required, and droop is eliminated. Figure 8.34(c) shows a flash converter chip. Note the resistor ladder and the comparators followed by the priority encoder. The MSB can be selectively inverted so that the device can be used either in offset binary or two's complement mode.

Reduction in component complexity can be achieved by quantizing serially. The most primitive method of generating different quantized voltages is to connect a counter to a DAC as in Figure 8.35. The resulting staircase voltage is compared with the input and used to stop the clock to the counter when the DAC output has just exceeded the input. This method is painfully slow, and is not used, as a much faster method exists which is only slightly more complex. Using successive approximation, each bit is tested in turn, starting with the MSB. If the input is greater than half range, the MSB will be retained and used as a base to test the next bit, which will be retained if the input exceeds three-quarters range and so on. The number of decisions is equal to the number of bits in the word, in contrast to the number of quantizing intervals which was the case in the previous example. A drawback of the successive

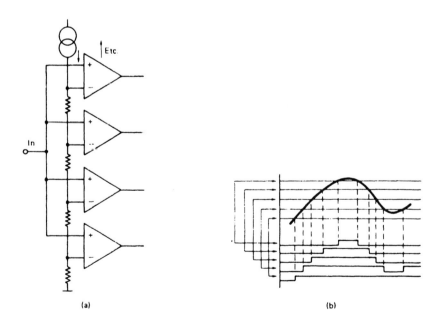

Figure 8.34 The flash converter. In (a) each quantizing interval has its own comparator, resulting in waveforms of (b). A priority encoder is necessary to convert the comparator outputs to a binary code

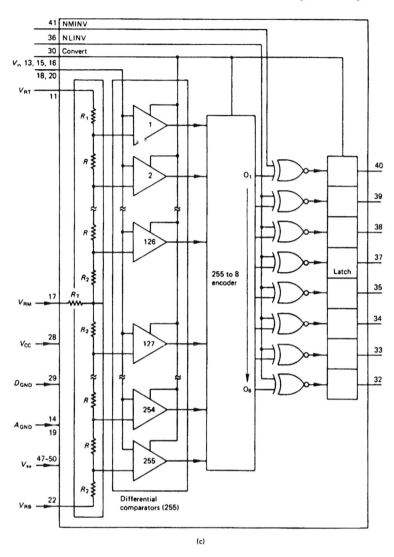

(c)

Figure 8.34 *(continued)* Shown in (c) is a typical eight-bit flash converter primarily intended for video applications. (Courtesy TRW.).

approximation converter is that the least significant bits are computed last, when droop is at its worst. Figure 8.36 and Figure 8.37 show that droop can cause a successive approximation converter to make a significant error under certain circumstances.

Analog-to-digital conversion can also be performed using the dual-current-source type DAC principle in a feedback system; the major difference is that the two current sources must work sequentially rather than concurrently. Figure 8.38 shows a 16-bit application in which the capacitor of the track/hold circuit is also used as the ramp integrator. The system operates as follows.

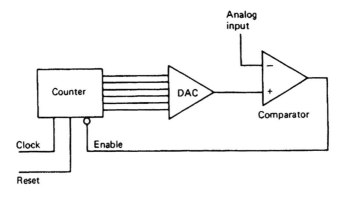

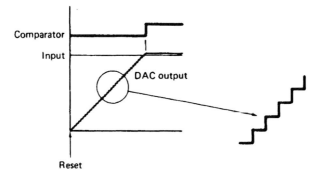

Figure 8.35 Simple ramp ADC compares output of DAC with input. Count is stopped when DAC output just exceeds input. This method, although potentially accurate, is much too slow for digital audio.

When the track/hold FET switches off, the capacitor C will be holding the sample voltage. Two currents of ratio 128:1 are capable of discharging the capacitor. Owing to this ratio, the smaller current will be used to determine the seven least significant bits, and the larger current will determine the nine most significant bits. The currents are provided by current sources of ratio 127:1. When both run together, the current produced is 128 times that from the smaller source alone. This approach means that the current can be changed simply by turning off the larger source, rather than by attempting a change-over.

With both current sources enabled, the high-order counter counts up until the capacitor voltage has fallen below the reference of $-128Q$ supplied to comparator 1. At the next clock edge, the larger current source is turned off. Waiting for the next clock edge is important, because it ensures that the larger source can only run for entire clock periods, which will discharge the integrator by integer multiples of $128Q$. The integrator voltage will overshoot the $128Q$ reference, and the remaining voltage on the integrator will be less than $128Q$ and will be measured by counting the number of clocks for which the smaller current source runs before the integrator voltage reaches zero. This process is termed residual expansion. The break in the slope of the integrator voltage

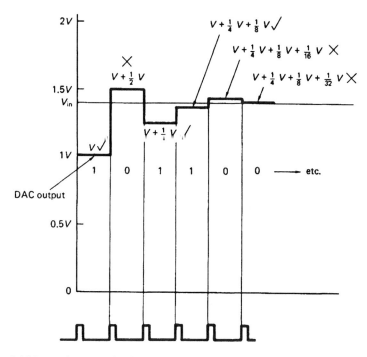

Figure 8.36 Successive approximation tests each bit in turn, starting with the most significant. The DAC output is compared with the input. If the DAC output is below the input ($\sqrt{}$) the bit is made 1; if the DAC output is above the input ($\times$) the bit is made zero.

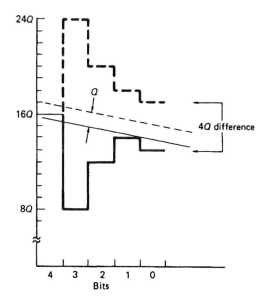

Figure 8.37 Two drooping track/hold signals (solid and dashed lines) which differ by one quantizing interval Q are shown here to result in conversions which are $4Q$ apart. Thus droop can destroy the monotonicity of a converter. Low-level signals (near the midrange of the number system) are especially vulnerable.

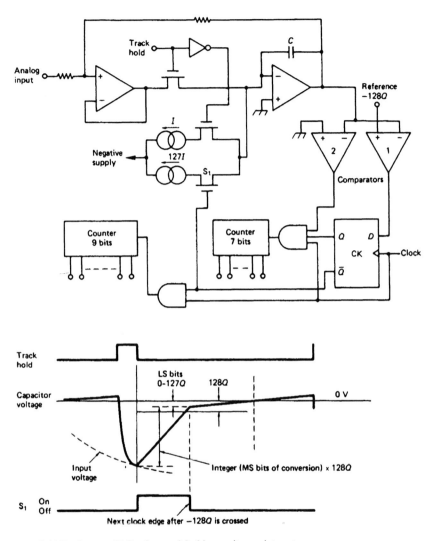

Figure 8.38 Dual-ramp ADC using track/hold capacitor as integrator.

gives rise to the alternative title of gear-change converter. Following ramping to ground in the conversion process, the track/hold circuit must settle in time for the next conversion. In this 16-bit example, the high-order conversion needs a maximum count of 512, and the low order needs 128: a total of 640. Allowing 25% of the sample period for the track/hold circuit to operate, a 48 kHz converter would need to be clocked at some 40 MHz. This is rather faster than the clock needed for the DAC using the same technology.

8.15 Alternative converters

Although PCM audio is universal because of the ease with which it can be recorded and processed numerically, there are several alternative related methods of converting an analog waveform to a bit stream. The output of these converter types is not Nyquist rate PCM, but this can be obtained from them by appropriate digital processing. In advanced conversion systems it is possible to adopt an alternative converter technique specifically to take advantage of a particular characteristic. The output is then digitally converted to Nyquist rate PCM in order to obtain the advantages of both.

Conventional PCM has already been introduced. In PCM, the amplitude of the signal only depends on the number range of the quantizer, and is independent of the frequency of the input. Similarly, the amplitude of the unwanted signals introduced by the quantizing process is also largely independent of input frequency.

Figure 8.39 introduces the alternative converter structures. The top half of the diagram shows converters which are differential. In differential coding the value of the output code represents the difference between the current sample voltage and that of the previous sample. The lower half of the diagram shows converters which are PCM. In addition, the left side of the diagram shows single-bit converters, whereas the right side shows multi-bit converters.

In differential pulse code modulation (DPCM), shown at top right, the difference between the previous absolute sample value and the current one is quantized into a multi-bit binary code. It is possible to produce a DPCM signal from a PCM signal simply by subtracting successive samples; this is digital differentiation. Similarly the reverse process is possible by using an accumulator or digital integrator (see Chapter 2) to compute sample values from the differences received. The problem with this approach is that it is very easy to lose the baseline of the signal if it commences at some arbitrary time. A digital high-pass filter can be used to prevent unwanted offsets.

Differential converters do not have an absolute amplitude limit. Instead there is a limit to the maximum rate at which the input signal voltage can change. They are said to be slew rate limited, and thus the permissible signal amplitude falls at 6 dB/octave. As the quantizing steps are still uniform, the quantizing error amplitude has the same limits as PCM. As input frequency rises, ultimately the signal amplitude available will fall down to it.

If DPCM is taken to the extreme case where only a binary output signal is available then the process is described as delta modulation (top left in Figure 8.39). The meaning of the binary output signal is that the current analog input is above or below the accumulation of all previous bits. The characteristics of the system show the same trends as DPCM, except that there is severe limiting of the rate of change of the input signal. A DPCM decoder must accumulate all the difference bits to provide a PCM output for conversion to analog, but with a one-bit signal the function of the accumulator can be performed by an analog integrator.

If an integrator is placed in the input to a delta modulator, the integrator's amplitude response loss of 6 dB/octave parallels the converter's amplitude limit of 6 dB/octave; thus the system amplitude limit becomes independent of frequency. This integration is responsible for the term sigma-delta modulation, since in mathematics sigma is used to denote summation. The input integrator can be combined with the integrator already present in a delta-modulator by a slight rearrangement of the components (bottom left in Figure 8.39). The transmitted signal is now the amplitude of the input, not the slope; thus the receiving integrator can be dispensed with, and all that is necessary after the

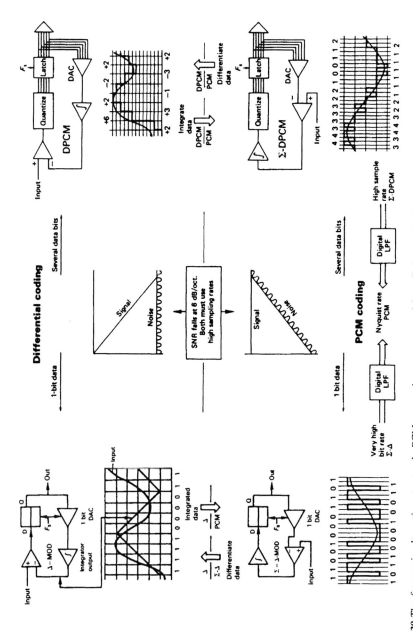

Figure 8.39 The four main alternatives to simple PCM conversion are compared here. Delta modulation is a 1-bit case of differential PCM, and conveys the slope of the signal. The digital output of both can be integrated to give PCM. Σ-Δ (sigma-delta) is a 1-bit case of Σ-DPCM. The application of integrator before differentiator makes the output true PCM, but tilts the noise floor; hence these can be referred to as 'noise-shaping' converters.

DAC is an LPF to smooth the bits. The removal of the integration stage at the decoder now means that the quantizing error amplitude rises at 6 dB/octave, ultimately meeting the level of the wanted signal.

The principle of using an input integrator can also be applied to a true DPCM system and the result should perhaps be called sigma-DPCM (bottom right in Figure 8.39). The dynamic range improvement over delta-sigma modulation is 6 dB for every extra bit in the code. Because the level of the quantizing error signal rises at 6 dB/octave in both delta-sigma modulation and sigma-DPCM, these systems are sometimes referred to as 'noise-shaping' converters, although the word 'noise' must be used with some caution. The output of a sigma-DPCM system is again PCM, and a DAC will be needed to receive it, because it is a binary code.

As the differential group of systems suffer from a wanted signal that converges with the unwanted signal as frequency rises, they must all use very high sampling rates.[17] It is possible to convert from sigma-DPCM to conventional PCM by reducing the sampling rate digitally. When the sampling rate is reduced in this way, the reduction of bandwidth excludes a disproportionate amount of noise because the noise shaping concentrated it at frequencies beyond the audio band. The use of noise shaping and oversampling is the key to the high resolution obtained in advanced converters.

8.16 Oversampling

Oversampling means using a sampling rate which is greater (generally substantially greater) than the Nyquist rate. Neither sampling theory nor quantizing theory *require* oversampling to be used to obtain a given signal quality, but Nyquist rate conversion places extremely high demands on component accuracy when a converter is implemented. Oversampling allows a given signal quality to be reached without requiring very close tolerance, and therefore expensive, components. Although it can be used alone, the advantages of oversampling are better realized when it is used in conjunction with noise shaping. Thus in practice the two processes are generally used together and the terms are often seen used in the loose sense as if they were synonymous. For a detailed and quantitative analysis of oversampling having exhaustive references the serious reader is referred to Hauser.[18]

Dynamic element matching trades component accuracy for accuracy in the time domain. Oversampling is another example of the same principle.

Figure 8.40 shows the main advantages of oversampling. At Figure 8.40(a) it will be seen that the use of a sampling rate considerably above the Nyquist rate allows the anti-aliasing and reconstruction filters to be realized with a much more gentle cut-off slope. There is then less likelihood of phase linearity and ripple problems in the audio passband.

Figure 8.40(b) shows that information in an analog signal is two dimensional and can be depicted as an area which is the product of bandwidth and the linearly expressed signal-to-noise ratio. The figure also shows that the same amount of information can be conveyed down a channel with an SNR of half as much (6 dB less) if the bandwidth used is doubled, with 12 dB less SNR if bandwidth is quadrupled, and so on, provided that the modulation scheme used is perfect.

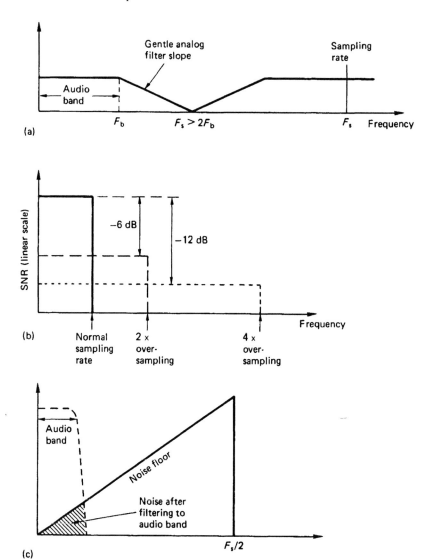

Figure 8.40 Oversampling has a number of advantages. In (a) it allows the slope of analog filters to be relaxed. In (b) it allows the resolution of converters to be extended. In (c) a *noise-shaped* converter allows a disproportionate improvement in resolution.

The information in an analog signal can be conveyed using some analog modulation scheme in any combination of bandwidth and SNR which yields the appropriate channel capacity. If bandwidth is replaced by sampling rate and SNR is replaced by a function of wordlength, the same must be true for a digital signal as it is no more than a numerical analog. Thus raising the sampling rate potentially allows the wordlength of each sample to be reduced without information loss.

Oversampling permits the use of a converter element of shorter wordlength, making it possible to use a flash converter. The flash converter is capable of working at very high frequency and so large oversampling factors are easily realized. The flash converter needs no track/hold system as it works instantaneously. The drawbacks of track/hold set out in Section 8.7 are thus eliminated. If the sigma-DPCM converter structure of Figure 8.39 is realized with a flash converter element, it can be used with a high oversampling factor. Figure 8.40(c) shows that this class of converter has a rising noise floor. If the highly oversampled output is fed to a digital low-pass filter which has the same frequency response as an analog anti-aliasing filter used for Nyquist rate sampling, the result is a disproportionate reduction in noise because the majority of the noise was outside the audio band. A high-resolution converter can be obtained using this technology without requiring unattainable component tolerances.

Information theory predicts that if an audio signal is spread over a much wider bandwidth by, for example, the use of an FM broadcast transmitter, the SNR of the demodulated signal can be higher than that of the channel it passes through, and this is also the case in digital systems. The concept is illustrated in Figure 8.41. At (a) four-bit samples are delivered at sampling rate F. As four bits have 16 combinations, the information rate is $16F$. At (b) the same information rate is obtained with three-bit samples by raising the sampling rate to $2F$ and at Figure 8.41(c) two-bit samples having four combinations require to be delivered at a rate of $4F$. Whilst the information rate has been maintained, it will be noticed that the bit rate of (c) is twice that of (a). The reason for this is shown in Figure 8.42. A single binary digit can only have two states; thus it can only convey two pieces of information, perhaps 'yes' or 'no'. Two binary digits together can have four states, and can thus convey four pieces of information, perhaps 'spring summer autumn or winter', which is two pieces of information per bit. Three binary digits grouped together

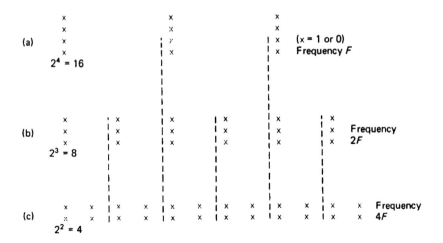

Figure 8.41 Information rate can be held constant when frequency doubles by removing 1 bit from each word. In all cases here it is $16F$. Note bit rate of (c) is double that of (a). Data storage in oversampled form is inefficient.

	0 = No 1 = Yes	00 = Spring 01 = Summer 10 = Autumn 11 = Winter	000 do 001 re 010 mi 011 fa 100 so 101 la 110 te 111 do	0000 0 0001 1 0010 2 0011 3 0100 4 0101 5 0110 6 0111 7 1000 8 1001 9 1010 A 1011 B 1100 C 1101 D 1110 E 1111 F		0000 FFFF	Digital audio sample values
No. of bits	1	2	3	4		16	
Information per word	2	4	8	16		65536	
Information per bit	2	2	≈3	4		4096	

Figure 8.42 The amount of information per bit increases disproportionately as wordlength increases. It is always more efficient to use the longest words possible at the lowest word rate. It will be evident that 16-bit PCM is 2048 times as efficient as delta modulation. Oversampled data are also inefficient for storage.

can have eight combinations, and convey eight pieces of information, perhaps 'doh re mi fah so lah te or doh', which is nearly three pieces of information per digit. Clearly the further this principle is taken, the greater the benefit. In a 16-bit system, each bit is worth 4K pieces of information. It is always more efficient, in information-capacity terms, to use the combinations of long binary words than to send single bits for every piece of information. The greatest efficiency is reached when the longest words are sent at the slowest rate which must be the Nyquist rate. This is one reason why PCM recording is more common than delta modulation, despite the simplicity of implementation of the latter type of converter. PCM simply makes more efficient use of the capacity of the binary channel.

As a result, oversampling is confined to converter technology where it gives specific advantages in implementation. The storage or transmission system will usually employ PCM, where the sampling rate is a little more than twice the audio bandwidth. Figure 8.43 shows a digital audio tape recorder which uses oversampling converters. The ADC runs at *n* times the Nyquist rate, but once in the digital domain the rate needs to be reduced in a type of digital filter called a *decimator*. The output of this is conventional Nyquist rate PCM, according to the tape format, which is then recorded. On replay the sampling rate is raised once more in a further type of digital filter called an *interpolator*. The

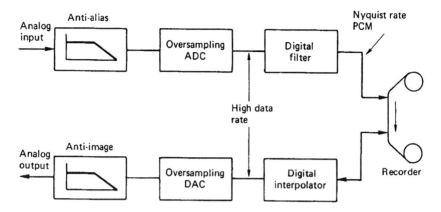

Figure 8.43 A recorder using oversampling in the converters overcomes the shortcomings of analog anti-aliasing and reconstruction filters and the converter elements are easier to construct; the recording is made with Nyquist rate PCM which minimizes tape consumption.

system now has the best of both worlds: using oversampling in the converters overcomes the shortcomings of analog anti-aliasing and reconstruction filters and the wordlength of the converter elements is reduced making them easier to construct; the recording is made with Nyquist rate PCM which minimizes tape consumption. Digital filters have the characteristic that their frequency response is proportional to the sampling rate. If a digital recorder is played at a reduced speed, the response of the digital filter will reduce automatically and prevent images passing the reconstruction process. If oversampling were to become universal, there would then be no need for the 48 kHz sampling rate.

Oversampling is a method of overcoming practical implementation problems by replacing a single critical element or bottleneck by a number of elements whose overall performance is what counts. As Hauser[18] properly observed, oversampling tends to overlap the operations which are quite distinct in a conventional converter. In earlier sections of this chapter, the vital subjects of filtering, sampling, quantizing and dither have been treated almost independently. Figure 8.44(a) shows that it is possible to construct an ADC of predictable performance by taking a suitable anti-aliasing filter, a sampler, a dither source and a quantizer and assembling them like building bricks. The bricks are effectively in series and so the performance of each stage can only limit the overall performance. In contrast Figure 8.44(b) shows that with oversampling the overlap of operations allows different processes to augment one another allowing a synergy which is absent in the conventional approach.

If the oversampling factor is n, the analog input must be bandwidth limited to $nF_s/2$ by the analog anti-aliasing filter. This unit need only have flat frequency response and phase linearity within the audio band. Analog dither of an amplitude compatible with the quantizing interval size is added prior to sampling at nF_s and quantizing.

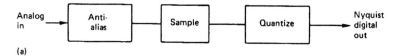

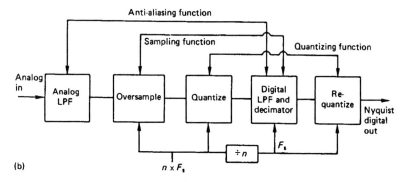

Figure 8.44 A conventional ADC performs each step in an identifiable location as in (a). With oversampling, many of the steps are distributed as shown in (b).

Next, the anti-aliasing function is completed in the digital domain by a low-pass filter which cuts off at $F_s/2$. Using an appropriate architecture this filter can be absolutely phase linear and implemented to arbitrary accuracy. Such filters are discussed in Chapter 4. The filter can be considered to be the demodulator of Figure 8.40 where the SNR improves as the bandwidth is reduced. The wordlength can be expected to increase. The multiplications taking place within the filter extend the wordlength considerably more than the bandwidth reduction alone would indicate. The analog filter serves only to prevent aliasing into the audio band at the oversampling rate; the audio spectrum is determined with greater precision by the digital filter.

With the audio information spectrum now Nyquist limited, the sampling process is completed when the rate is reduced in the decimator. One sample in n is retained.

The excess wordlength extension due to the anti-aliasing filter arithmetic must then be removed. Digital dither is added, completing the dither process, and the quantizing process is completed by requantizing the dithered samples to the appropriate wordlength which will be greater than the wordlength of the first quantizer. Alternatively noise shaping may be employed.

Figure 8.45(a) shows the building-brick approach of a conventional DAC. The Nyquist rate samples are converted to analog voltages and then a steep-cut analog low-pass filter is needed to reject the sidebands of the sampled spectrum.

Figure 8.45(b) shows the oversampling approach. The sampling rate is raised in an interpolator which contains a low-pass filter which restricts the baseband spectrum to the audio bandwidth shown. A large frequency gap now exists between the baseband and the lower sideband. The multiplications in the interpolator extend the wordlength considerably and this must be reduced

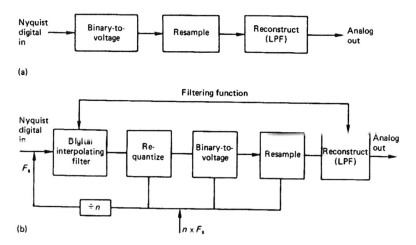

Figure 8.45 A conventional DAC in (a) is compared with the oversampling implementation in (b).

within the capacity of the DAC element by the addition of digital dither prior to requantizing. Again noise shaping may be used as an alternative.

8.17 Oversampling without noise shaping

If an oversampling converter is considered which makes no attempt to shape the noise spectrum, it will be clear that if it contains a perfect quantizer, no amount of oversampling will increase the resolution of the system, since a perfect quantizer is blind to all changes of input within one quantizing interval, and looking more often is of no help. It was shown earlier that the use of dither would linearize a quantizer, so that input changes much smaller than the quantizing interval would be reflected in the output and this remains true for this class of converter.

Figure 8.46 shows the example of a white-noise-dithered quantizer, oversampled by a factor of four. Since dither is correctly employed, it is valid to speak of the unwanted signal as noise. The noise power extends over the whole baseband up to the Nyquist limit. If the baseband width is reduced by the oversampling factor of four back to the bandwidth of the original analog input, the noise bandwidth will also be reduced by a factor of four, and the noise power will be one-quarter of that produced at the quantizer. One-quarter noise power implies one-half the noise voltage, so the SNR of this example has been increased by 6 dB, the equivalent of one extra bit in the quantizer. Information theory predicts that an oversampling factor of four would allow an extension by two bits. This method is suboptimal in that very large oversampling factors would be needed to obtain useful resolution extension, but it would still realize some advantages, particularly the elimination of the steep-cut analog filter.

The division of the noise by a larger factor is the only route left open, since all the other parameters are fixed by the signal bandwidth required.

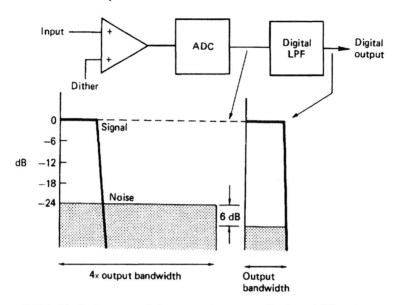

Figure 8.46 In this simple oversampled converter, 4× oversampling is used. When the converter output is low-pass filtered, the noise power is reduced to one-quarter, which in voltage terms is 6 dB. This is a suboptimal method and is not used.

The reduction of noise power resulting from a reduction in bandwidth is only proportional if the noise is white, i.e. it has uniform power spectral density (PSD). If the noise from the quantizer is made spectrally non-uniform, the oversampling factor will no longer be the factor by which the noise power is reduced. The goal is to concentrate noise power at high frequencies, so that after low-pass filtering in the digital domain down to the audio input bandwidth, the noise power will be reduced by more than the oversampling factor.

8.18 Noise shaping

Noise shaping dates from the work of Cutler[19] in the 1950s. It is a feedback technique applicable to quantizers and requantizers in which the quantizing process of the current sample is modified in some way by the quantizing error of the previous sample.

When used with requantizing, noise shaping is an entirely digital process which is used, for example, following word extension due to the arithmetic in digital mixers or filters in order to return to the required wordlength. It will be found in this form in oversampling DACs. When used with quantizing, part of the noise shaping circuitry will be analog. As the feedback loop is placed around an ADC it must contain a DAC. When used in converters, noise shaping is primarily an implementation technology. It allows processes which are conveniently available in integrated circuits to be put to use in audio conversion. Once integrated circuits can be employed, complexity ceases to be a drawback and low-cost mass production is possible.

It has been stressed throughout this chapter that a series of numerical values or samples is just another analog of an audio waveform. Chapter 2 showed that all analog processes such as mixing, attenuation or integration all have exact numerical parallels. It has been demonstrated that digitally dithered requantizing is no more than a digital simulation of analog quantizing. It should be no surprise that in this section noise shaping will be treated in the same way. Noise shaping can be performed by manipulating analog voltages or numbers representing them or both. If the reader is content to make a conceptual switch between the two, many obstacles to understanding fall, not just in this topic, but in digital audio in general.

The term noise shaping is idiomatic and in some respects unsatisfactory because not all devices which are called noise shapers produce true noise. The caution which was given when treating quantizing error as noise is also relevant in this context. Whilst 'quantizing-error-spectrum shaping' is a bit of a mouthful, it is useful to keep in mind that noise shaping means just that in order to avoid some pitfalls. Some noise shaper architectures do not produce a signal decorrelated quantizing error and need to be dithered.

Figure 8.47(a) shows a requantizer using a simple form of noise shaping. The low-order bits which are lost in requantizing are the quantizing error. If the value of these bits is added to the next sample before it is requantized, the quantizing error will be reduced. The process is somewhat like the use of negative feedback in an operational amplifier except that it is not instantaneous, but encounters a one sample delay. With a constant input, the mean or average quantizing error will be brought to zero over a number of samples, achieving one of the goals of additive dither. The more rapidly the input changes, the greater the effect of the delay and the less effective the error feedback will be. Figure 8.47(b) shows the equivalent circuit seen by the quantizing error, which is created at the requantizer and subtracted from itself one sample period later. As a result the quantizing error spectrum is not uniform, but has the shape of a raised sine wave shown at (c), hence the term noise shaping. The noise is very small at DC and rises with frequency, peaking at the Nyquist frequency at a level determined by the size of the quantizing step. If used with oversampling, the noise peak can be moved outside the audio band.

Figure 8.48 shows a simple example in which two low-order bits need to be removed from each sample. The accumulated error is controlled by using the bits which were neglected in the truncation, and adding them to the next sample. In this example, with a steady input, the roundoff mechanism will produce an output of 01110111.... If this is low-pass filtered, the three ones and one zero result in a level of three-quarters of a quantizing interval, which is precisely the level which would have been obtained by direct conversion of the full digital input. Thus the resolution is maintained even though two bits have been removed.

The noise shaping technique was used in the first generation Philips CD players which oversampled by a factor of four. Starting with 16-bit PCM from the disk, the 4× oversampling will in theory permit the use of an ideal 14-bit converter, but only if the wordlength is reduced optimally. The oversampling DAC system used is shown in Figure 8.49.[20] The interpolator arithmetic extends the wordlength to 28 bits, and this is reduced to 14 bits using the error feedback loop of Figure 8.47. The noise floor rises slightly towards the

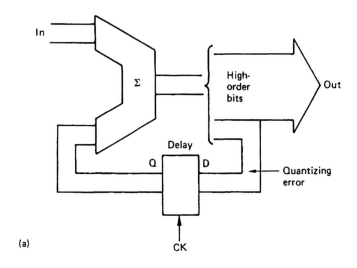

(a)

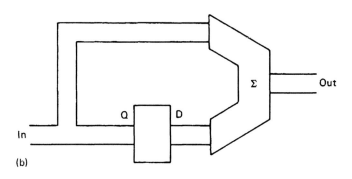

(b)

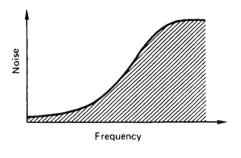

(c)

Figure 8.47 (a) A simple requantizer which feeds back the quantizing error to reduce the error of subsequent samples. The one-sample delay causes the quantizing error to see the equivalent circuit shown in (b) which results in a sinusoidal quantizing error spectrum shown in (c).

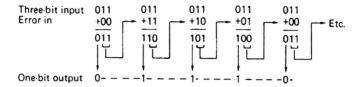

Figure 8.48 By adding the error caused by truncation to the next value, the resolution of the lost bits is maintained in the duty cycle of the output. Here, truncation of 011 by two bits would give continuous zeros, but the system repeats 0111, 0111, which, after filtering, will produce a level of three-quarters of a bit.

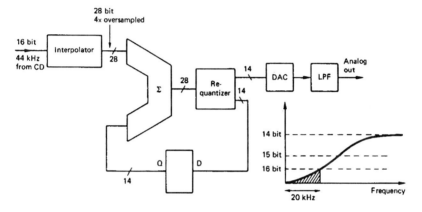

Figure 8.49 The noise-shaping system of the first generation of Philips CD players.

edge of the audio band, but remains below the noise level of a conventional 16-bit DAC which is shown for comparison.

The 14-bit samples then drive a DAC using dynamic element matching. The aperture effect in the DAC is used as part of the reconstruction filter response, in conjunction with a third-order Bessel filter which has a response 3 dB down at 30 kHz. Equalization of the aperture effect within the audio passband is achieved by giving the digital filter which produces the oversampled data a rising response. The use of a digital interpolator as part of the reconstruction filter results in extremely good phase linearity.

Noise shaping can also be used without oversampling. In this case the noise cannot be pushed outside the audio band. Instead the noise floor is shaped or weighted to complement the unequal spectral sensitivity of the ear to noise.[21,22] Unless we wish to violate Shannon's theory, this psychoacoustically optimal noise shaping can only reduce the noise power at certain frequencies by increasing it at others. Thus the average log PSD over the audio band remains the same, although it may be raised slightly by noise induced by imperfect processing.

Figure 8.50 shows noise shaping applied to a digitally dithered requantizer. Such a device might be used when, for example, making a CD master from a 20-bit recording format. The input to the dithered requantizer is subtracted from the output to give the error due to requantizing. This error is filtered

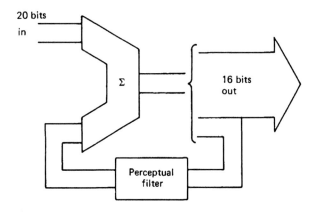

Figure 8.50 Perceptual filtering in a requantizer gives a subjectively improved SNR.

(and inevitably delayed) before being subtracted from the system input. The filter is not designed to be the exact inverse of the perceptual weighting curve because this would cause extreme noise levels at the ends of the band. Instead the perceptual curve is levelled off[23] such that it cannot fall more than, for example, 40 dB below the peak.

Psychoacoustically optimal noise shaping can offer nearly three bits of increased dynamic range when compared with optimal spectrally flat dither. Enhanced Compact Discs recorded using these techniques are now available.

8.19 Noise-shaping ADCs

The sigma-DPCM converter introduced in Figure 8.39 has a natural application here and is shown in more detail in Figure 8.51. The current digital sample from the quantizer is converted back to analog in the embedded DAC.

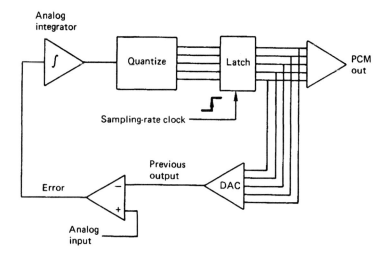

Figure 8.51 The sigma-DPCM converter of Figure 8.39 is shown here in more detail.

The DAC output differs from the ADC input by the quantizing error. The DAC output is subtracted from the analog input to produce an error which is integrated to drive the quantizer in such a way that the error is reduced. With a constant input voltage the average error will be zero because the loop gain is infinite at DC. If the average error is zero, the mean or average of the DAC outputs must be equal to the analog input. The instantaneous output will deviate from the average in what is called an idling pattern. The presence of the integrator in the error feedback loop makes the loop gain fall with rising frequency. With the feedback falling at 6 dB/octave, the noise floor will rise at the same rate.

Figure 8.52 shows a simple oversampling system using a sigma-DPCM converter and an oversampling factor of only four. The sampling spectrum shows that the noise is concentrated at frequencies outside the audio part of the oversampling baseband. Since the scale used here means that noise power is represented by the area under the graph, the area left under the graph after the filter shows the noise-power reduction. Using the relative areas of similar triangles shows that the reduction has been by a factor of 16. The corresponding noise-voltage reduction would be a factor of four, or 12 dB which corresponds to an additional two bits in wordlength. These bits will be available in the wordlength extension which takes place in the decimating filter. Owing to the rise of 6 dB/octave in the PSD of the noise, the SNR will be 3 dB worse at the edge of the audio band.

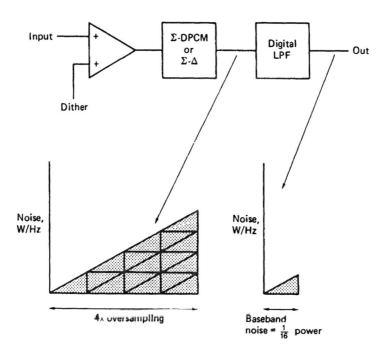

Figure 8.52 In a sigma-DPCM or Σ-Δ converter, noise amplitude increases by 6 dB/octave, noise power by 12 dB/octave. In this 4× oversampling converter, the digital filter reduces bandwidth by four, but noise power is reduced by a factor of 16. Noise voltage falls by a factor of four or 12 dB.

One way in which the operation of the system can be understood is to consider that the coarse DAC in the loop defines fixed points in the audio transfer function. The time averaging which takes place in the decimator then allows the transfer function to be interpolated between the fixed points. True signal-independent noise of sufficient amplitude will allow this to be done to infinite resolution, but by making the noise primarily outside the audio band the resolution is maintained but the audio band signal-to-noise ratio can be extended. A first-order noise-shaping ADC of the kind shown can produce signal-dependent quantizing error and requires analog dither. However, this can be outside the audio band and so need not reduce the SNR achieved.

A greater improvement in dynamic range can be obtained if the integrator is supplanted to realize a higher order filter.[24] The filter is in the feedback loop and so the noise will have the opposite response to the filter and will therefore rise more steeply to allow a greater SNR enhancement after decimation. Figure 8.53 shows the theoretical SNR enhancement possible for various loop filter orders and oversampling factors. A further advantage of high-order loop filters is that the quantizing noise can be decorrelated from the signal making dither unnecessary. High-order loop filters were at one time thought to be impossible to stabilize, but this is no longer the case, although care is necessary. One technique which may be used is to include some feedforward paths as shown in Figure 8.54.

An ADC with high-order noise shaping was disclosed by Adams[25] and a simplified diagram is shown in Figure 8.55. The comparator outputs of the 128 times oversampled four-bit flash ADC are directly fed to the DAC which consists of 15 equal resistors fed by CMOS switches. As with all feedback loops, the transfer characteristic cannot be more accurate than the feedback, and in this case the feedback accuracy is determined by the precision of the DAC.[26] Driving the DAC directly from the ADC comparators is more accurate

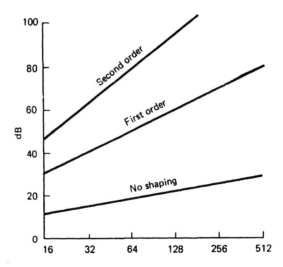

Figure 8.53 The enhancement of SNR possible with various filter orders and oversampling factors in noise-shaping converters.

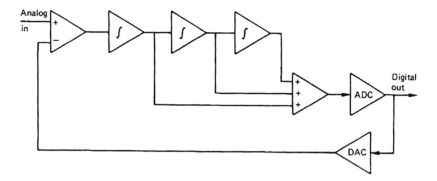

Figure 8.54 Stabilizing the loop filter in a noise-shaping converter can be assisted by the incorporation of feedforward paths as shown here.

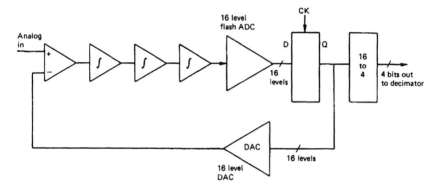

Figure 8.55 An example of a high-order noise-shaping ADC. See text for details.

because each input has equal weighting. The stringent MSB tolerance of the conventional binary weighted DAC is then avoided. The comparators also drive a 16 to 4 priority encoder to provide the four-bit PCM output to the decimator. The DAC output is subtracted from the analog input at the integrator. The integrator is followed by a pair of conventional analog operational amplifiers having frequency-dependent feedback and a passive network which gives the loop a fourth-order response overall. The noise floor is thus shaped to rise at 24 dB/octave beyond the audio band. The time constants of the loop filter are optimized to minimize the amplitude of the idling pattern as this is an indicator of the loop stability. The four-bit PCM output is low-pass filtered and decimated to the Nyquist frequency. The high oversampling factor and high-order noise shaping extend the dynamic range of the four-bit flash ADC to 108 dB at the output.

8.20 A one-bit DAC

It might be thought that the waveform from a one-bit DAC is simply the same as the digital input waveform. In practice this is not the case. The input signal is a logic signal which need only be above or below a threshold for its binary value to be correctly received. It may have a variety of waveform distortions and a duty cycle offset. The area under the pulses can vary enormously. In the DAC output the amplitude needs to be extremely accurate. A one-bit DAC uses only the binary information from the input, but reclocks to produce accurate timing and uses a reference voltage to produce accurate levels. The area of pulses produced is then constant. One-bit DACs will be found in noise-shaping ADCs as well as in the more obvious application of producing analog audio.

Figure 8.56(a) shows a one-bit DAC which is implemented with MOS field-effect switches and a pair of capacitors. Quanta of charge are driven into or out of a virtual earth amplifier configured as an integrator by the switched capacitor action. Figure 8.56(b) shows the associated waveforms. Each data bit period is divided into two equal portions; that for which the clock is high, and that for which it is low. During the first half of the bit period, pulse P+ is generated if the data bit is a 1, or pulse P− is generated if the data bit is a 0. The reference input is a clean voltage corresponding to the gain required.

C1 is *discharged* during the second half of every cycle by the switches driven from the complemented clock. If the next bit is a 1, during the next high period of the clock the capacitor will be connected between the reference and the virtual earth. Current will flow into the virtual earth until the capacitor is charged. If the next bit is not a 1, the current through C1 will flow to ground.

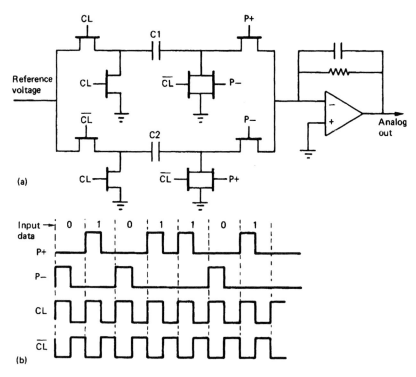

Figure 8.56 In (a) the operation of a 1-bit DAC relies on switched capacitors. The switching waveforms are shown in (b).

C2 is *charged* to reference voltage during the second half of every cycle by the switches driven from the complemented clock. On the next high period of the clock, the reference end of C2 will be grounded, and so the op-amp end will assume a negative reference voltage. If the next bit is a 0, this negative reference will be switched into the virtual earth, if not the capacitor will be discharged.

Thus on every cycle of the clock, a quantum of charge is either pumped into the integrator by C1 or pumped out by C2. The analog output therefore precisely reflects the ratio of ones to zeros.

8.21 One-bit noise-shaping ADCs

In order to overcome the DAC accuracy constraint of the sigma-DPCM converter, the sigma-delta converter can be used as it has only one bit internal resolution. A one-bit DAC cannot be non-linear by definition as it defines only two points on a transfer function. It can, however, suffer from other deficiencies such as DC offset and gain error although these are less offensive in audio. The one-bit ADC is a comparator.

As the sigma-delta converter is only a one-bit device, clearly it must use a high oversampling factor and high-order noise shaping in order to have sufficiently good SNR for audio.[27] In practice the oversampling factor is limited not so much by the converter technology as by the difficulty of computation in the decimator. A sigma-delta converter has the advantage that the filter input 'words' are one bit long and this simplifies the filter design as multiplications can be replaced by selection of constants.

Conventional analysis of loops falls down heavily in the one-bit case. In particular the gain of a comparator is difficult to quantify, and the loop is highly non-linear so that considering the quantizing error as additive white noise in order to use a linear loop model gives rather optimistic results. In the absence of an accurate mathematical model, progress has been made empirically, with listening tests and by using simulation.

Single-bit sigma-delta converters are prone to long idling patterns because the low resolution in the voltage domain requires more bits in the time domain to be integrated to cancel the error. Clearly the longer the period of an idling pattern the more likely it is to enter the audio band as an objectional whistle or 'birdie'. They also exhibit threshold effects or deadbands where the output fails to react to an input change at certain levels. The problem is reduced by the order of the filter and the wordlength of the embedded DAC. Second- and third-order feedback loops are still prone to audible idling patterns and threshold effect.[28] The traditional approach to linearizing sigma-delta converters is to use dither. Unlike conventional quantizers, the dither used was of a frequency outside the audio band and of considerable level. Square-wave dither has been used and it is advantageous to choose a frequency which is a multiple of the final output sampling rate as then the harmonics will coincide with the troughs in the stopband ripple of the decimator. Unfortunately the level of dither needed to linearize the converter is high enough to cause premature clipping of high-level signals, reducing the dynamic range. This problem is overcome by using in-band white noise dither at low level.[29]

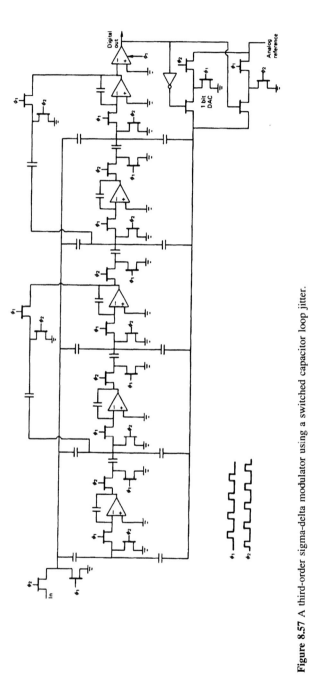

Figure 8.57 A third-order sigma-delta modulator using a switched capacitor loop jitter.

An advantage of the one-bit approach is that in the one-bit DAC, precision components are replaced by precise timing in switched capacitor networks. The same approach can be used to implement the loop filter in an ADC. Figure 8.57 shows a third-order sigma-delta modulator incorporating a DAC based on the principle of Figure 8.55. The loop filter is also implemented with switched capacitors.

8.22 Two's complement coding

In the two's complement system, the upper half of the pure binary number range has been redefined to represent negative quantities. If a pure binary counter is constantly incremented and allowed to overflow, it will produce all the numbers in the range permitted by the number of available bits, and these are shown for a four-bit example drawn around the circle in Figure 8.58. In two's complement, the quantizing range represented by the circle of numbers does not start at zero, but starts on the diametrically opposite side of the circle. Zero is mid-range, and all numbers with the MSB (most significant

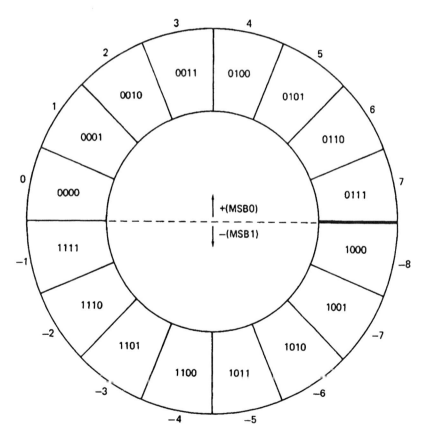

Figure 8.58 In this example of a 4-bit two's complement code, the number range is from −8 to +7. Note that the MSB determines polarity.

bit) set are considered negative. The MSB is thus the equivalent of a sign bit where 1 = minus. Two's complement notation differs from pure binary in that the most significant bit is inverted in order to achieve the half circle rotation.

Figure 8.59 shows how a real ADC is configured to produce two's complement output. At (a) an analog offset voltage equal to one half the quantizing range is added to the bipolar analog signal in order to make it unipolar as at (b). The ADC produces positive-only numbers at (c) which are proportional to the input voltage. The MSB is then inverted at (d) so that the all-zeros code moves to the centre of the quantizing range.

Figure 8.60 shows how the two's complement system allows two sample values to be added, or mixed in audio parlance, in a manner analogous to adding analog signals in an operational amplifier. The waveform of input A is depicted by solid black samples, and that of B by samples with a solid outline. The result of mixing is the linear sum of the two waveforms obtained by adding pairs of sample values. The dashed lines depict the output values. Beneath each set of samples is the calculation which will be seen to give the correct result. Note that the calculations are pure binary. No special arithmetic is needed to handle two's complement numbers.

It is often necessary to phase reverse or invert an audio signal, for example a microphone input to a mixer. The process of inversion in two's complement is

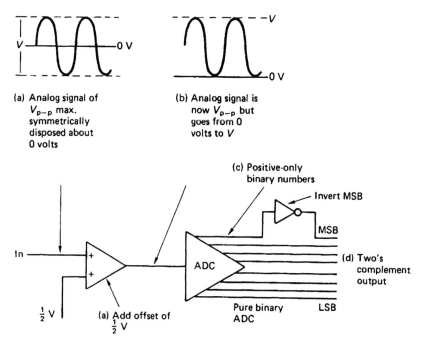

(a) Analog signal of V_{p-p} max. symmetrically disposed about 0 volts

(b) Analog signal is now V_{p-p} but goes from 0 volts to V

(c) Positive-only binary numbers

Invert MSB

MSB

In

ADC

$\frac{1}{2}$ V

(a) Add offset of $\frac{1}{2}$ V

Pure binary ADC

LSB

(d) Two's complement output

Figure 8.59 A two's complement ADC. At (a) an analog offset voltage equal to one-half the quantizing range is added to the bipolar analog signal in order to make it unipolar as at (b). The ADC produces positive-only numbers at (c), but the MSB is then inverted at (d) to give a two's complement output.

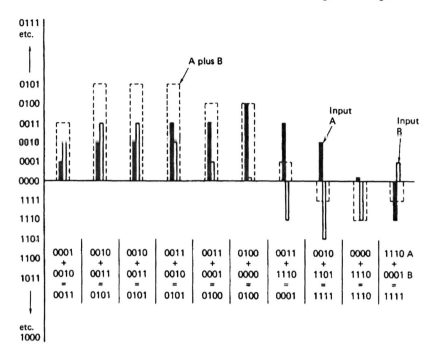

Figure 8.60 Using two's complement arithmetic, single values from two waveforms are added together with respect to midrange to give a correct mixing function.

simple. All bits of the sample value are inverted to form the one's complement, and one is added. This can be checked by mentally inverting some of the values in Figure 8.58. The inversion is transparent and performing a second inversion gives the original sample values. Using inversion, signal subtraction can be performed using only adding logic. The inverted input is added to perform a subtraction, just as in the analog domain.

8.23 Level in digital audio

Analog tape recorders use operating levels which are some way below saturation. The range between the operating level and saturation is called the headroom. In this range, distortion becomes progressively worse and sustained recording in the headroom is avoided. However, transients may be recorded in the headroom as the ear cannot respond to distortion products unless they are sustained. The PPM level meter has an attack time constant which simulates the temporal distortion sensitivity of the ear. If a transient is too brief to deflect a PPM into the headroom, it will not be heard either

Operating levels are used in two ways. On making a recording from a microphone, the gain is increased until distortion is just avoided, thereby obtaining a recording having the best SNR. In post production the gain will be set to whatever level is required to obtain the desired subjective effect in the context of the program material. This is particularly important to broadcasters who require the relative loudness of different material to be controlled so that the listener does not need to make continuous adjustments to the volume control.

In order to maintain level accuracy, analog recordings are traditionally preceded by line-up tones at standard operating level. These are used to adjust the gain in various stages of dubbing and transfer along land lines so that no level changes occur to the program material.

Figure 8.61 shows some audio waveforms at various levels with respect to the coding values. Where an audio waveform just fits into the quantizing range without clipping it has a level which is defined as 0 dBFs where Fs indicates *full scale*. Reducing the level by 6.02 dB makes the signal half as large and results in the second bit in the sample becoming the same as the sign bit. Reducing the level by a further 6.02 dB to −12 dBFs will make the second and third bits the same as the sign bit and so on.

Unlike analog recorders, digital recorders do not have headroom, as there is no progressive onset of distortion until converter clipping, the equivalent of saturation, occurs at 0 dBFs. Accordingly many digital recorders have level meters which read in dBFs. The scales are marked with 0 at the clipping level and all operating levels are below that. This causes no difficulty provided the user is aware of the consequences.

However, in the situation where a digital copy of an analog tape is to be made, it is very easy to set the input gain of the digital recorder so that line-up tone from the analog tape reads 0 dB. This lines up digital clipping with the analog operating level. When the tape is dubbed, all signals in the headroom suffer converter clipping.

In order to prevent such problems, manufacturers and broadcasters have introduced artificial headroom on digital level meters, simply by calibrating the scale and changing the analog input sensitivity so that 0 dB analog is some way below clipping. Unfortunately there has been little agreement on how much artificial headroom should be provided, and machines which have it are seldom labelled with the amount. There is an argument which suggests that the amount of headroom should be a function of the sample wordlength, but this causes difficulties when transferring from one wordlength to another. The EBU[30]

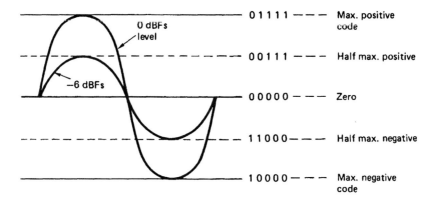

Figure 8.61 0 dBFs is defined as the level of the largest sinusoid which will fit into the quantizing range without clipping.

concluded that a single relationship between analog and digital level was desirable. In 16-bit working, 12 dB of headroom is a useful figure, but now that 18- and 20-bit converters are available, the new EBU draft recommendation specifies 18 dB.

8.24 The AES/EBU interface

The AES/EBU digital audio interface, originally published in 1985,[31] was proposed to embrace all the functions of existing formats in one standard. The goal was to ensure interconnection of professional digital audio equipment irrespective of origin. Alongside the professional format, Sony and Philips developed a similar format now known as SPDIF (Sony Philips Digital Interface) intended for consumer use. This offers different facilities to suit the application, yet retains sufficient compatibility with the professional interface so that, for many purposes, consumer and professional machines can be connected together.[32,33]

It was desired to use the existing analog audio cabling in such installations, which would be 600 Ω balanced line screened, with one cable per audio channel, or in some cases one twisted pair per channel with a common screen. At audio frequency the impedance of cable is high and the 600 Ω figure is that of the source and termination. If a single serial channel is to be used, the interconnect has to be self-clocking and self-synchronizing, i.e. the single signal must carry enough information to allow the boundaries between individual bits, words and blocks to be detected reliably. To fulfil these requirements, the AES/EBU and SPDIF interfaces use FM channel code which is DC free, strongly self-clocking and capable of working with a changing sampling rate. FM code generation is described in Section 10.5. Synchronization of the deserialization process is achieved by violating the usual encoding rules.

The use of FM means that the channel frequency is the same as the bit rate when sending data ones. Tests showed that in typical analog audio-cabling installations, sufficient bandwidth was available to convey two digital audio channels in one twisted pair. The standard driver and receiver chips for RS-422A[34] data communication (or the equivalent CCITT-V.11) are employed for professional use, but work by the BBC[35] suggested that equalization and transformer coupling are desirable for longer cable runs, particularly if several twisted pairs occupy a common shield. Successful transmission up to 350 m has been achieved with these techniques.[36] Figure 8.62 shows the standard

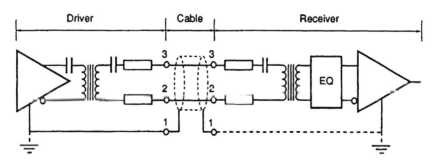

Figure 8.62 Recommended electrical circuit for use with the standard two-channel interface.

configuration. The output impedance of the drivers will be about 110 Ω, and the impedance of the cable used should be similar at the frequencies of interest. The driver was specified in AES-3-1985 to produce between 3 and 10 V_{p-p} into such an impedance but this was changed to between 2 and 7 volts in AES-3-1992 to better reflect the characteristics of actual RS-422 driver chips.

The original receiver impedance was set at a high 250 Ω, with the intention that up to four receivers could be driven from one source. This has been found to be inadvisable on long cables because of reflections caused by impedance mismatches and AES-3-1992 is now a point-to-point interface with source, cable and load impedance all set at 110 Ω.

In Figure 8.63, the specification of the receiver is shown in terms of the minimum eye pattern which can be detected without error. It will be noted that the voltage of 200 mV specifies the height of the eye opening at a width of half a channel bit period. The actual signal amplitude will need to be larger than this, and even larger if the signal contains noise. Figure 8.64 shows

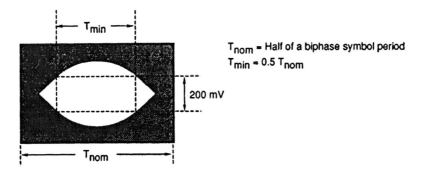

T_{nom} = Half of a biphase symbol period
T_{min} = 0.5 T_{nom}

Figure 8.63 The minimum eye pattern acceptable for correct decoding of standard two-channel data.

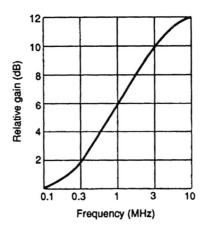

Figure 8.64 EQ characteristic recommended by the AES to improve reception in the case of long lines.

the recommended equalization characteristic which can be applied to signals received over long lines.

The purpose of the standard is to allow the use of existing analog cabling, and as an adequate connector in the shape of the XLR is already in wide service, the connector made to IEC 268 Part 12 has been adopted for digital audio use. Effectively, existing analog audio cables having XLR connectors can be used without alteration for digital connections. The AES/EBU standard does, however, require that suitable labelling should be used so that it is clear that the connections on a particular unit are digital.

The need to drive long cables does not generally arise in the domestic environment, and so a low-impedance balanced signal is not necessary. The electrical interface of the consumer format uses a 0.5 V peak single-ended signal, which can be conveyed down conventional audio-grade coaxial cable connected with RCA 'phono' plugs. Figure 8.65 shows the resulting consumer interface as specified by IEC 958.

There is a separate proposal[37] for a professional interface using coaxial cable for distances of around 1000 m. This is simply the AES/EBU protocol but with a 75 Ω coaxial cable carrying a 1 V signal so that it can be handled by analog video distribution amplifiers. Impedance converting transformers allow balanced 110 Ω to unbalanced 75 Ω matching.

In Figure 8.66 the basic structure of the professional and consumer formats can be seen. One subframe consists of 32 bit-cells, of which four will be used by a synchronizing pattern. Subframes from the two audio channels, A and B, alternate on a time division basis. Up to 24-bit sample wordlength can be used, which should cater for all conceivable future developments, but normally 20-bit maximum length samples will be available with four auxiliary data bits, which can be used for a voice-grade channel in a professional application. In a consumer RDAT machine, subcode can be transmitted in bits 4–11, and the 16-bit audio in bits 12–27.

Preceding formats sent the most significant bit first. Since this was the order in which bits were available in successive approximation converters it has become a *de facto* standard for inter-chip transmission inside equipment. In contrast, this format sends the least significant bit first. One advantage of this approach is that simple serial arithmetic is then possible on the samples

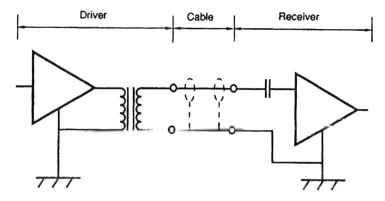

Figure 8.65 The consumer electrical interface.

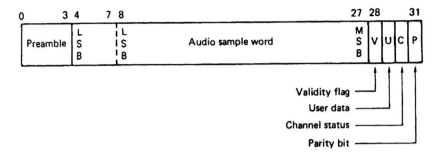

Figure 8.66 The basic subframe structure of the AES/EBU format. Sample can be 20 bits with four auxiliary bits, or 24 bits. LSB is transmitted first.

because the carries produced by the operation on a given bit can be delayed by one bit period and then included in the operation on the next higher order bit.

The format specifies that audio data must be in two's complement coding. Whilst pure binary could accept various alignments of different wordlengths with only a level change, this is not true of two's complement. If different wordlengths are used, the MSBs must always be in the same bit position otherwise the polarity will be misinterpreted. Thus the MSB has to be in bit 27 irrespective of wordlength. Shorter words are leading zero filled up to the 20-bit capacity. The channel status data included from AES-3-1992 signalling of the actual audio wordlength used so that receiving devices could adjust the digital dithering level needed to shorten a received word which is too long or pack samples onto a disk more efficiently.

Four status bits accompany each subframe. The validity flag will be reset if the associated sample is reliable. Whilst there have been many aspirations regarding what the V bit could be used for, in practice a single bit cannot specify much, and if combined with other V bits to make a word, the time resolution is lost. AES-3-1992 described the V bit as indicating that the information in the associated subframe is 'suitable for conversion to an analog signal'. Thus it might be reset if the interface was being used for non-audio data as is done, for example, in CD-I players.

The parity bit produces even parity over the subframe, such that the total number of ones in the subframe is even. This allows for simple detection of an odd number of bits in error, but its main purpose is that it makes successive sync patterns have the same polarity, which can be used to improve the probability of detection of sync. The user and channel-status bits are discussed later.

Two of the subframes described above make one frame, which repeats at the sampling rate in use. The first subframe will contain the sample from channel A, or from the left channel in stereo working. The second subframe will contain the sample from channel B, or the right channel in stereo. At 48 kHz, the bit rate will be 3.072 MHz, but as the sampling rate can vary, the clock rate will vary in proportion.

In order to separate the audio channels on receipt the synchronizing patterns for the two subframes are different as Figure 8.67 shows. These sync patterns begin with a run length of 1.5 bits which violates the FM channel coding rules and so cannot occur due to any data combination. The type of sync pattern is denoted by the position of the second transition which can be 0.5, 1.0 or

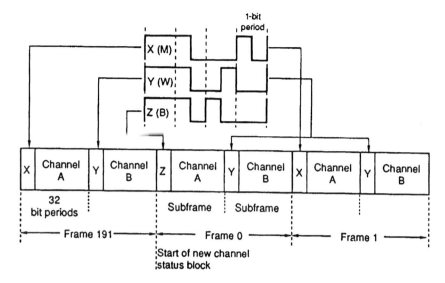

Figure 8.67 Three different preambles (X, Y and Z) are used to synchronize a receiver at the start of subframes.

1.5 bits away from the first. The third transition is designed to make the sync patterns DC free.

The channel-status and user bits in each subframe form serial data streams with one bit of each per audio channel per frame. The channel-status bits are given a block structure and synchronized every 192 frames, which at 48 kHz gives a block rate of 250 Hz, corresponding to a period of 4 ms. In order to synchronize the channel-status blocks, the channel A sync pattern is replaced for one frame only by a third sync pattern which is also shown in Figure 8.67. The AES standard refers to these as X, Y and Z whereas IEC 958 calls them M, W and B. As stated, there is a parity bit in each subframe, which means that the binary level at the end of a subframe will always be the same as at the beginning. Since the sync patterns have the same characteristic, the effect is that sync patterns always have the same polarity and the receiver can use that information to reject noise. The polarity of transmission is not specified, and indeed an accidental inversion in a twisted pair is of no consequence, since it is only the transition that is of importance, not the direction.

When 24-bit resolution is not required, which is most of the time, the four auxiliary bits can be used to provide talkback. This was proposed by broadcasters[38] to allow voice coordination between studios as well as program exchange on the same cables. Twelve-bit samples of the talkback signal are taken at one-third the main sampling rate. Each 12-bit sample is then split into three nibbles (half a byte, for gastronomers) which can be sent in the auxiliary data slot of three successive samples in the same audio channel. As there are 192 nibbles per channel-status block period, there will be exactly 64 talkback samples in that period. The reassembly of the nibbles can be synchronized by the channel-status sync pattern as shown in Figure 8.68. Channel-status byte 2 reflects the use of auxiliary data in this way.

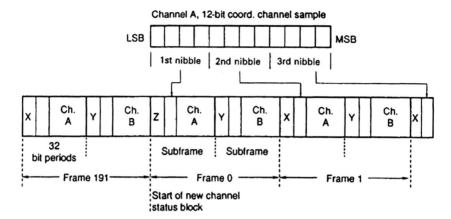

Figure 8.68 The coordination signal is of a lower bit rate to the main audio and thus may be inserted in the auxiliary nibble of the interface subframe, taking three subframes per coordination sample.

In the both the professional and consumer formats, the sequence of channel-status bits over 192 subframes builds up a 24-byte channel-status block. However, the contents of the channel-status data is completely different between the two applications. The professional channel-status structure is shown in Figure 8.69. Byte 0 determines the use of emphasis and the sampling rate, with details in Figure 8.70. Byte 1 determines the channel usage mode, i.e. whether the data transmitted are a stereo pair, two unrelated mono signals or a single mono signal, and details the user bit handling. Figure 8.71 gives details. Byte 2 determines wordlength as in Figure 8.72. This was made more comprehensive in AES-3-1992. Byte 3 is applicable only to multichannel applications. Byte 4 indicates the suitability of the signal as a sampling rate reference and will be discussed in more detail later in this chapter.

There are two slots of four bytes each which are used for alphanumeric source and destination codes. These can be used for routing. The bytes contain seven-bit ASCII characters (printable characters only) sent LSB first with the eighth bit set to zero according to AES-3-1992. The destination code can be used to operate an automatic router, and the source code will allow the origin of the audio and other remarks to be displayed at the destination.

Bytes 14–17 convey a 32-bit sample address which increments every channel-status frame. It effectively numbers the samples in a relative manner from an arbitrary starting point. Bytes 18–21 convey a similar number, but this is a time-of-day count, which starts from zero at midnight. As many digital audio devices do not have real-time clocks built in, this cannot be relied upon. AES-3-92 specified that the time-of-day bytes should convey the real time at which a recording was made, making it rather like timecode. There are enough combinations in 32 bits to allow a sample count over 24 hours at 48 kHz. The sample count has the advantage that it is universal and independent of local supply frequency. In theory if the sampling rate is known, conventional hours, minutes, seconds, frames timecode can be calculated from the sample count, but in practice it is a lengthy computation and users have proposed alternative formats in which the data from EBU or SMPTE timecode

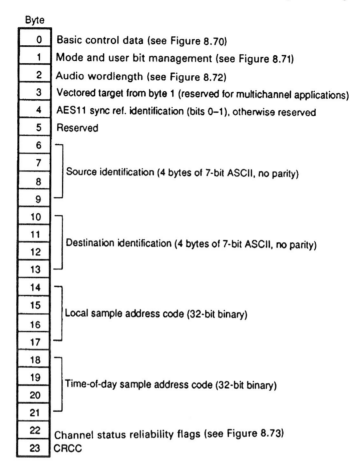

Figure 8.69 Overall format of the professional channel-status block.

is transmitted directly in these bytes. Some of these proposals are in service as *de facto* standards.

The penultimate byte contains four flags which indicate that certain sections of the channel-status information are unreliable (see Figure 8.73). This allows the transmission of an incomplete channel-status block where the entire structure is not needed or where the information is not available. For example, setting bit 5 to a logical one would mean that no origin or destination data would be interpreted by the receiver, and so it need not be sent.

The final byte in the message is a CRCC (cyclic redundancy check character) which converts the entire channel-status block into a codeword (see Chapter 10). The channel-status message takes 4 ms at 48 kHz and in this time a router could have switched to another signal source. This would damage the transmission, but will also result in a CRCC failure so the corrupt block is not used. Error correction is not necessary, as the channel-status data are either stationary, i.e. they stay the same, or change at a predictable rate, e.g. timecode. Stationary data will only change at the receiver if a good CRCC is obtained.

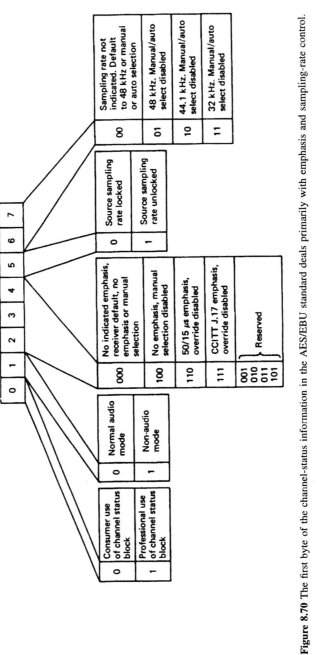

Figure 8.70 The first byte of the channel-status information in the AES/EBU standard deals primarily with emphasis and sampling-rate control.

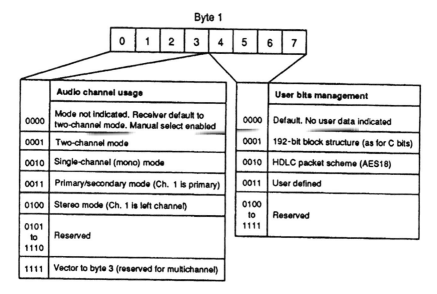

Figure 8.71 Format of byte 1 of professional channel status.

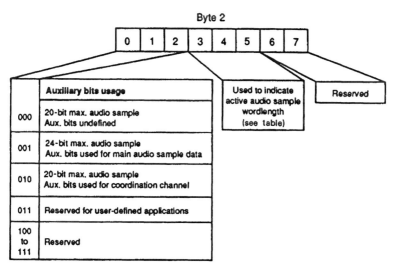

Bits states 3 4 5	Audio wordlength (24-bit mode)	Audio wordlength (20-bit mode)
0 0 0	Not indicated	Not indicated
0 0 1	23 bits	19 bits
0 1 0	22 bits	18 bits
0 1 1	21 bits	17 bits
1 0 0	20 bits	16 bits
1 0 1	24 bits	20 bits

Figure 8.72 Format of byte 2 of professional channel status.

Channel status byte 22

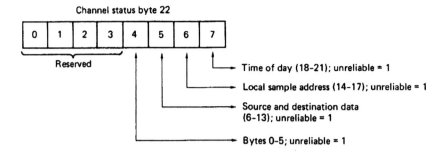

Reserved

Time of day (18-21); unreliable = 1

Local sample address (14-17); unreliable = 1

Source and destination data
(6-13); unreliable = 1

Bytes 0-5; unreliable = 1

Figure 8.73 Byte 22 of channel status indicates if some of the information in the block is unreliable.

For consumer use, a different version of the channel-status specification is used. As new products come along, the consumer subcode expands its scope. Figure 8.74 shows that the serial data bits are assembled into 12 words of 16 bits each. In the general format, the first six bits of the first word form a control code, and the next two bits permit a mode select for future expansion. At the moment only mode zero is standardized, and the three remaining codes are reserved.

Figure 8.75 shows the bit allocations for mode zero. In addition to the control bits, there are a category code, a simplified version of the AES/EBU source field, a field which specifies the audio channel number for multichannel working, a sampling-rate field, and a sampling-rate tolerance field.

The category code specifies the type of equipment which is transmitting, and its characteristics. In fact each category of device can output one of two

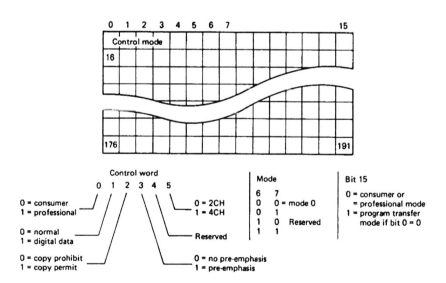

Figure 8.74 The general format of the consumer version of channel status. Bit 0 has the same meaning as in the professional format for compatibility. Bits 6−7 determine the consumer format mode, and presently only mode 0 is defined (see Figure 8.75).

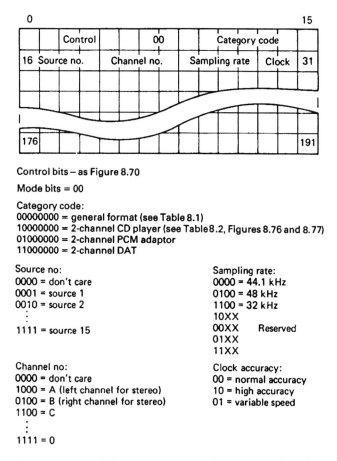

Control bits – as Figure 8.70

Mode bits = 00

Category code:
00000000 = general format (see Table 8.1)
10000000 = 2-channel CD player (see Table 8.2, Figures 8.76 and 8.77)
01000000 = 2-channel PCM adaptor
11000000 = 2-channel DAT

Source no:
0000 = don't care
0001 = source 1
0010 = source 2
⋮
1111 = source 15

Channel no:
0000 = don't care
1000 = A (left channel for stereo)
0100 = B (right channel for stereo)
1100 = C
⋮
1111 = O

Sampling rate:
0000 = 44.1 kHz
0100 = 48 kHz
1100 = 32 kHz
10XX
00XX   Reserved
01XX
11XX

Clock accuracy:
00 = normal accuracy
10 = high accuracy
01 = variable speed

Figure 8.75 In consumer mode 0, the significance of the first two 16-bit channel-status words is shown here. The category codes are expanded in Tables 8.1 and 8.2.

category codes, depending on whether bit 15 is or is not set. Bit 15 is the 'L-bit' and indicates whether the signal is from an original recording (0) or from a first generation copy (1) as part of the SCMS (serial copying management system) first implemented to resolve the stalemate over the sale of consumer DAT machines. In conjunction with the copyright flag, a receiving device can determine whether to allow or disallow recording. There were originally four categories; general purpose, two-channel CD player, two-channel PCM adapter and two-channel digital tape recorder (RDAT or SDAT), but the list has now extended as Figure 8.75 shows.

Table 8.1 illustrates the format of the subframes in the general purpose category. When used with CD players, Table 8.2 applies. In this application, the extensive subcode data of a CD recording (see Chapter 11) can be conveyed down the interface. In every CD sync block, there are 12 audio samples, and eight bits of subcode, P–W. The P flag is not transmitted, since it is solely positioning information for the player; thus only Q–W are sent. Since the interface can carry one user bit for every sample, there is surplus capacity in the user-bit channel for subcode. A CD subcode block is built up over 98 sync blocks, and has a repetition rate of 75 Hz. The start of the subcode data in the user bit stream will be seen in Figure 8.76 to be denoted by a minimum of 16 zeros, followed by a start bit which is always set to one. Immediately after the start bit, the receiver expects to see seven subcode bits, Q–W. Following these, another start bit and another seven bits may follow immediately, or a space of up to eight zeros may be left before the next start bit. This sequence repeats 98 times, when another sync pattern will be expected. The ability to leave zeros between the subcode symbols simplifies the handling of the disparity between user bit capacity and subcode bit rate. Figure 8.77 shows a representative example of a transmission from a CD player.

Table 8.1 The general category code causes the subframe structure of the transmission to be interpreted as below and the stated channel-status bits are valid.

Category code
00000000 = two-channel general format

Subframe structure

Two's complement, MSB in position 27, max 20 bits/sample
User-bit channel = not used
V bit optional
Channel status left = Channel status right, unless channel number (Figure 8.75) is non-zero

Control bits in channel status
Emphasis = bit 3
Copy permit = bit 2

Sampling-rate bits in channel status
Bits 4–27 = according to rate in use

Clock-accuracy bits in channel status
Bits 28–29 = according to source accuracy

Table 8.2 In the CD category, the meaning below is placed on the transmission. The main difference from the general category is the use of user bits for subcode as specified in Figure 8.76.

Category code
10000000 = two-channel CD player

Subframe structure

Two's complement MSB in position 27, 16 bits/sample
User-bit channel = CD subcode (see Figure 8.76)
V bit optional

Control bits in channel status
Derived from Q subcode control bits (see Chapter 12)

Sampling-rate bits in channel status
Bits 24–27 = 0000 = 44.1 kHz

Clock-accuracy bits in channel status
Bits 28–29 = according to source accuracy and use of variable speed

In a PCM adapter, there is no subcode, and the only ancillary information available from the recording consists of copy-protect and emphasis bits. In other respects, the format is the same as the general-purpose format.

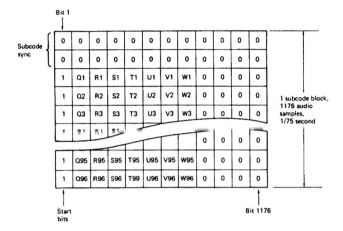

Figure 8.76 In CD, one subcode block is built up over 98 sync blocks. In this period there will be 1176 audio samples, and so there are 1176 user bits available to carry the subcode. There is insufficient subcode information to fill this capacity, and zero packing is used.

Subframe no.	Pre-amble SYNC	Aux	L S B	Audio samples			M S B	V	U	C	P
Channel status — 1	CS	0000	0000	XXXX	XXXX	XXXX	XXXX	0	0	C0L	P
block sync 2	B	0000	0000	XXXX	XXXX	XXXX	XXXX	0	0	C0R	P
3	A	0000	0000	XXXX	XXXX	XXXX	XXXX	0	0	C1L	P
4	B	0000	0000	XXXX	XXXX	XXXX	XXXX	0	0	C1R	P
A = left — 5	A	0000	0000	XXXX	XXXX	XXXX	XXXX	0	0	C2L	P
channel sample 6	B	0000	0000	XXXX	XXXX	XXXX	XXXX	0	0	C2R	P
7	A	0000	0000	XXXX	XXXX	XXXX	XXXX	0	0	C3L	P
8	B	0000	0000	XXXX	XXXX	XXXX	XXXX	0	0	C3R	P
9	A	0000	0000	XXXX	XXXX	XXXX	XXXX	0	0	C4L	P
B = right —10	B	0000	0000	XXXX	XXXX	XXXX	XXXX	0	0	C4R	P
channel sample 11	A	0000	0000	XXXX	XXXX	XXXX	XXXX	0	0	C5L	P
12	B	0000	0000	XXXX	XXXX	XXXX	XXXX	0	0	C5R	P
13	A	0000	0000	XXXX	XXXX	XXXX	XXXX	0	0	C6L	P
14	B	0000	0000	XXXX	XXXX	XXXX	XXXX	0	0	C6R	P
15	A	0000	0000	XXXX	XXXX	XXXX	XXXX	0	0	C7L	P
16	B	0000	0000	XXXX	XXXX	XXXX	XXXX	0	0	C7R	P
17	A	0000	0000	XXXX	XXXX	XXXX	XXXX	0	0	C8L	P
16 zeros 18	B	0000	0000	XXXX	XXXX	XXXX	XXXX	0	0	C8R	P
in user bits 19	A	0000	0000	XXXX	XXXX	XXXX	XXXX	0	0	C9L	P
= subcode 20	B	0000	0000	XXXX	XXXX	XXXX	XXXX	0	0	C9R	P
sync word 21	A	0000	0000	XXXX	XXXX	XXXX	XXXX	0	0	C10L	P
22	B	0000	0000	XXXX	XXXX	XXXX	XXXX	0	0	C10R	P
23	A	0000	0000	XXXX	XXXX	XXXX	XXXX	0	0	C11L	P
24	B	0000	0000	XXXX	XXXX	XXXX	XXXX	0	0	C11R	P
1 in user — 25	A	0000	0000	XXXX	XXXX	XXXX	XXXX	0	1	C12L	P
bits = subcode 26	B	0000	0000	XXXX	XXXX	XXXX	XXXX	0	Q1	C12R	P
start bit 27	B	0000	0000	XXXX	XXXX	XXXX	XXXX	0	R1	C13L	P
28	B	0000	0000	XXXX	XXXX	XXXX	XXXX	0	S1	C13R	P
U = Subcode 29	A	0000	0000	XXXX	XXXX	XXXX	XXXX	0	T1	C14L	P
30	B	0000	0000	XXXX	XXXX	XXXX	XXXX	0	U1	C14R	P
31	A	0000	0000	XXXX	XXXX	XXXX	XXXX	0	V1	C15L	P
32	B	0000	0000	XXXX	XXXX	XXXX	XXXX	0	W1	C15R	P
33	A	0000	0000	XXXX	XXXX	XXXX	XXXX	0	0	C16L	P
Subcode 34	B	0000	0000	XXXX	XXXX	XXXX	XXXX	0	0	C16R	P
space 35	A	0000	0000	XXXX	XXXX	XXXX	XXXX	0	0	C17L	P
36	B	0000	0000	XXXX	XXXX	XXXX	XXXX	0	0	C17R	P
Start bit 37	A	0000	0000	XXXX	XXXX	XXXX	XXXX	0	1	C18L	P
38	B	0000	0000	XXXX	XXXX	XXXX	XXXX	0	Q2	C18R	P
39	A	0000	0000	XXXX	XXXX	XXXX	XXXX	0	R2	C19L	P
40	B	0000	0000	XXXX	XXXX	XXXX	XXXX	0	S2	C19R	P
U = Subcode 41	A	0000	0000	XXXX	XXXX	XXXX	XXXX	0	T2	C20L	P
42	B	0000	0000	XXXX	XXXX	XXXX	XXXX	0	U2	C20R	P
43	A	0000	0000	XXXX	XXXX	XXXX	XXXX	0	V2	C21L	P
44	B	0000	0000	XXXX	XXXX	XXXX	XXXX	0	W2	C21R	P
45	A	0000	0000	XXXX	XXXX	XXXX	XXXX	0	0	C22L	P
46	B	0000	0000	XXXX	XXXX	XXXX	XXXX	0	0	C22R	P
47	A	0000	0000	XXXX	XXXX	XXXX	XXXX	0	0	C23L	P
48	B	0000	0000	XXXX	XXXX	XXXX	XXXX	0	0	C23R	P

Figure 8.77 Compact Disc subcode transmitted in user bits of serial interface.

When a DAT player is used with the interface, the user bits carry several items of information.[39] Once per drum revolution, the user bit in one subframe is raised when that subframe contains the first sample of the interleave block (see Chapter 10). This can be used to synchronize several DAT machines together for editing purposes. Immediately following the sync bit, start ID will be transmitted when the player has found the code on the tape. This must be asserted for 300 ±30 drum revolutions, or about 10 seconds. In the third bit position the skip ID is transmitted when the player detects a skip command on the tape. This indicates that the player will go into fast forward until it detects the next start ID. The skip ID must be transmitted for 33 ±3 drum rotations. Finally DAT supports an end-of-skip command which terminates a skip when it is detected. This allows jump editing to omit short sections of the recording. DAT can also transmit the track number (TNO) of the track being played down the user bit stream.

The user channel consists of one bit per audio channel per sample period. Unlike channel status, which only has a 192-bit frame structure, the user channel can have a flexible frame length. Figure 8.71 showed how byte 1 of the channel-status frame describes the state of the user channel. Many professional devices do not use the user channel at all and would set the all zeros code. If the user-channel frame has the same length as the channel-status frame then code 0001 can be set. One user-channel format which is standardized is the data-packet scheme of AES18-1992.[40,41] This was developed from proposals to employ the user channel for labelling in an asynchronous format.[42] A computer industry standard protocol known as HDLC (high-level data link control)[43] is employed in order to take advantage of readily available integrated circuits.

The frame length of the user channel can be conveniently made equal to the frame period of an associated device. For example, it may be locked to Film, TV or DAT frames. The frame length may vary in NTSC as there are not an integer number of samples in a frame.

References

1. Moore, B.C.J., *An Introduction to the Psychology of Hearing*, London: Academic Press (1989)
2. Muraoka, T., Iwahara, M. and Yamada, Y., Examination of audio bandwidth requirements for optimum sound signal transmission. *J. Audio Eng. Soc.*, **29**, 2–9 (1982)
3. Muraoka, T., Yamada, Y. and Yamazaki, M., Sampling frequency considerations in digital audio. *J. Audio Eng. Soc.*, **26**, 252–256 (1978)
4. Widrow, B., Statistical analysis of amplitude quantized sampled-data systems. *Trans. AIEE*, Part II, **79**, 555–568 (1961)
5. Lipshitz, S.P., Wannamaker, R.A. and Vanderkooy, J., Quantization and dither: a theoretical survey. *J. Audio Eng. Soc.*, **40**, 355–375 (1992)
6. Maher, R.C., On the nature of granulation noise in uniform quantization systems. *J. Audio Eng. Soc.*, **40**, 12–20 (1992)
7. Roberts, L.G., Picture coding using pseudo-random noise. *IRE Trans. Inform. Theory*, **IT-8**, 145–154 (1962)
8. Schuchman, L., Dither signals and their effect on quantizing noise. *Trans. Commun. Technol.*, **COM-12**, 162–165 (1964)
9. Sherwood, D.T., Some theorems on quantization and an example using dither. In *Conf. Rec., 19th Asilomar Conf. on circuits, systems and computers*, Pacific Grove, CA (1985)
10. Gerzon, M. and Craven, P.G., Optimal noise shaping and dither of digital signals. Presented at 87th Audio Eng. Soc. Conv. (New York, 1989), Preprint No. 2822 (J-1)

11. Vanderkooy, J. and Lipshitz, S.P., Resolution below the least significant bit in digital systems with dither. *J. Audio Eng. Soc.*, **32**, 106–113 (1984)
12. Blesser, B., Advanced A-D conversion and filtering: data conversion. In *Digital Audio*, edited by B.A. Blesser, B. Locanthi, and T.G. Stockham Jr, 37–53. New York: Audio Engineering Society (1983)
13. Vanderkooy, J. and Lipshitz, S.P., Digital dither. Presented at 81st Audio Eng. Soc. Conv. (Los Angeles, 1986), preprint 2412 (C-8)
14. Vanderkooy, J. and Lipshitz, S.P., Digital dither. In *Audio in Digital Times*, New York: Audio Engineering Society (1989)
15. v.d. Plassche, R.J., Dynamic element matching puts trimless converters on chip. *Electronics* (16 June 1983)
16. v.d. Plassche, R.J. and Goedhart, D., A monolithic 14-bit D/A converter. *IEEE J. Solid-State Circuits*, **SC-14**, 552–556 (1979)
17. Adams, R.W., Companded predictive delta modulation: a low-cost technique for digital recording. *J. Audio Eng. Soc.*, **32**, 659–672 (1984)
18. Hauser, M.W., Principles of oversampling A/D conversion. *J. Audio Eng. Soc.*, **39**, 3–26 (1991)
19. Cutler, C.C., Transmission systems employing quantization. US Pat. No. 2.927,962 (1960)
20. v.d. Plassche, R.J. and Dijkmans, E.C., A monolithic 16-bit D/A conversion system for digital audio. In *Digital Audio*, B.A. Blesser, B. Locanthi and T.G. Stockham Jr (eds), 54–60, New York: Audio Engineering Society (1983)
21. Fielder, L.D., Human auditory capabilities and their consequences in digital audio converter design. In *Audio in Digital Times*, New York: Audio Engineering Society (1989)
22. Wannamaker, R.A., Psychoacoustically optimal noise shaping. *J. Audio Eng. Soc.*, **40**, 611–620 (1992)
23. Lipshitz, S.P., Wannamaker, R.A. and Vanderkooy, J., Minimally audible noise shaping. *J. Audio Eng. Soc.*, **39**, 836–852 (1991)
24. Adams, R.W., Design and implementation of an audio 18-bit A/D converter using oversampling techniques. Presented at the 77th Audio Engineering Society Convention, Hamburg, preprint 2182
25. Adams, R.W., An IC chip set for 20-bit A/D conversion. In *Audio in Digital Times*, New York: Audio Engineering Society (1989)
26. Richards, M., Improvements in oversampling analogue to digital converters. Presented at 84th Audio Eng. Soc. Conv. (Paris, 1988), preprint 2588 (D-8)
27. Inose, H. and Yasuda, Y., A unity bit coding method by negative feedback. *Proc. IEEE*, **51**, 1524–1535 (1963)
28. Naus, P.J. et al., Low signal level distortion in sigma-delta modulators. Presented at 84th Audio Eng. Soc. Conv. (Paris 1988), preprint 2584
29. Stikvoort, E., High order one bit coder for audio applications. Presented at 84th Audio Eng. Soc. Conv. (Paris, 1988), preprint 2583(D-3)
30. Moller, L., Signal levels across the EBU/AES digital audio interface. In *Proc. 1st NAB Radio Montreux Symp.* (Montreux, 1992), 16–28
31. Audio Engineering Society, AES recommended practice for digital audio engineering–serial transmission format for linearly represented digital audio data. *J. Audio Eng. Soc.* **33**, 975–984 (1985)
32. EIAJ CP-340 *A Digital Audio Interface*, EIAJ Tokyo (1987)
33. EIAJ CP-1201 *A Digital Audio Interface* (revised), EIAJ Tokyo (1992)
34. EIA RS-422A. Electronic Industries Association, 2001 Eye St N.W., Washington, DC 20006, USA
35. Smart, D.L., Transmission performance of digital audio serial interface on audio tie lines. *BBC Designs Dept Technical Memorandum*, 3.296/84
36. European Broadcasting Union, Specification of the digital audio interface. *EBU Doc. Tech.*, 3250
37. Rorden, B. and Graham M., A proposal for integrating digital audio distribution into TV production. *J. SMPTE*, 606–608 (Sept 1992)
38. Gilchrist, N., Co-ordination signals in the professional digital audio interface. In *Proc. AES/EBU Interface Conf.* 13–15. Burnham: Audio Eng. Soc. (1989)
39. Digital audio tape recorder system (RDAT). Recommended design standard. DAT Conference, Part V (1986)

40. AES18-1992 Format for the user data channel of the AES digital audio interface. *J. Audio Eng. Soc.*, **40**, 167–183 (1992)
41. Nunn, J.P., Ancillary data in the AES/EBU digital audio interface. In *Proc. 1st NAB Radio Montreux Symp.*, 29–41 (1992)
42. Komly, A. and Viallevieille, A., Programme labelling in the user channel. In *Proc. AES/EBU Interface Conf.*, 28–51. Burnham: Audio Eng. Soc. (1989)
43. ISO 3309 *Information processing systems–data communications–high-level data link frame structure* (1984)

Chapter 9

Analog audio recording

Analog audio recording has a long history which was outlined in Chapter 1. Today analog recording is somewhat in retreat owing to the progress made in the digital domain. Nevertheless a significant amount of analog recording equipment will remain in service and a proper treatment is justified here. The audio recording systems considered here are magnetic tape and vinyl disk.

9.1 Introduction to analog recording

As transducers and media cannot know the meaning of the waveforms which they handle, they cannot be analog or digital; instead the term describes the way in which the replayed signals are *interpreted*. When discrete decisions are made on the replay signal, the machine is digital. It is a fundamental of analog recording that the medium carries a direct *analog* of the audio waveform and the accuracy to which this is retrieved determines the sound quality.

In all analog media the analog of time is the distance along the track on the medium. Consequently the average and instantaneous speed of the medium on reproduction must be exactly what it was during recording otherwise the analog of time will be incorrect. In practice recorders attempt to drive the medium at a constant speed in both record and reproduce. If the reproduce speed is constant but incorrect, the result will be a pitch change. Speed instabilities are divided into two categories, *wow* and *flutter*. These names are well chosen because they are almost self-explanatory. Wow is generally the result of eccentricity in some rotating component causing slow cyclic speed variations, whereas flutter is a faster speed variation which may be due to vibrations in the mechanism.

The analog of the audio waveform voltage may be the strength of a magnetic field, the transmissivity of light or the transverse velocity of a mechanical groove. The voltage analog must be precisely linear as shown in Section 2.2 otherwise the audio waveform will be distorted and harmonics will be created. In practical analog recorders this aspect of performance is generally not as good as is necessary with distortion figures often being one or two orders of magnitude greater than the distortion performance of good electronic circuitry.

With a linear transfer function, the waveform can still be distorted if different frequencies travel through the recorder at different speeds. This causes lack of phase linearity which can easily be revealed with a square-wave test.

In practice the reproduced waveform will have *noise* superimposed on it. Noise can be random due to thermal or statistical effects, electromagnetic interference, or a function of the recorded signal.

The development of successful high-quality analog audio recorders has basically been a constant battle against the above-mentioned problems. Unfortunately the battle will never be won because the enemy is logarithmic. Steady progress results in improvement increments which become smaller until diminishing returns call a halt.

As analog recording is imperfect, copying from one tape to another introduces *generation loss*. Bandwidth will be reduced, distortion and noise will rise. Professional recorders must have the highest possible first-generation quality so that copying in the production process does not cause excessive loss. Consumer recorders are generally only used to make first-generation recordings and tape consumption is more of an issue.

9.2 Tape recording

The basic principle of the tape recorder is very simple. Figure 9.1(a) shows that the tape medium is transported at a steady speed by pressing it against a rotating *capstan* with a compliant *pressure roller* also called a pinch wheel or puck. The tape comes from a *supply reel* and after passing through the machine is wound onto the *take-up reel*. Clearly the speed of the reels will change as the diameter of the tape pack changes and the reels must be driven in a way which keeps the tape at a reasonably constant tension despite that.

The tape carries a magnetic coating which stores the audio signal as variations in magnetic flux strength and direction. The tape from the supply reel first meets an *erase head* which removes any previous recordings. The input audio signal is impressed upon the tape by the *record head*. High-grade machines have a separate *playback head* so that the recording can be monitored as it is being made, giving the earliest warning of a problem. The record and playback heads are transducers converting audio signals between electrical and magnetic forms.

After recording, the tape can be rewound by retracting the pressure roller and applying power to the supply reel. The tape can also be driven forward at speed in a similar way to locate a desired recording. Some form of *counter* will be provided to indicate how far the tape has been wound. The recording can be played by driving the tape through the machine again but with the erase and record heads disabled. Clearly if the machine is accidentally set into record a valuable recording may be destroyed. Safety switches and interlocks are provided so that a deliberate act is required to make the machine record.

Tape may be used on open reels, or in cassettes or cartridges. Open-reel recorders intended for consumer use have hubs which accept ciné spools. These were originally intended for 8 mm film, but are ideal for $\frac{1}{4}$-inch tape. Professional recorders usually use NAB spools which have a three-inch central hole. Adapters are available to convert ciné hubs to mount NAB spools. Sometimes the tape is carried on spools called pancakes which only have a lower flange. In all cases the central holes have slots which engage with matching dogs on the hub to give positive torque transmission.

Figure 9.2 shows how tape should be wound on a spool. The tape is supported by the hub only and does not touch the flanges, avoiding edge

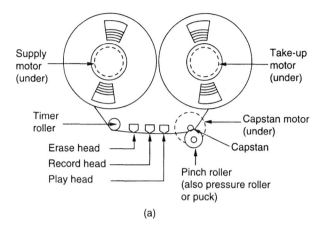

(a)

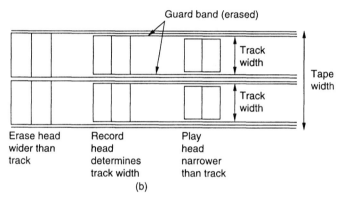

(b)

Figure 9.1 (a) An analog tape deck. The capstan and pressure roller provide constant tape speed and the supply and take-up reels apply tape tension. The supply reel back tension is obtained by driving the motor lightly against its rotation which causes the tape to be held against the heads. (b) The heads have different track widths to ensure that slight vertical misalignment does not affect recording or replay.

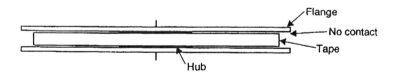

Figure 9.2 The flanges of a spool are for protection, not support, and the tape should not touch them.

damage. The flanges are to protect the tape, not to support it. Tape reels should be stored vertically to avoid sagging. When not on a machine, a tape reel should always be in a sealed box to prevent evaporation of plasticizers and lubricants. High-quality tape containers are of rigid plastics material with a sealing gasket, but a tape in a polyethylene bag inside a cardboard box is almost as well protected.

In cassettes, the tape runs between two spools in a container which protects against contamination and makes threading easier. The ends of the tape are firmly attached to prevent run-off. In the Compact Cassette the spools are flangeless hubs and the edge of the tape is guided by lubricated liner sheets. The advantage of this approach is that the cassette can be made smaller because the flangeless hubs can be put closer together. All other aspects of this construction are disadvantages.

In cartridges the tape is in the form of an endless loop which is wound in a single pack. A capstan and pressure-roller system pull tape from the centre of the pack which rotates to take it up on the outside. The tape is made with a special lubricant to allow this to happen. The tape can only be transported forwards.

One of the advantages of tape over other analog media is that it is easy to divide the width of the tape into separate *tracks* with *guard bands* between them to prevent crosstalk. Any suitable compromise can be made between a machine which can record several audio signals simultaneously or one which can record for a longer time.

Figure 9.3 shows how tape tracks are organized. In monophonic *half-track* consumer recorders and monophonic Compact Cassettes, the recorded track uses a little less than one-half of the tape width (a). When the end of the tape is reached the tape is physically turned over so that the other part of the tape surface can be used travelling in the opposite direction. Confusingly, the two directions are referred to as *sides* which have nothing to do with the physical sides of the tape.

In the *quarter-track* monophonic recorder the playing time was doubled by using narrower tracks (b). The heads had two independent magnetic circuits and the record and playback circuitry could be connected to one or the other. The design of the heads was eased by putting the two magnetic circuits

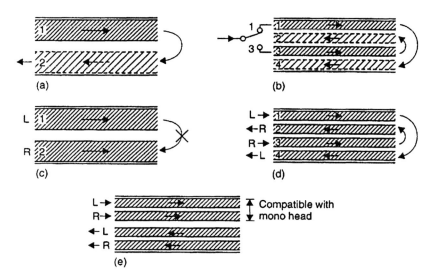

Figure 9.3 Track organization on tape. (a) Half-track. (b) Quarter-track. (c) Half-track stereo. (d) Quarter-track stereo. (e) Stereo Compact Cassette.

further apart. As a result the two tracks which can be recorded in a forward direction are not adjacent. As before the tape has to be turned over to use the other two tracks.

In a half-track stereo recorder, the same track layout as mono is used, but the machine has duplicated circuitry and double heads which can record both tracks simultaneously (c). Clearly if a half-track stereo tape is inadvertently turned over it will play backwards with the channels reversed.

In the quarter-track stereo recorder the same principle is used. In this case the tape can be turned over (d) to record the second stereo pair in the reverse direction.

In the stereo Compact Cassette there is a requirement that stereo tapes can be reproduced in mono players. The arrangement of Figure 9.3(d) cannot be used. Instead the stereo track pair are designed to occupy the same space as the mono track as in (e). When played on a mono machine the two tracks are added magnetically in the head.

Professional machines do not use formats which require the tape to be turned over as splice editing will destroy the recording on the other 'side'. In consumer products the need to turn over the tape can be eliminated by using a tape transport which can transport the tape in either direction, generally achieved by using an additional capstan near the supply reel. Further magnetic circuits are required in the replay heads to reproduce the tracks appropriate to the direction. Auto-reverse machines change direction and select the alternative heads automatically at the end of the tape. In some machines the heads would be turned over mechanically so that additional magnetic circuits were not required. Some quality consumer machines have additional heads so that both quarter- and half-track tapes could be played.

In the most common consumer cartridge format there are eight tape tracks arranged as four stereo pairs. The track pair is selected by mechanically moving the head across the tape. The head could be motorized or solenoid indexed so that it would move to a new position every time the join in the tape loop passed.

Multitrack recorders evolved for popular record production using tape up to two inches wide having 16 or 24 tracks.

9.3 Magnetic tape

Magnetic tape[1] consists of a substrate, nowadays almost universally polyethylene terephthalate film, known as Mylar (DuPont) or Melinex (ICI), which carries the magnetic layer. The film will be stretched during manufacture to obtain the smoothest possible finish. The surface may be passed through a high-voltage corona to aid the adhesion of the active layers. The tape will be produced in widths of a metre or more and slit into required widths quite late in the process. The quality of the slitting process has a bearing on the sound quality as will be discussed. Figure 9.4 shows common tape widths and their applications.

The substrate comprises about two-thirds of the thickness of the tape. Overall thickness is a compromise between long playing time, which requires a thin tape, and robustness and low print-through, which require a thick tape. Figure 9.5 shows the various thicknesses available in open-reel and cassette tapes. Standard play tape is in almost universal use for professional

Width		Application	No. of tracks
0.15 in	3.81 mm	Compact Cassette	2 (mono) 4 (stereo)
0.25 in	6.3 mm	Most open reel	1–4
0.5 in	12.7 mm	Pro open reel	1–8
1.0 in	25.4 mm	Multitrack	8–16
2.0 in	50.8 mm	Multitrack	16–24

Figure 9.4 Tape widths in common use and their applications.

Standard play	52 μm	Professional
Long play	35 μm	Consumer. Thinner grades liable to break and give poor print-through
Double play	26 μm	
Triple play	18 μm	
Quad play	12 μm	
Sextuple play	9 μm	

Figure 9.5 Tape thicknesses used for various purposes.

applications because the extra cost is a small part of the overall process and the best quality is essential. In the Compact Cassette the tape is somewhat thinner as the quality expectations are not so high.

The magnetic layer consists of a polymeric binder material into which is mixed the magnetic particles and various lubricants and mild abrasives which are designed to aid the smooth passage of the tape across the heads and to keep them clean. An introduction to magnetic principles may be found in Chapter 2. The magnetic particles may be gamma ferric oxide or chromium dioxide. The coating materials are suspended in a solvent which evaporates after application to the substrate. The back of the tape may also have a matt coating which helps to stabilize the packing of tape during winding. This back coating may also be weakly electrically conductive in order to prevent the build-up of static electricity.

The magnetic particles are *acicular* or needle-like and when initially applied to the substrate their orientation is random. Before the coating has dried the acicular particles are *oriented* by passing the tape over a powerful magnet. This improves the signal-to-noise ratio. In audio tapes the orientation is as parallel to the tape motion as possible. In practice the particles are left at a slight angle to the tape motion resulting in a slight change of equalization depending on the direction in which the tape is run. This is known as the velour effect. The tape will be erased before shipping to remove the magnetization due to the orientation process.

Only about 70 % of the volume of the tape coating is magnetic because of the need for other compounds and the binder to hold the whole together. The coating thickness and density of magnetic particles must be precisely uniform otherwise the gain of the tape fluctuates dynamically. Variations in particle

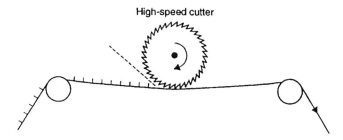

Figure 9.6 A tape-linishing machine uses a cutter to remove asperities.

density cause the reproduced signal to be amplitude modulated resulting in modulation sidebands called *modulation noise*. A local deficiency in particle density may cause brief signal loss known as *dropout*.

Dropout may also be due to the tape being lifted out of close head contact by tape *asperities* which are bristle-like contamination protruding from the binder. Asperities are removed by rollers or by *linishing* the tape surface on a rotary cutter as shown in Figure 9.6.

The choice of magnetic material is a compromise between performance and cost. Gamma ferric oxide is brown in appearance and is extremely cheap, but has poor short wavelength performance and a high noise level. Chromium dioxide[2] is almost black and has higher *coercivity*, i.e. it is capable of storing more magnetic energy. This produces a larger playback signal if a machine having more powerful record and bias drive is used. This improves the signal-to-noise ratio. Chromium dioxide tape also has smaller particles improving the short wavelength response.

In professional machines higher speeds are used so that wavelengths are longer. The quantity of tape used is much higher, and the cost of chromium tape is prohibitive. In Compact Cassette, the speed is very low and wavelengths are very short. The quantity of tape is quite small and additional cost of chromium tape is not so serious especially as it gives a tangible improvement to the marginal quality.

Ferric tapes have been improved by various techniques. These include doping with cobalt ions, or placing a thin layer of chromium dioxide on the surface of a ferric tape to produce what is known as ferrichrome tape. Both of these techniques improve short wavelength performance and noise.

9.4 Heads

Figure 9.7 shows the construction of a magnetic tape record head intended for analog use. Heads designed for use with tape work in actual contact with the magnetic coating. The tape is tensioned to pull it against the head. There will be a wear mechanism and need for periodic cleaning. A magnetic circuit carries a coil through which the record current passes and generates flux. The magnetic circuit may be made from permalloy, which is electrically conductive. It will need to be laminated to reduce eddy currents which cause losses at high frequencies. An alternative is ferrite which is an insulator or sendust which is a material made by hot-pressing powder.

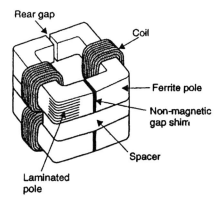

Figure 9.7 An analog magnetic head showing major components.

A non-magnetic gap at the front of the head forces the flux to leave the magnetic circuit of the head and penetrate the medium. In analog record heads a second gap may be used at the rear of the head. The reluctance of the magnetic circuit is dominated by that of the rear gap so that variations in tape contact do not modulate the recorded level so much.

Figure 9.8 shows that the recording is actually made just after the trailing pole of the record head where the flux strength from the gap is falling. The width of the gap is generally made quite large to ensure that the full thickness of the magnetic coating is recorded. A gap width similar to the coating thickness is typical. The recording process does not have a flat frequency response and a recording with constant head drive will result in a tape magnetization shown in Figure 9.9 in which a 6 dB/octave roll-off will occur after the turnover frequency. As the loss is wavelength related, the turnover frequency is a function of the tape speed used. Unfortunately this loss cannot be corrected by equalization of the record amplifier because the result would be distortion.

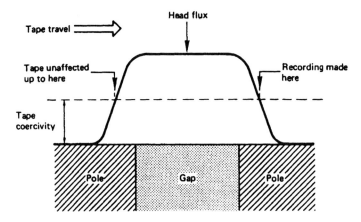

Figure 9.8 The recording is actually made near the trailing pole of the head where the head flux falls below the coercivity of the tape.

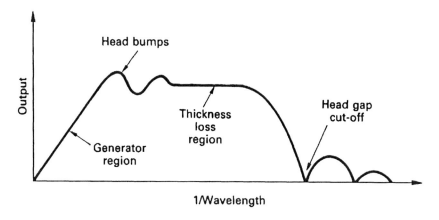

Figure 9.9 The frequency response of a conventional inductive head. See text for details.

Instead the correction for this loss has to be done on replay. In order to permit interchange, standards have been set which define what the tape magnetization with respect to frequency shall be. These are close to what naturally occurs, but any variations due to specific record head designs will have to be eliminated by the use of equalization in the record amplifier. In this way, irrespective of the actual record head performance, the tape has the standard characteristic so that a standard equalization curve can be applied on replay.

A conventional inductive replay head has a frequency response shown in Figure 9.9. At zero Hertz there is no change of flux and no output. As a result inductive heads are at a disadvantage at very low speeds. The output rises with frequency until the rise is halted by the onset of thickness loss which is actually a form of spacing loss. As the recorded wavelength falls, flux from the shorter magnetic patterns cannot be picked up so far away. At some point, the wavelength becomes so short that flux from the back of the tape coating cannot reach the head and a decreasing thickness of tape contributes to the replay signal.[3] Initially thickness loss causes a 6 dB/octave loss which levels the rising inductive response. However, at shorter wavelengths thickness loss increases to 12 dB/octave because the recording is not made uniformly through the tape thickness.

As wavelength reduces, gap loss also occurs, where the head gap is too big to resolve detail on the track. The result is an aperture effect (see Section 8.7) where the response has nulls where flux cancellation takes place across the gap. Clearly the smaller the gap the shorter the wavelength of the first null. This contradicts the requirement of the record head to have a large gap. In quality analog recorders, it is the norm to have different record and replay heads for this reason. Replay heads have no need for a rear gap as this reduces their sensitivity. Clearly where the same head is used for record and play, as in many Compact Cassette recorders, the head gap size will be determined by the playback requirement.

Whereas head losses are a function of frequency, thickness loss and gap effect are wavelength related and their severity is a function of tape speed. Figure 9.10 shows the effect of increasing the tape speed. The onset frequency

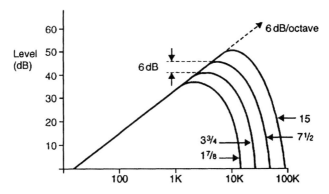

Figure 9.10 The effect of raising tape speed on the replay frequency response.

of thickness loss rises allowing the signal amplitude to increase, improving the noise performance. The cut-off frequency rises, improving the frequency response. Naturally raising the tape speed raises the cost of the tape consumed. It is obvious that professional recorders will use high tape speeds for the best quality. Figure 9.10 shows some common speeds. Note the very low speed of the Compact Cassette which fundamentally restricts high frequency and noise performance.

It will also be seen from Figure 9.9 that there are some irregularities in the low-frequency response region which are known as head bumps. These irregularities come about because of interactions between longer wavelengths on the medium and the width of the head poles. Analog audio recorders must use physically wide heads to drive the head bumps down to low frequencies, although at high tape speeds, such as 30 inches per second, this is impossible and the head bumps become audible. Compact Cassette heads must also be small as otherwise they would not fit the cassette apertures and again head bumps are a problem. Large head poles do, however, result in low contact pressure which makes the reproduction of high frequencies difficult.

As can be seen, the frequency response of a magnetic recorder is far from ideal, and equalization will be needed both on record and replay to compensate for the sum of these effects to give an overall flat response.

9.5 Biased recording

Magnetic recording relies on the hysteresis of magnetically hard media as described in Section 2.20. By definition the transfer function is non-linear, and so the recorded signal will be a distorted version of the input. Analog magnetic recorders have to use bias[4] to linearize the process whereas digital recorders are not concerned with the non-linearity, and bias is unnecessary.

Bias consists of a sine-wave signal well above the audible frequency range. Figure 9.11 shows that when applied to a tape head such a signal produces a magnetic field strength which diminishes with distance away from the head gap. A particle of tape coating experiences a bias frequency waveform decaying with respect to time as it recedes from the trailing pole of the head. In the case of an erase head, only the high-frequency oscillation is supplied.

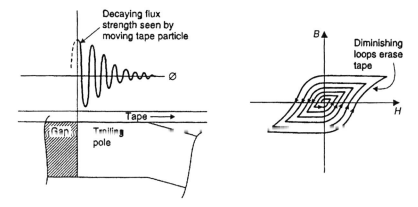

Figure 9.11 A high-frequency signal produces a field which appears to decay with time to a tape receding from the gap. This sweeps the tape around a set of diminishing loops until it is erased.

As the tape recedes from the head it will be swept around a set of diminishing hysteresis loops until it is demagnetized. The same technique is used when demagnetizing heads. A degausser is a solenoid connected to the AC supply via a switch. An alternating field from a degausser is applied, and the degausser is physically removed to a distance before being switched off, resulting in exactly the same decaying waveform.

In the case of the record head, the bias is linearly added to the audio waveform to be recorded. Now as the tape recedes from the trailing pole it will experience a decaying oscillation superimposed upon the audio signal. Figure 9.12 shows that the result is that the tape is swept through a series of

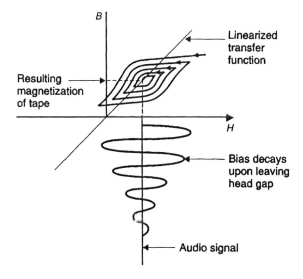

Figure 9.12 When bias is added to the audio signal the tape is swept through a series of loops which converge upon a straight transfer function.

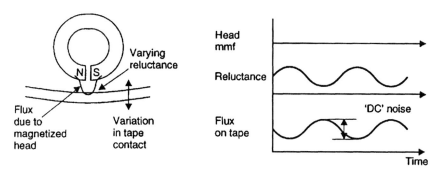

Figure 9.13 Origin of DC noise is the variation in reluctance due to unstable head/tape contact. Any head magnetization or DC component of the record signal will cause modulation of the tape.

minor loops which converge upon the magnetization determined by the audio waveform, linearizing the transfer function. The purity of the bias signal is extremely important. As the bias amplitude is considerably larger than that of the signal, noise in the bias signal should be extremely low.

The bias signal must also have perfect symmetry, i.e. it must be DC free. If this is not the case a noise source exists. Figure 9.13 shows that the record head and the tape surface form a magnetic circuit. The flux in this circuit for a constant current is a function of the total reluctance. Practical tape coatings are not homogeneous and the contact between the head and the tape is variable. Both of these factors cause the reluctance to vary. In the presence of a steady magnetomotive force due to a DC component of the bias, the resultant flux will be modulated by the reluctance variations and the bias field will cause the flux variations to be recorded on the tape. The same effect will occur if the head is magnetized, hence the requirement for regular head degaussing.

In the presence of bias the normal B-H loop is not exhibited and instead the tape exhibits what is known as an *anhysteresis curve* shown in Figure 9.14 for low frequencies.[5] Recording is carried out in the linear portion of the correctly biased curve. Note that underbiasing fails to remove the non-linearity of hysteresis whereas overbiasing reduces the sensitivity of the tape, raising the noise level.

The correctly biased anhysteresis curve flattens at high levels.[6] Distortion begins when the level of signal plus bias reaches tape saturation. Consequently analog tape has a maximum operating level (MOL) often defined as the level above which distortion exceeds, for example, 3%. Gross distortion occurs at saturation and between MOL and saturation is a region known as headroom in which transients can be recorded because the hearing mechanism cannot register the distortion unless it is sustained for more than about a millisecond.

In very short wavelength recording, such as in Compact Cassettes, the anhysteresis curve of Figure 9.14 is not valid. High audio frequencies can add to the bias and reduce the signal level by effectively overbiasing. This results in dynamic range compression which is colloquially known as high-frequency squashing. In the Dolby HX (headroom extension) system, the high-frequency content of the input audio is monitored and used to modulate the bias level.

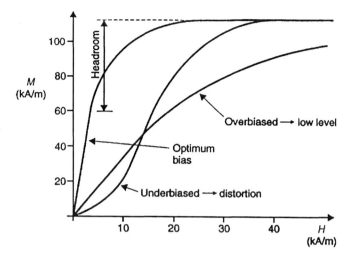

Figure 9.14 Anhysteresis curve shows response of magnetic tape in the presence of bias.

Unfortunately the intensity of the bias field falls with increasing distance from the head. This means that only one layer of the tape coating will be correctly biased. Layers closer to the head will be overbiased and layers deeper in the coating will be underbiased. High frequencies are reproduced by the surface layer of the tape because deeper layers suffer spacing loss. Consequently biasing for the best high-frequency output results in distortion at low frequencies.

As there is no optimum amplitude of the bias signal a practical compromise must be made between distortion, replay level and frequency response which will need to be individually assessed for each type of tape used. As overbiasing is less damaging than underbiasing, often the bias level is increased until the replay output has reduced by 2–3 dB.

9.6 Cross-field bias

The system of adding bias to the record head signal is non-optimal because the bias field decays with distance from the head whereas the opposite is required. If the bias is supplied from an additional head on the opposite side of the tape using an arrangement shown in Figure 9.15 this problem can be overcome. In practice the cross-field head is mounted on an arm so that it can be retracted for easy tape threading. The head can be engaged by the same mechanism which operates the pressure roller.

In cross-field biasing, pioneered by the Tandberg company, the intensity of the bias field is greatest deep in the tape coating where low frequencies are recorded, but reduced at the tape surface where the high frequencies are recorded. Placing the cross-field head slightly ahead of the record head causes the bias field to decay at the point where the record head creates the strongest field so that the recording is made at that point.

As the bias level is optimized over a wider range of frequencies, the result is that the tape has a higher sensitivity at high frequency before low-frequency

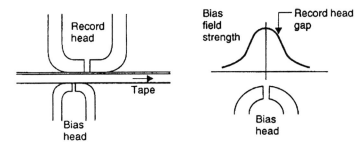

Figure 9.15 Cross-field bias uses a bias head on the opposite side of the tape.

distortion occurs. This results in a larger high-frequency replay signal needing less equalization and consequently having lower noise.

The cross-field mechanism would work well at the low speeds used in Compact Cassette but unfortunately the design of the cassette precludes the use of head on the back of the tape.

9.7 Pre-emphasis and equalization

Both of these are processes in which frequency-dependent filtering is applied. They differ in principle and achieve different purposes. Pre-emphasis is part of the interchange standard for the tape format and is designed to reduce hum by boosting low frequencies in the record process and applying an equal and opposite attenuation on playback. Equalization is necessary to compensate for losses and deficiencies in the record and playback processes whilst maintaining interchange. Thus equalization is necessarily adjustable whereas pre-emphasis is fixed. Different tape formats and speeds use different pre-emphasis and equalization, and a switch may be provided to select the correct response so that one machine may be used with several formats.

Figure 9.16 shows the various frequency-dependent processes in an analog tape recorder. The trick is to get everything to cancel out so that the overall frequency response is flat and the response on the tape adheres to the standard. In the absence of low-frequency pre-emphasis the actual magnetization of the tape should be flat up to the turnover frequency and then roll off as defined by the high-frequency time constant of the standard in use, e.g. 70 μs. Any departure from the standard response on tape in the record process is compensated by record equalization. These compensations are highly machine dependent and vary with head wear.

Figure 9.16(a) shows the optional low-frequency pre-emphasis and (b) the adjustable record equalization which is used to compensate for high-frequency losses in the record head. The combination of the record-head loss and the record equalization serves to put the standard frequency response on the tape.

As the tape playback process is a generator process with output rising at 6 dB/octave, the replay equalization begins with an integrator (c) having a response falling at 6 dB/octave. Because of the turnover in the recorded characteristic, the integrator response is designed to level out at the same frequency (d). If low-frequency pre-emphasis has been used, the integrator

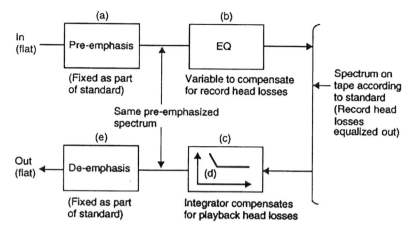

Figure 9.16 The frequency response corrections needed in an analog tape recorder. See text for details.

response will also be modified at low frequency to give the required roll-off (e) of 1590 or 3180 μs.

In the absence of replay-head losses this will give a flat output frequency response and in 30 ips machines this is very nearly the case. At lower speeds replay losses such as gap and thickness loss are significant and further equalization in the form of a high-frequency boost (f) is needed. This will need to be adjustable as it will vary with head wear.

Consequently the replay circuits of a typical tape deck contain fixed low-frequency de-emphasis, fixed integration which levels out according to the high-frequency time constant and adjustable equalization to counteract the effects of replay loss. Where a severe high-frequency loss is caused by the use of a very low tape speed and a large head gap, exact equalization cannot be used because so much gain would be needed that the high-frequency noise level would be unacceptable. In practice the bandwidth of the reproduced signal will be limited to keep the noise at a moderate level.

As the record and replay equalizations are in series, then in theory it would be possible to make all of the corrections on, say, replay. However, although the machine would be able to play its own tapes, there would be a serious dynamic range problem. Furthermore these tapes would display response irregularities when played elsewhere. Similarly tapes from elsewhere would have an irregular response when played on such a machine. Thus it is the requirements of interchange which demand separate record and replay equalization and standardized pre-emphasis.

Pre-emphasis takes advantage of the fact that most audio program material has lower signal levels at the extremes of the audible range. Low frequencies can be boosted on record and given a corresponding reduction on replay which will reduce hum. This is used in the NAB and Compact Cassette standards. Clearly in some cases audio signals have exceptional levels at frequency extremes, for example organ music and percussion. In these cases pre-emphasis is not beneficial as it could result in overloading. Recording level meters should

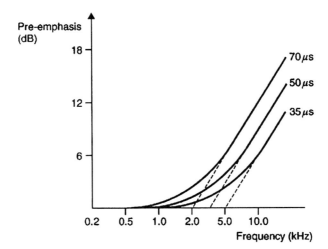

Figure 9.17 Common recording characteristics.

be connected after the pre-emphasis stage so that this condition will be recognized. However, it might be better to use a standard without low-frequency equalization.

Figure 9.17 shows the time constants and responses used in a number of different recording standards. The turnover frequency is a function of tape speed. With low tape speeds the highest frequencies may not be reproduced at all and greater benefit is obtained by using a lower turnover frequency. Note that when chromium tape is used in Compact Cassette the improved frequency response means that a higher turnover frequency can be used in the pre-emphasis.

The replay equalization is adjusted using a test tape containing tones at a wide range of frequencies recorded at an identical level except for pre-emphasis. The equalization is adjusted so that all of the tones playback at the same level. The record equalization is then adjusted by recording onto a blank tape which uses the pre-adjusted replay system as a reference.

Care should be taken at low frequency that one of the test tones does not coincide with a head bump as this could result in a false adjustment. Where head bumps are suspected, low-frequency equalization should be adjusted to give a good average response over a range of low frequencies.

The main disadvantage of high-quality analog recorders is that they have to be cleaned and adjusted frequently in order to maintain performance. As there are so many adjustments this is quite a labour intensive process and in a commercial enterprise can be costly.

9.8 Sources of noise

Unfortunately analog tape recording suffers from a large number of different noise mechanisms making high quality difficult to achieve. In high-quality equipment electronic and head noise is engineered to be negligible compared with the noises of magnetic origin.

The lowest noise floor is that due to the intrinsic noise of the tape which occurs because in a given tape area there is a finite number of particles with a finite number of domains within them giving a limit to the resolution of the recording. Traditionalists who like to attack digital recording for its finite resolution tend not to be aware that analog recording has such a limit.

Clearly the resolution of a recording can be increased by allotting a larger area of tape to it. Linsley Hood[7] rightly draws a parallel with the professional use of large format film to reduce the effect of grain. Doubling the track width doubles the replay voltage but doubles the noise power, resulting in a 3 dB improvement in SNR. This is why professional recorders have such wide tracks. When it is considered that doubling the tape consumption of a digital recorder would allow the sample wordlength to be doubled it should be clear that in comparison analog recording is not very efficient.

The noise floor of the tape can only be reached by a bulk eraser, but when such a tape is recorded, the bias field has the effect of increasing the noise. Tape noise produces noise flux in the record head and the bias field causes it to be recorded, doubling the noise flux power and causing a loss of 3 dB SNR. Tapes which have been used before need to be effectively erased. The erase depth figure of a recorder is a measure of the amount of a previous signal which is left after the tape passes the erase head.

A mechanism similar to that responsible for bias noise occurs in the erase head where some of the recorded flux penetrates the head and is weakly recorded by the biasing effect of the erase waveform. Many erase heads have double gaps so that the tape passes through two erase processes which effectively eradicate the re-recording phenomenon.

Tapes which have been in storage for some time may be more difficult to erase because the magnetic material has annealed and become magnetically hard to some extent. Such recordings appear to bulk erase successfully, but when used again the bias field causes the original recording to reappear. Mee and Daniel[5] quote a worst case 20 dB above noise. For this reason professionals prefer to use new tapes for important projects and security organizations physically destroy unwanted recordings instead of simply erasing them.

Tape noise is random and uncorrelated with the audio signal. However, there are other noise mechanisms which are signal related. Surface noise is due to irregularities on the tape surface modulating the intimacy of head-tape contact. Modulation noise is due to inhomogeneities in the tape coating which vary the reluctance of the magnetic circuit during recording. This results in the audio waveform being amplitude modulated. Any head magnetization or DC component of the bias waveform will also be modulated by this mechanism to produce noise.

In multitrack formats crosstalk between tracks can be a problem. This is generally avoided by leaving wide guard bands between the tracks on tape and by installing magnetic shields between the magnetic circuits in the head. Where one track of a multitrack tape is used for timecode or console automation data this should be put on an edge track. If crosstalk is experienced the adjacent track may have to remain unused.

Print-through is a phenomenon where the recorded signal affects adjacent layers of tape on the reel and is normally evident as a series of pre- and post-echoes of a transient. Cobalt-doped tapes are significantly worse offenders especially at high storage temperatures. Print-through is greater to layers in the direction of the tape coating, leading to stronger pre-echoes in a conventionally wound open-reel recorder. In the Compact Cassette the tape is wound coating outwards, leading to stronger post-echoes. Post-echoes are less obvious which is just as well given the very thin substrates of cassette media.

Print-through is reduced in professional tapes by using thick substrates. Print-through takes time to build up and is reduced if the tape is rewound. Consequently it is good practice to store tapes tail-out so that they have to be rewound immediately before playing.

9.9 Head alignment

Figure 9.1(b) shows a typical analog head configuration.[5] Tape is located with respect to the heads by *guides* which are hardened to resist wear. As tape varies in width it is important to prevent tape weave which changes the alignment with the heads. This can be done by using guides with sprung flanges. The fixed flange runs along the reference edge of the tape and the other edge is allowed to wander by the sprung flange. It is important that tape guides are kept free from debris as this will spoil the alignment. Care should be taken when cleaning that the sprung flange is not tilted as it may jam out of contact with the tape.

There are three heads, erase, record and replay. The erase head erases a wider path on the tape than the track width to ensure that new recordings are always made in fully erased areas even if there is slight head misalignment. Similarly the replay head is slightly smaller than the track width to prevent lateral alignment variation altering the amplitude of the replay signal. The height at which the heads are installed must be correct or the above processes will not work properly.

Head height can be adjusted using special tapes which have tones recorded in the areas *outside* the tape track. The height of the head is adjusted to minimize the replayed signal. Alternatively the record head can be checked by making a test recording and developing the tape to make the tracks visible. This is done using a suspension of magnetic powder in a volatile liquid which evaporates after application.

Figure 9.18 shows the three axes in which a head can be aligned. If the *wrap* adjustment is incorrect, the maximum pressure due to the tape tension will not act at the gap and high-frequency response will be impaired. If the *zenith* angle is incorrect, tape tension will be greater at one edge than the other leading to wear problems and differing frequency responses in analog multitrack machines.

The azimuth angle is critical because if it is incorrect it makes the effective head gap larger, increasing the aperture effect of the head and impairing high-frequency response. It will be clear from Figure 9.19(a) that azimuth error is more significant in the wide tracks of professional analog recorders than it is with the narrow tracks of a digital recorder.

Azimuth is adjusted using a reference tape containing transitions recorded at 90° to the tape edge. The replay head is adjusted for maximum output on a high frequency (b), or if the head has more than one track it may be adjusted to align the phases of two replay signals on a dual trace oscilloscope (c). Caution is needed to ensure that alignment is not achieved with exactly one cycle of phase error.

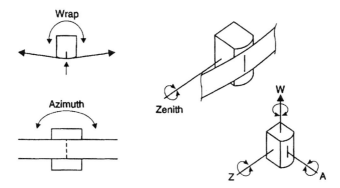

Figure 9.18 The three axes in which heads can be aligned. Wrap centres the tape tension pressure on the gap. Zenith equalizes the pressure across the tape and azimuth sets the gap at 90° to the tape path.

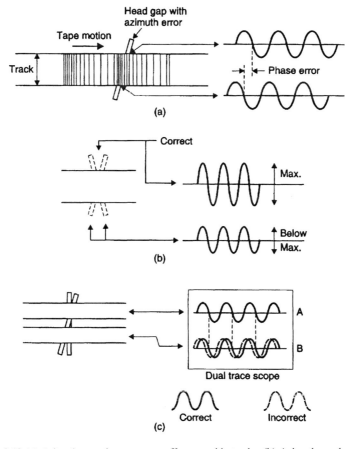

Figure 9.19 (a) Azimuth error has a greater effect on wide tracks. (b) Azimuth can be adjusted for maximum output with a high-frequency test tape or (c) on a stereo machine for minimum phase error between the tracks.

The record head azimuth is adjusted using a blank tape which is played back at the previously adjusted replay head.

In stereo analog machines, it is important that the record azimuth adjustment is not made when the bias levels in the two channels are unknown. As recording actually takes place at some point after the trailing pole where the bias field decays, a different bias level in the two channels has the same effect as an azimuth error. Where a machine is being extensively overhauled, the azimuth may need to be adjusted before a bias adjustment can be performed. In this case it will be necessary to repeat the record azimuth adjustment.

9.10 Capstan motors

The job of the tape transport is to move tape linearly past the head at constant speed whilst exerting constant pressure on the head surface. In addition it must perform various other conveniences such as winding and rewinding, measuring the amount of tape used and locking to another recorder.

The basic tape drive mechanism is the capstan which is a precision shaft rotating at constant speed driven by a motor. In quality machines the reel turntables or hubs are driven by separate motors in preference to a dubious arrangement of belts, pulleys and felt-slipping clutches. In low-cost machines the motor would be a shaded-pole squirrel-cage induction type shown in Figure 9.20. The rotor of such a motor consists of a laminated soft iron stack into which is set a peripheral array of copper bars which are parallel to the shaft and joined by shorting rings, hence the term squirrel cage. When such a rotor is subject to a rotating magnetic field, the rotating flux cuts the cage bars and generates an emf. As the series resistance of the cage is so low, large

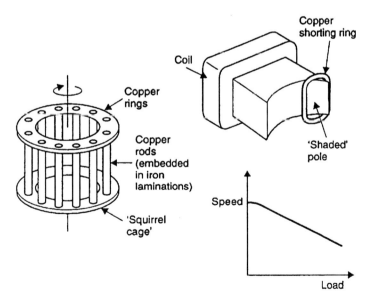

Figure 9.20 Squirrel-cage induction motor used in low-cost tape decks has poor speed stability due to slip.

currents flow which produce a further field which interacts with the applied field to produce torque. The currents are induced in the rotor which needs no brushes, hence the term induction motor.

The rotor turns to follow the rotating field, but it can never catch up because once it does so there will be no field cutting and the induced currents and torque will fall to zero. As a result the motor suffers from *slip* which is a load-dependent disparity between shaft speed and field rotation speed. The rotating field is obtained by winding short-circuited turns of copper around one side of the stator poles. This causes a phase lag in the field from that part of the pole. A more sophisticated alternative is to have additional windings on displaced poles which are fed by a series capacitor which introduces a phase shift.

Shaded pole motors generally run far too fast for direct drive and have no speed change ability. Consequently they will be fitted with a stepped shaft which drives the capstan flywheel via a rubber roller. Speed change is by selecting a different diameter.

Capacitor-run squirrel-cage motors are also found built inside-out with stationary inner windings and an outer rotor. These are capable of high torque for direct drive applications and have low leakage fields.

Greater speed stability can be obtained with a synchronous motor, which is a relative of the stepping motor and of motors found in AC driven clocks. In synchronous motors the rotor is a multipole permanent magnet. The poles of the rotor are attracted to opposite poles on the stator and so try to align themselves with the applied rotating field. The configuration will produce torque with a small load-dependent phase lag, but there is no slip so the speed is exactly determined by the supply frequency. Synchronous motors suffer from *cogging* which is a cyclic torque variation that has to be eliminated with a large flywheel. They also produce very little torque when starting.

Synchronous motors can have their speed doubled by pole switching. Pairs of stator poles are commoned by connecting their windings so that the field rotates twice as fast. This allows a two-speed tape deck with a direct drive motor. A direct drive synchronous capstan gives a very stable result but can only be use in countries having a suitable power frequency.

9.11 Servo-controlled capstans

The stability of the power frequency is not very good, especially in cold weather and this and the increase in international travel led to the demise of the synchronous motor. Modern machines use DC motors whose speed is controlled by a feedback servo. Figure 9.21 shows that the motor whose speed is to be controlled is fitted with a toothed wheel or slotted disk. For convenience, the number of slots will usually be some power of two. A sensor, magnetic or optical, will produce one pulse per slot, and these will be counted by a binary divider. A similar counter is driven by a reference frequency which will usually be derived from a crystal oscillator.

The outputs of the two counters are taken to a full adder, whose output drives a DAC which in turn drives the motor. The bias of the motor amplifier is arranged so that a DAC code of one half of the quantizing range results in zero drive to the motor, and smaller or larger codes will result in forward or reverse drive.

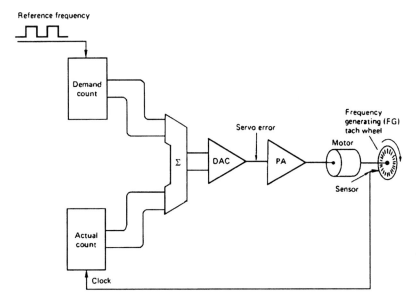

Figure 9.21 Motor speed control using frequency generator on motor shaft. Pulses from FG are compared with pulses from reference to derive servo error.

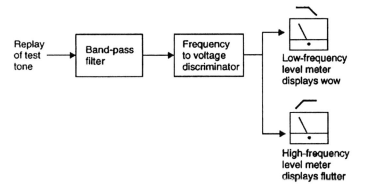

Figure 9.22 Wow and flutter meter uses frequency discriminator to measure speed irregularities.

If the count in the tacho divider lags the count in the reference divider, the motor will receive increased power, whereas if the count in the tacho divider leads the count in the reference divider, the motor will experience reverse drive, which slows it down. The result is that the speed of the motor is exactly proportional to the reference frequency. In principle the system is a phase-locked loop, where the voltage-controlled oscillator has been replaced by a motor and a frequency generating (FG) wheel. Feedback systems of this kind are very stable and independent of AC supply frequency and voltage. DC motors can produce torque over a wide speed range so that tape speed changes can simply be performed by changing the reference frequency in a divider chain.

The performance of the tape drive reflects in the wow and flutter figures. A wow/flutter meter is shown in Figure 9.22. A test tape is recorded with a fixed frequency and this is played back through a narrow band filter and a limiter to prevent level fluctuations affecting the reading. The filtered signal is fed to a frequency discriminator whose output drives a level meter. Wow and flutter cause frequency modulation of the sinusoidal test signal and the amount can be read directly.

9.12 Brushless motors

The weak point of the DC motor is the brush/commutator system, which serves to distribute current to the windings which are in the best position relative to the field to generate torque. There is a physical wear mechanism, and the best brush material developed will still produce a conductive powder which will be distributed far and wide by the need to have airflow through the motor for cooling. The interruption of current in an inductive circuit results in sparking, which can cause electrical interference and erosion of the brushes and commutator. The brushless motor allows all of the benefits of the DC motor, without the drawbacks of commutation.

Figure 9.23 shows that a brushless motor is not unlike a normal brush motor turned inside-out. In a normal motor, the field magnet is stationary, and the windings rotate. In a brushless motor, the windings are stationary, and the magnet rotates. The stationary windings eliminate the need to supply current to the rotor, and with it the need for brushes. The commutating action is replaced by electronic switches. These need to be told the rotational angle of the shaft, a function which is intrinsic in the conventional commutator. A rotation sensor performs this function.

Figure 9.24 shows the circuit of a typical brushless motor. The three-phase winding is driven by six switching devices, usually power FETs. By switching

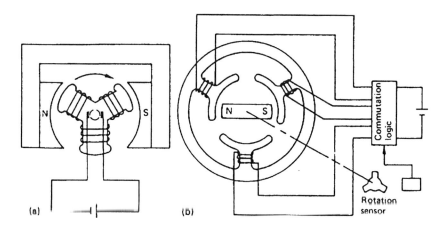

Figure 9.23 In (a) a conventional brush motor has rotating windings. Current is fed through stationary bushes and commutation is automatic. In (b) the motor has been turned inside-out: the magnet revolves and the windings are now stationary, so they can be directly connected to an electronic commutation circuit. The rotation sensor on the magnet shaft tells the commutator when to switch.

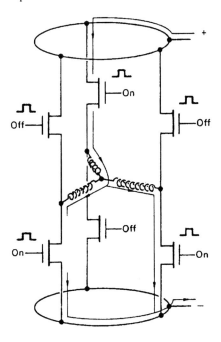

Figure 9.24 Circuit of a brushless DC motor is basically a three-phase bridge. Six different flux directions can be obtained by switching combinations of FETs. One example is shown.

these on in various combinations, it is possible to produce six resultant field directions in the windings. The switching is synchronized to the rotation of the magnet, such that the magnetic field of the rotor always finds itself at right angles to the field from the windings, and so produces the most torque. In this condition the motor will produce the most back emf, and the normal way of adjusting the rotation sensor is to allow the motor to run on no load and to mechanically adjust the angular sensor position until the motor consumes minimum current.

The rotating magnet has to be sufficiently powerful to provide the necessary field, and in early brushless motors was heavy, which gave the motor a high inertial time constant. Whilst this was of no consequence in steady speed applications, it caused problems in servos. The brushless servo motor was basically waiting for high-energy permanent magnets, which could produce the necessary field without excessive mass. These rare-earth magnets are currently more expensive than the common ferrite magnet, and so the first applications of brushless motors have been in machines where the motor cost is a relatively small proportion of the high overall cost, and where reliability is paramount.

The development of the power FET also facilitated brushless motors, since all DC motors regenerate, which is to say that when the torque opposes the direction of rotation, the current reverses. FETs are not perturbed by current direction, whereas bipolar devices were at best inefficient or would at worst fail in the presence of reverse current.

The number of motor poles directly influences the complexity of the switching circuitry, and brushless motors tend to have relatively few poles

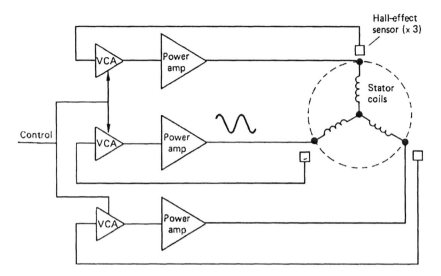

Figure 9.25 A brushless three-phase motor is constructed with a stationary coil set, driving a rotating magnet which produces three feedback sinusoids at 120° phase from Hall-effect magnetic sensors. These signals drive *analog* amplifiers to generate the power waveforms for the motor coils. Motor torque is controlled with VCAs (voltage-controlled amplifiers). The three-phase system makes the motor torque independent of shaft angle, and reduces cogging.

and can suffer from cogging. For direct drive capstan applications, it is possible to construct a cogging-free motor. The commutation circuit is no longer switching, but consists of a number of analog amplifiers which will feed the motor with sinusoidal waveforms in fixed phase relationships. In the case of a two-phase motor, this will be two waveforms at 90°, for a three-phase motor, shown in Figure 9.25, it will be three waveforms at 120°.

The drive signals for the amplifiers are produced by a number of rotation sensors, which directly produce analog sinusoids. These can conveniently be Hall-effect sensors which detect the fields from rotating magnets.

9.13 Reel motors

Whilst the slip of a squirrel-cage motor makes accurate capstan drive difficult, it is an ideal characteristic for a reel motor which has to turn at a range of speeds as the radius of the tape pack on the spool changes. Outer rotor capacitor-run induction motors are used for reel drive in semi-professional recorders such as the Revox. When supplied with a voltage somewhat below their free running requirements these motors work in high slip mode and effectively are torque generators where their speed is determined by the tape. The supply reel motor is driven in reverse to provide back tension and it is actually forced backwards by the tape being pulled from the reel. In some machines a switch allows the motor torque to be reduced when small reels are in use.

In order to wind the tape the reel which is to wind in is fed with a high power level and the reel which is to pay out is fed with a low power drive to give some back tension. Winding speed has to be limited because the tape entrains air as it moves and leads this in to the spool which is winding in. This air forms a lubricating layer between the pack and the latest layer of tape until it escapes. If there is insufficient time for this to happen the tape layers can slip axially leading to a poor pack which could cause tape damage in storage. Some machines offer a slow wind facility to pack tapes neatly prior to archiving.

When braking, the braking effort on the wind out reel must exceed that on the wind in reel slightly to avoid a slack loop of tape. However, the braking difference must not be too great as this will result in high tape tension. In some machines the reel motors carry out all braking operations and mechanical brakes are only fitted in case of a power failure. In other machines the motors are switched off to stop and all braking is done by the friction brakes. In direct drive reel motors the brake drums will be fitted directly below the reel turntable. Mechanical brakes are rendered failsafe because they are applied by coil springs. The brakes have to be released by solenoids so that they are automatically applied if the power fails.

The brakes are differential which means that they are designed so that the braking torque is higher in the wind out direction than in the wind in direction in order to prevent tape slack.

In low-cost machines a resettable mechanical counter is driven from one of the reels. As the reel speed varies the count is a somewhat non-linear function of playing time and is specific to that machine. In professional machines the tape passes around a timer roller generally between the supply reel and the head block. This produces a count which is linear. Early timer rollers drove mechanical counters but in recent machines the timer roller is a pulse generator and the counting is electronic. Such counters are accurate enough to count in units of minutes and seconds.

9.14 Tension servos

In professional machines head/tape contact is obtained by tape tension and it is important that this is kept constant. Tension control will also be necessary for shuttle to ensure correct tape packing. These requirements can be met using a tension servo with a DC motor.

Figure 9.26 shows a typical back tension control system. A swinging guide called a tension arm is mounted on a position sensor which can be optical or magnetic. A spring applies a force to the tension arm in a direction which would make the tape path longer. The position sensor output is connected to the supply reel motor amplifier which is biased in such a way that when the sensor is in the centre of its travel the motor will receive no drive. The polarity of the system is arranged so that the reel motor will be driven in reverse when the spring contracts.

The result is that the reel motor will apply a reverse torque which will extend the spring as it attempts to return the tension arm to the neutral position. If the system is given adequate loop gain, a minute deflection of the arm will result in full motor torque, so the arm will be kept in an essentially constant position. This means that the spring will have a constant length, and so the tape tension will be constant. Different tensions can be obtained by switching

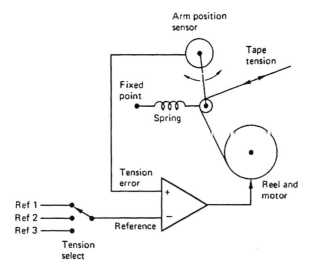

Figure 9.26 Tension servo controls reel motor current so that motor extends sprung arm. Sensor determines extension of spring and thus tape tension.

offsets into the feedback loop, which result in the system null being at a different extension of the spring.

9.15 The analog multitrack recorder

The analog multitrack recorder has evolved to suit a particular form of music production in which the recording balance can be changed after the recording because every source has an individual microphone and tape track. Analog recorders have a maximum of 24 tracks on 2-inch tape but can be locked together to allow more tracks to be used.

It is important that each track should be quite independent so that some may record whilst others do not. Separate *record standby* switches are used for each channel and these determine which channels will record when the master record interlock is operated. Crosstalk between the tracks has to be minimized and this is done using wide guard bands which are often nearly the same width as the track.

In many productions the tracks are recorded sequentially because this avoids microphone spill and gives better isolation of the instruments. A rhythm track will be recorded first and all subsequent recording is carried out by musicians who listen to the rhythm track on headphones and play in time with it. In order to avoid a time displacement on the tape between the rhythm track and the new track, the rhythm track must be played back with the record head as shown in Figure 9.27. All multitrack recorders have the facility to switch heads in this way for synchronous recording. Although this does not achieve the highest quality it is sufficient for timing purposes.

In sequential recording and in mix-down it is required to return to the beginning of the recording time after time. This procedure is simplified using an *autolocator* which returns the tape to a predetermined place automatically.

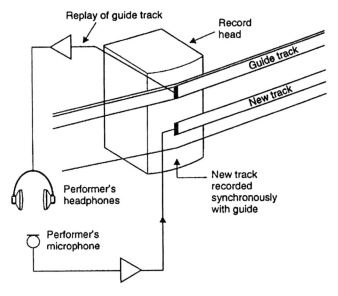

Figure 9.27 In synchronous recording it is essential to play back the guide track using the record head to avoid a time delay.

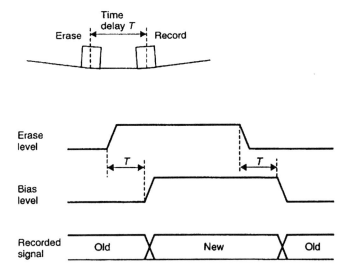

Figure 9.28 Quiet punch-in/out requires careful synchronization and ramping of the erase and bias signals.

Tape position is measured using a timer roller which generates a fixed number of pulses per length of tape.

In many productions a single wrong note may occur in an otherwise good track. This can be remedied by a form of editing called punch-in. The band play once more using headphones, and at appropriate points the machine is set into record briefly to update the defective area of the track. Machines generally have in- and out-point memories which store the timer roller count so that the punch-in can be automatic. Generally the punch-in will be rehearsed with the engineer listening to a simulation a few times before it is committed to tape.

Successful punch-in requires careful control of the erase and bias signals to prevent clicks. Figure 9.28 shows the sequence. As the punch-in begins, the erase current is ramped up gradually. Just as the erased part of the tape reaches the record head, the audio signal and the bias are also ramped. At the out-point the erase drive is ramped down and just as the original recording reaches the record head the audio and bias signals are ramped down. Ramping the bias gives an effect which is nearly a crossfade on the recording.

A multitrack recorder is a sizeable item and operation is eased by the provision of a remote control unit which can be placed near the mixing console. The remote control will have the main tape function buttons, the auto-locator, the record and synchronous record function switches and the punch-in controls.

9.16 The Compact Cassette

The compact cassette was introduced in 1963 as a dictating machine format and this humble origin dictated many of the design parameters which are not what one would choose if designing a quality audio cassette from the outset. The slow speed of $1\frac{7}{8}$ ips (47.5 mm/s) makes high-frequency response difficult to achieve and the narrow tracks of only 0.61 mm make noise a problem. There are no spools as such and the edge of the tape pack rubs against the liner sheets causing friction. Tape guidance in the cassette is by a couple of plastic rollers in the corners.

The tape manufacturers have worked wonders in developing better media which have allowed better quality within the restrictions of the format. Chromium and ferrichrome tapes give much better results on machines with suitable equalization and bias settings. However, probably the single largest improvement to the Compact Cassette was the adoption of the Dolby noise reducing system.

Mass-produced cassette mechanisms can be made at very low cost and their poor quality is masked by the equally poor loudspeakers fitted in portable consumer equipment, most of which is not Dolby equipped. However, if one of these decks is connected to a loudspeaker system of reasonable quality the shortcomings are readily apparent.

When the cassette is placed in the machine it is located by tapered pegs which engage a pair of holes. The capstan passes through a small hole and the reel turntables engage dogs on the hubs. There are three large apertures on the tape face and two small ones. The central aperture accepts the record/replay head and a felt pad and hum shield are fitted opposite this aperture. The outer apertures accept the erase head and the pressure roller. These exchange when the tape is turned over.

A small number of high-quality Compact Cassette decks have been made and these have had to pay fanatical attention to almost every detail in addition to having to use the most expensive cassettes available. One of the biggest problems of the cassette for quality applications is accurate tape guidance. If

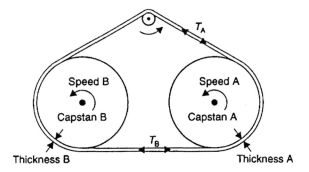

$T_A > T_B$ ∴ Thickness A < Thickness B
∴ Speed A > Speed B
∴ Tape between capstans A and B is in tension

Figure 9.29 Twin capstan system maintains tape tension through minute difference in capstan speeds.

not given careful attention tape weave causes stereo image shift and variable azimuth error. Friction in the cassette causes wow.

One solution to this problem has been the use of dual capstan transports which completely isolate the goings on in the cassette from the head area. Figure 9.29 shows a dual capstan system. The two capstans are driven by an elastomeric belt from a single motor. The second pressure roller occupies the aperture intended for the erase head and this has to be relocated. The section of belt between the motor and the right-hand capstan is under more tension than that between the capstans and consequently is slightly thinner. This slightly reduces its effective radius with the result that the right-hand capstan goes minutely faster. This slight speed difference causes the tape between the capstans to be put under tension. Tape guides adjacent to the pressure rollers ensure that the tape height is set by the deck and not the cassette so that weave is reduced as much as possible.

The original cassette design did not anticipate separate record and replay heads and a lot of effort has gone into making separate record and play heads which will fit the standard aperture. The adoption of separate heads makes a big difference to the sound quality because the slow tape speed and short wavelengths make a single head too much of a compromise.

9.17 Additional quality criteria in stereo

In stereo recorders it is important that the two channels are not reversed. There is a universal standard which states that the left channel should be recorded on the track which is further from the deckplate, i.e. the upper track on a horizontal machine. Inputs and outputs should be identified in plain text or colour coded. Professional machines use the navigational colour code of red for left (port) and green for right (starboard). In consumer equipment a number of standards exist, including red for right and white for left.

In stereo working the quality criteria for recorders are more strict than in multitrack or mono working because of the need to consider the spatial information in a stereo signal. Intensity stereo signals encode positional information in the relative level of the two channels and there are no timing differences. Inadvertent level differences or timing errors will disturb the stereo imaging.

It is not practicable to build analog tape recorders with an exactly flat frequency response, but what is needed in stereo is for the response of the two channels to be as identical as possible. If this is not the case the apparent position of virtual sound sources will become frequency dependent. In the case of a recorder in which the left channel suffers a drooping high-frequency response, low-frequency sources will be correctly placed, but high-frequency sources will be pulled to the right. Sources with a harmonic content will be *smeared* so that they appear wider than they should.

There are a number of ways in which timing errors can occur between the channels. Figure 9.30 shows that tape weave produces varying interchannel timing which causes imaging to wander. This can be due to dirty guides, poor mechanics or inferior tape with poor edge straightness.

A fixed timing error can be due to azimuth error in record or replay heads. An azimuth error which causes almost negligible frequency response loss will cause an image shift. Both tape weave and azimuth error get worse at low tape speeds and are particularly problematical in Compact Cassette and in video cassette recorders.

When adjusting azimuth in a stereo machine with a high frequency it is better to adjust for minimum phase error between the channels using a double-beam oscilloscope than to adjust for maximum output.

The bias level can also affect the stereo image. As the point at which the recording is made moves away from the head as the bias level rises, the bias control affects the timing. It is important that the bias level in the two tracks should be identical. Because of this interaction the correct order in which to make adjustments is not obvious. In practice some iteration will be needed.

Unfortunately to maintain the highest standards in stereo this time-consuming alignment procedure will have to be carried out regularly, resulting

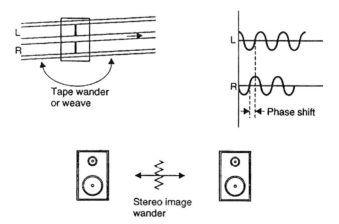

Figure 9.30 Tape wander causes relative timing errors and stereo image shift.

in a high operating cost. It is easy to see how digital machines have become popular for stereo as they have negligible interchannel differences yet need no adjustment.

9.18 Analog noise reduction

The random noise caused by the finite number of particles in tape limits the performance of analog recorders. As has been seen, the only way in which the noise can be actually reduced is to make the tracks wider, but only 3 dB improvement is obtained for every doubling of track width, so this is not a practical solution. As the noise of tape recording is significantly higher in level than noise from any other component in an analog production system, efforts have been made to reduce its effect. Of these the best known and most successful have been the various systems produced by Dolby Laboratories.[8]

Noise reducers (NRs) are essentially reversible analog compressors and function by compressing the dynamic range of the signal at the encoder prior to recording in a way which can be reversed at the decoder on replay. The expansion process reduces the effect of tape noise.

As noise reduction is designed to work with unmodified analog recorders the decoder must work on the audio signal only. Figure 9.31 shows that this is overcome by using a recursive type compressor which operates based on its own output. Clearly when this is done, the decoder and the encoder can be made complementary because both work from the same signal provided the recorder has no gain error and a reasonably flat frequency response. In fact a recursive decoder can be used instead with equal success.

Static gain error is overcome by generating an NR line-up tone in the encoder which is put on tape before the program. On replay the input level of the decoder is adjusted during the line-up tone to give a particular level indication on the decoder meter. Companders cannot handle dynamic gain errors very well as the decoder is misled. The result is called mistracking because the decoder fails to track the encoder.

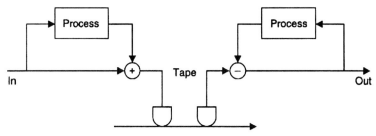

Tape = In + Process × In = In (1 + Process)
Out = Tape − Process × Out
∴ Tape = Out + Process × Out = (1 + Process)
∴ In (1 + Process) = Out (1 + Process)
∴ In = Out

Figure 9.31 In order to obtain complementary expansion of the compressed signal, noise reducers work by analysing their own output. The decoder then has access to the same signal as the encoder (assuming the tape deck has unity gain).

In a compression system the gain variation could result in the noise floor rising and falling. If audible this will result in annoying breathing or pumping effects. The noise floor will be audible whenever it is not masked. Consequently practical NR systems must restrict the degree of compression or divide the audio spectrum into sub-bands in which the gain is individually controlled. The performance of this approach increases with the number of bands. In the Dolby A system the audio input is divided into four bands at 80 Hz, 3 kHz and 9 kHz.

The approach of the Dolby systems is to modify the signal at low levels whilst largely leaving high-level signals alone. As the input in a particular band falls in level, gain is used so that a −40 dB input is recorded at −30 dB. On replay that band will receive a 10 dB attenuation giving a corresponding improvement in apparent signal-to-noise ratio.

In the Dolby B system, designed for the Compact Cassette, the goal is to reduce the high-frequency hiss. This is done using a level-dependent pre-emphasis. At high levels the response of the compressor is flat, but as level goes down the amount of pre-emphasis increases as shown in Figure 9.32 to 10 dB at −45 dB. This is done by a level sensor which controls a programmable filter in a side chain. Encode level is increased by adding the side chain signal to the input. The overshoot clipper prevents high-level transients which are faster than the compressor time constant from incorrectly increasing the recorded level.

The decoder incorporates an inverter so that the side chain subtracts from the main path to return the signal to the correct level. About 10 dB of noise reduction is obtained with the appropriate decoder. The Dolby B system was designed so that reasonable results would be obtained in the absence of a decoder such as when a B-coded cassette is played on a cheap machine. The result is that the high frequencies appear compressed, i.e. at low signal levels the sound appears brighter than it should. As very cheap cassette players tend to have a fairly miserable high-frequency response this is not as bad as it sounds.

The Dolby C system has been used in video cassette recorders, including the popular Betacam professional VCR. Dolby C uses two cascaded stages in

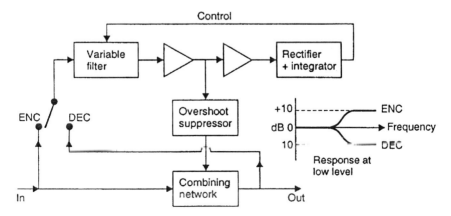

Figure 9.32 Dolby B NR system.

both encoder and the mirrored decoder to offer up to 20 dB of noise reduction. Figure 9.33 shows that the stages are known as the high-level and low-level stages. The stages are similar in concept to those used in Dolby B except that the variable pre-emphasis is extended down to around 100 Hz. The low level stage handles levels 20 dB below those of the first stage.

As a high degree of companding is used, it is important to prevent audible artifacts due to mistracking. In cassettes the biggest problem is inconsistency of high-frequency response due to tape weave and variable head contact. A decoder interprets these response changes as level changes and expands them. The non-obvious solution adopted by Dolby is known as 'spectral skewing' and acts to de-emphasize the high frequency prior to encoding and equalize again on replay. The result is that variable high-frequency response causes significantly less mistracking modulation elsewhere in the audio band. The penalty is that the noise reduction is only 10 dB at the highest frequency but this is a small price to pay as the ear is not so sensitive to noise here.

The Dolby SR system has a novel approach in which the sub-band boundaries can move. The spectrum of the signal is analysed and a sub-band boundary is positioned just above the significant energy in the signal. This allows the maximum masking effect.

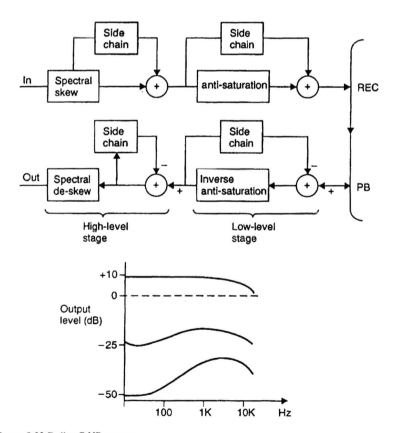

Figure 9.33 Dolby C NR system.

Correct replay of an NR-encoded tape can only be achieved with the correct decoder. Consequently it is important that tapes are marked with the type of NR used. A drawback of analog NR is that errors in replay level caused by dropouts cause mistracking meaning that the dropouts are expanded. High-quality professional analog tapes are specifically designed to minimize dropout frequency.

9.19 Audio in analog VTRs

In most analog video recorders, the audio tracks are longitudinal and are placed at one edge of the tape. Figure 9.34 shows a typical analog tape track arrangement. The audio performance of video recorders has tradition-ally lagged behind that of audio-only recorders. In video recorders, the use of rotary heads to obtain sufficient bandwidth results in a wide tape which travels relatively slowly by professional audio standards. The audio performance is limited by the format itself and by the nature of video recording. In all rotary-head recorders, the intermittent head contact causes shock-wave patterns to propagate down the tape, making low flutter figures difficult to achieve. This is compounded by the action of the capstan servo which has to change tape speed to maintain control-track phase if the video heads are to track properly.

The requirements of lip-sync dictate that the same head must be used for both recording and playback, when the optimum head design for these two functions is different. When dubbing from one track to the next, one head gap will be recording the signal played back by the adjacent magnetic circuit in the head, and mutual inductance can cause an oscillatory loop if extensive anti-phase crosstalk cancelling is not employed. Placing the tracks on opposite sides of the tape would help this problem, but gross phase errors between the channels would then be introduced by tape weave. This can mean the difference between a two-channel recorder and a stereo recorder. Even when the audio channels are on the same edge of the tape, most analog video recorders have marginal interchannel phase accuracy. As the linear tape speed is low, recorded wavelengths are short and a given amount of tape weave results in greater phase error. Crosstalk between the timecode and audio tracks can also restrict performance. The introduction of stereo audio with television

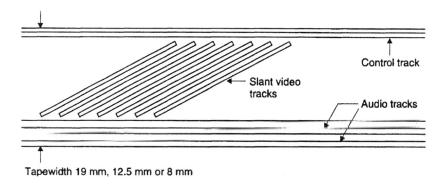

Figure 9.34 Track layout of an analog VTR. Note the linear audio tracks are relatively narrow. The linear tape speed is quite low by audio recording standards.

and the modern trend toward extensive post-production makes sound quality and the ability to support multigeneration work essential.

In consumer recorders, economy of tape consumption is paramount. It is also important in portable professional recorders used for electronic news gathering (ENG). In video cassette recorders such as VHS and Betamax, and in the professional Betacam and M-II formats, the rotary head combined with azimuth recording allowed extremely narrow tape tracks, with the result that linear tape speed was very low. The analog audio tracks of these systems gave marginal performance.

One solution to the audio quality issue was to frequency-modulate the audio onto a carrier which was incorporated in the spectrum of the signals recorded by the rotary heads. The audio quality of such systems was very much better than that offered by the analog tracks, but there was a tremendous drawback for production applications that the audio could not be recorded independently of the video, or vice versa, as they were both combined into one signal.

9.20 The vinyl disk

The vinyl disk is now obsolete having been replaced by CD. However, this section is included for completeness and for the benefit of those such as archivists who still have to reproduce vinyl disks. The vinyl microgroove disk is the descendant of the shellac disk and shares with it the fact that it can easily and cheaply be duplicated by pressing. Disks are available in 12-inch and 7-inch diameters which turn at $33\frac{1}{3}$ and 45 rpm respectively. The basic principle of the audio disk is simple and is shown in Figure 9.35. The spiral groove is modulated by vibrating a cutting stylus so that the stylus velocity is proportional to the audio waveform voltage. In mono disks the stylus moves from side to side whereas in stereo disks the groove walls are at 90° so that they can move orthogonally to replicate the two audio channels.

Bearing in mind the M-S stereo principles explained in Chapter 7, the stylus in an orthogonal groove behaves like a vectorscope display. The phasing is arranged such that the M signal $(L + R)$ causes motion in the plane of the disk and the S signal causes motion along the disk axis. This approach gives a degree of mono compatibility. Provided a monophonic pickup allows vertical stylus movement, the normal lateral movement will produce an output which is $L + R$. As the largest amplitudes are associated with low frequencies which are generally nearly in-phase in stereo, this arrangement allows better tracking

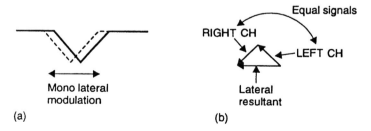

(a) Mono lateral modulation

(b) Lateral resultant — Equal signals, RIGHT CH, LEFT CH

Figure 9.35 (a) Mono groove modulation is lateral. (b) Stereo modulates groove walls independently so that in-phase signals cause lateral motion.

on replay than the alternative phasing arrangement which would cause low-frequency transients to throw the stylus from the groove.

The disk cutter is a massive device in order to suppress vibrations reaching it through the floor which would cause noise on the recording. It has a heavy turntable to ensure constant speed and a cutter head which is driven slowly in from the perimeter towards the centre along an exact radius. The speed of the cutter head is variable so that the groove pitch can be changed. Coarser pitch is needed at the lead-in and lead-out grooves as well as between the tracks on long-playing albums. Early cutters used constant pitch in the tracks, but an increase in quality and playing time was obtained by making the pitch vary dynamically as a function of the audio signal amplitude. The grooves would be close together on quiet passages, extending the recording time, but coarser spacing would be used for loud passages to prevent one groove breaking into another.

Figure 9.36 shows the components of a variable pitch disk cutter. As shown at (a) the groove pitch must begin to coarsen somewhat in advance of a loud passage to give the cutter head time to move away from the previous track. In early equipment (b) this was done using a special tape deck with an advanced replay head in addition to the normal one. The signal picked up by the advanced head was analysed and used to produce a velocity command signal to the cutter. When a loud passage was encountered by the advanced head it would speed up the cutter head so that the groove pitch was already coarsened by the time the signal from the normal head reached the cutter.

In later systems it was possible to use an unmodified tape deck because the audio signal to the cutter was delayed whereas the signal to the groove pitch

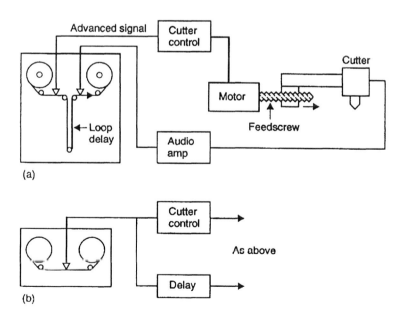

Figure 9.36 (a) Variable pitch disk cutter uses advanced playback head to control lathe pitch. (b) Later system using digital delay and standard tape deck.

control was not. The delay was implemented by digitizing the input signal and temporarily storing the data in RAM (random access memory) prior to outputting it to a DAC. Naturally the delay unit had to be of high quality as it was in series with the main recorded signal. As the quality of the vinyl disk is fairly poor this was not difficult.

In the early days of the Compact Disc it was normal to hear the misinformed criticizing the alleged shortcomings of digital audio and singing the praises of the analog disk. It was a matter of great hilarity that the 'sonically superior' signal on the treasured analog vinyl disks belonging to these people would have been through the digital domain during the cutting process.

The cutter head contains two orthogonal moving coil actuators which drive the cutting stylus. In order to obtain the best linearity velocity feedback is used, with sense coils producing signals proportional to the cutter velocity which are compared with the original waveform.

The original disk-cutting process used a soft wax disk which was then coated in a conductive layer so that it could be electroplated. The electroplated layer could be used directly as a stamper, or it could be used to make more stampers via a pair of electroplating steps for large production runs. Later a technique called direct metal mastering (DMM) was developed in which the stylus cut directly into a copper master disk. The greater rigidity of the copper improved the sound quality obtained. DMM was basically the swan song of the vinyl disk before it was eclipsed by CD.

The basic modulation method of the vinyl disk is that the stylus velocity is proportional to the audio waveform voltage. When played with a magnetic pickup having voltage proportional to velocity this gives the correct waveform directly. However, some modifications are necessary. First with a precise velocity characteristic the amplitude of groove deflection becomes too high at low frequencies, restricting playing time. As there is little disk surface noise at low frequency this problem is solved by de-emphasizing the low frequencies on recording by 6 dB/octave so that the groove has constant amplitude below about 600 Hz. A corresponding low-frequency boost is required on replay.

As surface noise increases with frequency, whereas the signal amplitude generally falls on most program material, a high-frequency pre-emphasis is also applied. The overall replay equalization required is known as the RIAA (Radio Industries Association of America) characteristic which is specified by the time constants of the response curve which are 3180, 318 and 75 µs.

9.21 Vinyl disk players

The disk player requires a turntable which is designed to rotate at constant speed to avoid wow and flutter. Early designs used shaded pole induction motors which drove the inside of the turntable rim by a rubber faced idler wheel as shown in Figure 9.37(a). These motors would vibrate because of their high speed and the vibrations would be picked up by the stylus as *rumble*. This approach gave way to low-speed synchronous motors driving a sub-platter through a flexible belt (b). The use of the belt allowed the motor and the turntable to be suspended independently and improved the rumble figures provided a precision spindle bearing was used. However, the stylus drag is a function of the amplitude of groove modulation and this could cause belt stretch and signal-induced wow on some belt driven turntables. Finally direct drive brushless motors (c) were developed incorporating velocity feedback and these allowed low rumble as well as exceptionally good speed stability. It was common for quality turntables to have a stroboscopic pattern on the rim so that the speed could be checked.

The disk is traced with a stylus carried on a cantilever which is compliantly mounted in the pickup. The pickup is carried in an arm. The majority of players use a radially swinging arm about nine inches long for economy, even though the disks were cut with a parallel tracking cutter. Figure 9.38 shows that by bending the end of the arm the tracking error could be minimized although the stylus drag then caused a moment which tended to pull the arm in towards the disk centre. Quality arms would have an *anti-skating* mechanism which would apply a compensating outward force to the arm. Unfortunately the stylus drag is not constant but is a function of the modulation. As a result of varying stylus drag the arm would move in and out as shown in Figure 9.39, displacing the stylus arm in the pickup and adding low-frequency noise as well as a component of motion along the groove which causes flutter.

For the finest quality, a parallel tracking straight arm was used in which the arm pivot was on a motorized carriage which moved to keep the arm aligned.

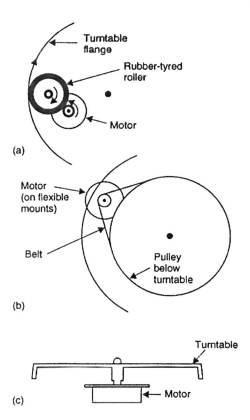

Figure 9.37 (a) Simple vinyl turntable using induction motor. (b) Belt drive reduces motor vibrations. (c) Direct drive servo motor.

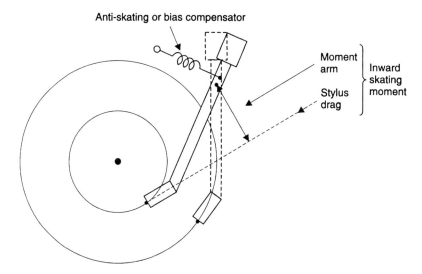

Figure 9.38 Bent arm reduces tracking error as arm swings.

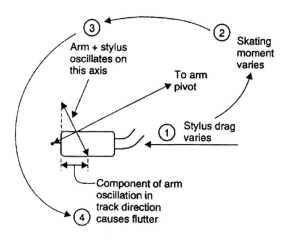

Figure 9.39 Bent arm has complex motion as stylus drag varies.

The straight arm means that there is no tracking error, no skating force and no modulation-dependent effects. The audible superiority of parallel tracking arms led to their availability from a number of manufacturers.

Figure 9.40 shows that the effective mass of the arm and pickup forms a resonant system with the stylus cantilever compliance. Below resonance the arm moves in sympathy with the stylus and there is little pickup output. Above resonance the arm remains fixed due to its mass and the pickup works normally. The resonant frequency is optimally set at around 20 Hz so that lower frequencies due to disk warps and rumble are not reproduced, but genuine low frequency on the disk is. Unfortunately most hi-fi pickup arms have detachable headshells so that pickups can be changed. The mass of these arms is so great

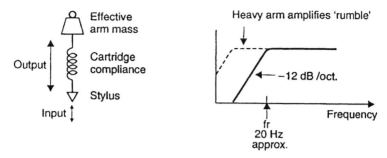

Figure 9.40 Cartridge compliance resonates with arm mass to produce low-pass filter. With heavy arm the turnover frequency would be too low.

that the resonant frequency is too low. The situation is made even worse in 12-inch arms designed to reduce tracking error.

The best arms have no headshell, but the pickup plugs directly into the arm. This allows the resonance to be set at the optimal frequency even with a high compliance pickup. Parallel tracking turntables can have very short arms indeed because there is no tracking error.

The groove is cut with a sharp stylus shown in Figure 9.41(a) having a very small tip radius and an included angle of nearly 90°. The depth of cut is adjusted to give an average groove width of 0.0025 inch. The replay stylus has to be rounded to prevent the contact pressure being too high. The tip radius is larger than the cutter tip radius so that it does not drag in the bottom of the groove. If the end of the replay stylus is spherical, with a radius of about 18 μm, optimal contact is made half-way up the groove walls as shown in (b);

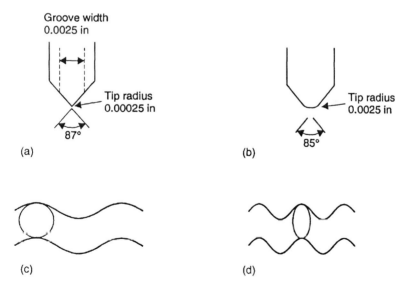

Figure 9.41 (a) Dimensions of cutter. (b) Spherical replay stylus. (c) Poor high-frequency response of spherical stylus. (d) Biradial stylus follows short wavelengths better.

however, the spherical shape is not optimal for high-frequency reproduction as can be seen in (c) because the stylus is too big to follow short wavelength groove detail. Even if the spherical stylus can follow the groove wall, the reproduced waveform will not be the original because the cutter had sharp edges. The result is *tracing distortion*. The solution is the biradial or elliptical stylus which is spherical in the plane of groove contact, but elliptical in plan so that the radius along the groove is smaller; typically 7 µm. The smaller radius reduces tracing distortion and allows shorter wavelengths to be reproduced.

The mass of the stylus must be accelerated to make it follow the groove. The groove can supply an upward force, but the stylus will only return if a downward force is available and this comes from the tracking weight which is the static downward force the stylus exerts on the disk. The lighter the stylus the less force is needed to accelerate it and the lower the tracking force can be. This is beneficial because it reduces groove and stylus wear. A further issue is that if large forces are needed to accelerate the stylus the groove wall will flex and distortion will result. In practice the distortion from even a high-quality vinyl disk pickup is relatively poor compared to an open-reel tape and dismal compared to digital systems. If the tracking force is insufficient the stylus will not fall as the groove recedes and it will lose contact resulting in mistracking. Figure 9.42 shows that this is most common at the inner grooves of a disk where the linear speed is lowest and the modulation is steeper.

Disk pickups or cartridges have been built using crystal and ceramic piezo-electric elements, strain gauges, moving magnets, moving coils and variable reluctance. Piezo-electric pickups are only used in the cheapest equipment as they exert large dynamic forces on the groove wall and need a heavy tracking force. The magnetic cartridge is capable of better performance because it can have a low moving mass which allows a lower tracking force. Figure 9.43 shows the three types of magnetic cartridge. In the variable reluctance

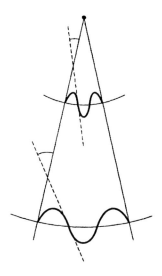

Figure 9.42 Groove modulation gets steeper near centre of disk where linear speed is lowest.

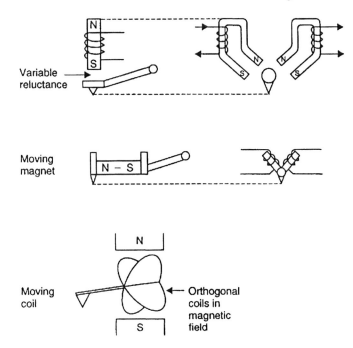

Figure 9.43 Variable reluctance, moving magnet and moving coil cartridges.

approach the cantilever drives a ferrous plate which changes the reluctance of a magnetic circuit as it moves. The variations in flux induce voltages in a coil. The moving coil pickup works like a moving coil microphone with the diaphragm replaced by a stylus whereas the moving magnet system holds the coil still and allows the magnet to move with the stylus. With the use of rare-earth magnets the moving mass can be very small.

The moving coil pickup is considered to have the lowest distortion but it suffers from a very small output signal requiring special low noise pre-amplifiers or step-up transformers.

References

1. Mallinson, J.C., *The Foundations of Magnetic Recording*, San Diego: Academic Press (1987)
2. Naumann, K.E. and Daniel, E.D., Audio cassette chromium dioxide tape. *J. Audio Eng. Soc.* **19**, 822 (1971)
3. Mee, C.D., *The Physics of Magnetic Recording*, Amsterdam and New York: Elsevier–North Holland Publishing (1978)
4. Bertram, H.N., Long wavelength AC bias recording theory. *IEEE Trans. Magn.* **MAG 10**, 1039 (1974)
5. Mee, C.D. and Daniel, E.D., (eds) *Magnetic Recording Vol. III*, Chapter 3, McGraw-Hill (1988)
6. Köster, E., A contribution to anhysteretic remanence and AC bias recording, *IEEE Trans. Magn.*, **MAG-II**, 1185 (1975)
7. Linsley Hood, J., *Audio Electronics*, Chapter 1, Oxford: Butterworth-Heinemann (1995)
8. Dolby, R.M., An audio noise reduction system. *JAES*, **15**, 383 (1967)

Chapter 10

Digital recording

Unlike analog recording, digital recording reproduces discrete numbers, and coding techniques can be used to ensure that the occurrence of bit errors is negligible. When this is done, digital recording has no sound quality because the presence of the recorder makes no difference to the data. This is advantageous because more effort can then be applied to the design of converters which do affect the quality.

10.1 Introduction

In digital tape recording the transport may resemble the analog transports shown in Chapter 9 except that the tape tracks are narrower. The difference is in the way the head signals are processed. Figure 10.1 shows the block diagram of a machine of this type. Analog inputs are converted to the digital domain by converters. Clearly there will be one converter for every audio channel to be recorded. Unlike an analog machine, there is not necessarily one tape track per audio channel. In stereo machines the two channels of audio samples may be distributed over a number of tracks each in order to reduce the tape speed and extend the playing time.

The samples from the converter will be separated into odd and even for concealment purposes (see Section 10.7) and usually one set of samples will be delayed with respect to the other before recording. The continuous stream of samples from the converter will be broken into blocks by time compression prior to recording. As Section 10.2 shows, time compression allows the insertion of edit gaps, addresses and redundancy into the data stream. An interleaving process is also necessary to reorder the samples prior to recording. As explained above, the subsequent deinterleaving breaks up the effects of burst errors on replay.

The result of the processes so far is still raw data, and these will need to be channel coded (see Section 10.5) before they can be recorded on the medium. On replay a data separator reverses the channel coding to give the original raw data with the addition of some errors. Following deinterleave, the errors are reduced in size and are more readily correctable. The memory required for deinterleave may double as the timebase correction memory, so that variations in the speed of the tape are rendered undetectable. Any errors which are beyond the power of the correction system will be concealed after the odd-even shift is reversed. Following conversion in the DAC an analog output emerges.

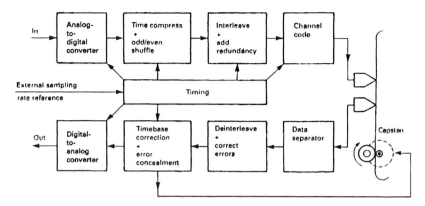

Figure 10.1 Block diagram of one channel of a stationary-head digital audio recorder. See text for details of the function of each block. Note the connection from the timebase corrector to the capstan motor so that the tape is played at such a speed that the TBC memory neither underflows nor overflows.

In a digital recorder the amplitude of the record current is constant, and recording is performed by reversing the direction of the current with respect to time. As the track passes the head, this is converted to the reversal of the magnetic field left on the tape with respect to distance. The magnetic recording is therefore bipolar.[1]

The record current is selected to be as large as possible to obtain the best SNR without resulting in transition spreading which occurs as saturation is approached. Note that in a digital machine the SNR of the replay signal is *not* the SNR of the audio. In practice the best value for the record current may be that which minimizes the error rate. As no attempt is made to record a varying signal, bias is quite unnecessary and is not used.

Figure 10.2 shows what happens when a conventional inductive head, i.e. one having a normal winding, is used to replay the bipolar track made by reversing the record current. The head output is proportional to the rate of change of flux and so only occurs at flux reversals. In other words, the replay head differentiates the flux on the track.

The amplitude of the replay signal is of no consequence and often an automatic gain control (AGC) system is used. What matters is the time at which the write current, and hence the flux stored on the medium, reversed. This can be determined by locating the peaks of the replay impulses, which can conveniently be done by differentiating the signal and looking for zero crossings. Figure 10.3 shows that this results in noise between the peaks. This problem is overcome by the gated peak detector, where only zero crossings from a pulse which exceeds the threshold will be counted. The AGC system allows the thresholds to be fixed. As an alternative, the record waveform can also be restored by integration, which opposes the differentiation of the head as in Figure 10.4.[2]

A more recent development is the magneto-resistive (M-R) head. This is a head which measures the flux on the tape rather than using it to generate a signal directly. Flux measurement works down to DC and so offers advantages at low tape speeds. Recorders which have low head-to-medium speed, such as

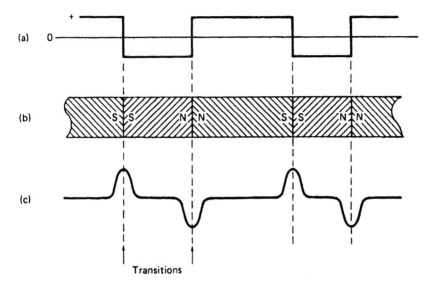

Figure 10.2 Basic digital recording. At (a) the write current in the head is reversed from time to time, leaving a binary magnetization pattern shown at (b). When replayed, the waveform at (c) results because an output is only produced when flux in the head changes. Changes are referred to as transitions.

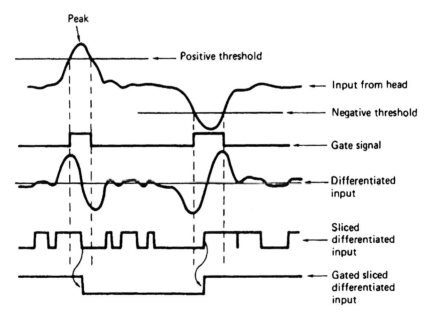

Figure 10.3 Gated peak detection rejects noise by disabling the differentiated output between transitions.

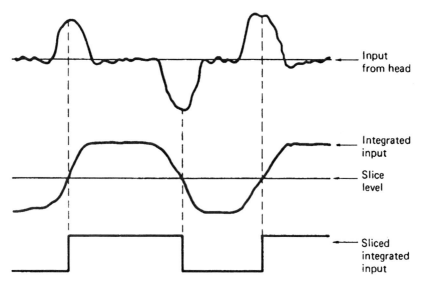

Figure 10.4 Integration method for recreating write-current waveform.

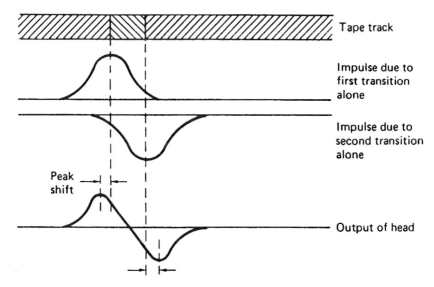

Figure 10.5 Readout pulses from two closely recorded transitions are summed in the head and the effect is that the peaks of the waveform are moved outwards. This is known as peak-shift distortion and equalization is necessary to reduce the effect.

DCC (Digital Compact Cassette) tend to use M-R heads, whereas recorders with high speeds, such as digital VTRs, DAT and magnetic disk drives tend to use inductive heads.

Digital recorders are sold into a highly competitive market and must operate at high density in order to be commercially viable. As a result the shortest possible wavelengths will be used. Figure 10.5 shows that when two flux changes, or transitions, are recorded close together, they affect each other on replay. The amplitude of the composite signal is reduced, and the position of the peaks is pushed outwards. This is known as inter-symbol interference, or peak-shift distortion and it occurs in all magnetic media. The effect is primarily due to high-frequency loss and it can be reduced by equalization on replay.

In digital recording the goal is not to reproduce the recorded waveform, but to reproduce the data it represents. Small irregularities are of no consequence, especially as an error-correction system will be fitted. Consequently digital recorders have fewer adjustments. Azimuth adjustment is generally only necessary if a head is replaced, sometimes not even then. There is no bias to adjust and some machines automatically adjust the replay equalization to optimize the error rate. Digital machines require much less maintenance and if the tape is in a cassette, much less operator skill.

10.2 Time compression

When samples are converted, the ADC must run at a constant clock rate and it outputs an unbroken stream of samples. Time compression allows the sample stream to be broken into blocks for convenient handling.

Figure 10.6 shows an ADC feeding a pair of RAMs. When one is being written by the ADC, the other can be read, and vice versa. As soon as the first RAM is full, the ADC output switches to the input of the other RAM so that there is no loss of samples. The first RAM can then be read at a higher clock rate than the sampling rate. As a result the RAM is read in less time than it took to write it, and the output from the system then pauses until the second RAM is full. The samples are now time compressed. Instead of being an unbroken stream which is difficult to handle, the samples are now arranged in blocks with convenient pauses in between them. In these pauses numerous processes can take place. A rotary head recorder might switch heads; a hard disk might move to another track. On a tape recording, the time compression of the audio samples allows time for synchronizing patterns, subcode and error-correction words to be recorded. In digital video recorders, time compression allows the continuous audio samples to be placed in blocks recorded at the same bit rate as the video data.

Subsequently, any time compression can be reversed by time expansion. Samples are written into a RAM at the incoming clock rate, but read out at the standard sampling rate. Unless there is a design fault, time compression is totally inaudible. In a recorder, the time expansion stage can be combined with the timebase correction stage so that speed variations in the medium can be eliminated at the same time. The use of time compression is universal in digital audio recording. In general the *instantaneous* data rate at the medium is not the same as the rate at the converters, although clearly the *average* rate must be the same.

Another application of time compression is to allow more than one channel of audio to be carried on a single cable. In the AES/EBU interface described in Chapter 8, for example, audio samples are time-compressed by a factor of two, so it is possible to carry samples from a stereo source in one cable.

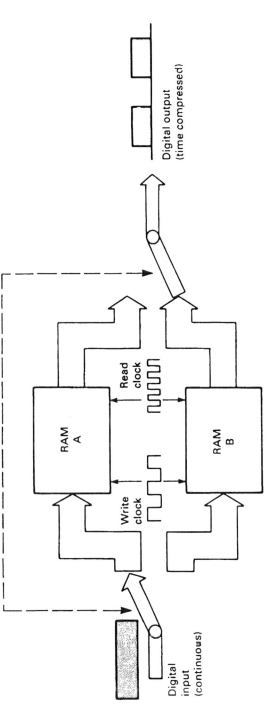

Figure 10.6 In time compression, the unbroken real-time stream of samples from an ADC is broken up into discrete blocks. This is accomplished by the configuration shown here. Samples are written into one RAM at the sampling rate by the write clock. When the first RAM is full, the switches change over, and writing continues into the second RAM whilst the first is read using a higher frequency clock. The RAM is read faster than it was written and so all the data will be output before the other RAM is full. This opens spaces in the data flow which are used as described in the text.

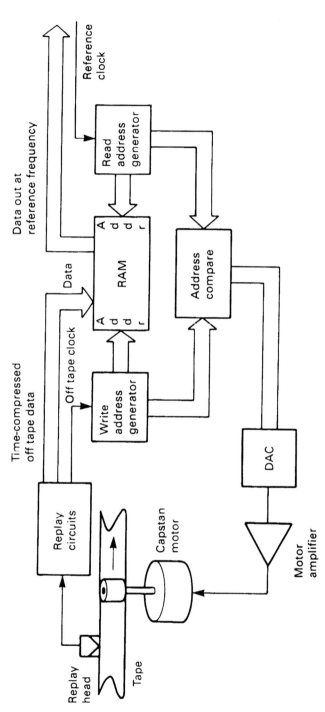

Figure 10.7 In a recorder using time compression, the samples can be returned to a continuous stream using RAM as a timebase corrector (TBC). The long-term data rate has to be the same on the input and output of the TBC or it will lose data. This is accomplished by comparing the read and write addresses and using the difference to control the tape speed. In this way the tape speed will automatically adjust to provide data as fast as the reference clock takes it from the TBC

10.3 Replay synchronization

Transfer of samples between digital audio devices in real time is only possible if both use a common sampling rate and they are synchronized. A digital audio recorder must be able to synchronize to the sampling rate of a digital input in order to record the samples. It is frequently necessary for such a recorder to be able to play back locked to an external sampling rate reference so that it can be connected to, for example, a digital mixer. The process is already common in video systems but now extends to digital audio.

Figure 10.7 shows how the external reference locking process works. The timebase expansion is controlled by the external reference which becomes the read clock for the RAM and so determines the rate at which the RAM address changes. In the case of a digital tape deck, the write clock for the RAM would be proportional to the tape speed. If the tape is going too fast, the write address will catch up with the read address in the memory, whereas if the tape is going too slow the read address will catch up with the write address. The tape speed is controlled by subtracting the read address from the write address. The address difference is used to control the tape speed. Thus if the tape speed is too high, the memory will fill faster than it is being emptied, and the address difference will grow larger than normal. This slows down the tape.

Thus in a digital recorder the speed of the medium is constantly changing to keep the data rate correct. Clearly this is inaudible as properly engineered timebase correction totally isolates any instabilities on the medium from the data fed to the converter.

In multitrack recorders, the various tracks can be synchronized to sample accuracy so that no timing errors can exist between the tracks. Extra transports can be slaved to the first to the same degree of accuracy if more tracks are required. In stereo recorders, image shift due to phase errors is eliminated.

In order to replay without a reference, perhaps to provide an analog output, a digital recorder generates a sampling clock locally by means of a crystal oscillator. Provision will be made on professional machines to switch between internal and external references.

10.4 Practical digital recorders

As the process of timebase correction can be used to eliminate irregularities in data flow, digital recorders do not need to record continuously as analog recorders do and so there is much more design freedom. Recorders can be made which have discontinuous tracks or tracks which are subdivided into blocks. Rotary-head recorders and disk drives have many advantages over the stationary head approach of the traditional analog tape recorder.

In a rotary-head recorder, the heads are mounted in a revolving drum and the tape is wrapped around the surface of the drum in a helix as can be seen in Figure 10.8. The helical tape path results in the heads traversing the tape in a series of diagonal or slanting tracks. The space between the tracks is controlled not by head design but by the speed of the tape and in modern recorders this space is reduced to zero with corresponding improvement in packing density.

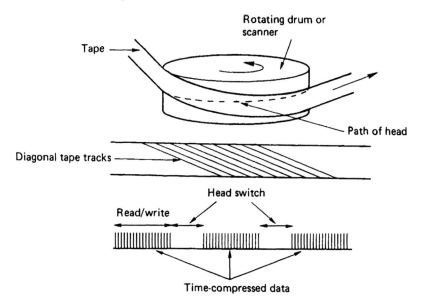

Figure 10.8 In a rotary-head recorder, the helical tape path around a rotating head results in a series of diagonal or slanting tracks across the tape. Time compression is used to create gaps in the recorded data which coincide with the switching between tracks.

The added complexity of the rotating heads and the circuitry necessary to control them is offset by the improvement in density. The discontinuous tracks of the rotary-head recorder are naturally compatible with time-compressed data. As Figure 10.8 illustrates, the audio samples are time-compressed into blocks each of which can be contained in one slant track.

Rotary-head recording is naturally complemented by azimuth recording. Figure 10.9(a) shows that in azimuth recording, the transitions are laid down at an angle to the track by using a head which is tilted. Machines using azimuth recording must always have an even number of heads, so that adjacent tracks can be recorded with opposite azimuth angle. The two track types are usually referred to as A and B. Figure 10.9(b) shows the effect of playing a track with the wrong type of head. The playback process suffers from an enormous azimuth error. The effect of azimuth error can be understood by imagining the tape track to be made from many identical parallel strips. In the presence of azimuth error, the strips at one edge of the track are played back with a phase shift relative to strips at the other side. At some wavelengths, the phase shift will be 180°, and there will be no output; at other wavelengths, especially long wavelengths, some output will reappear. The effect is rather like that of a comb filter, and serves to attenuate crosstalk due to adjacent tracks so that no guard bands are required. Since no tape is wasted between the tracks, more efficient use is made of the tape. The term guard-band-less recording is often used instead of, or in addition to, the term azimuth recording. The failure of the azimuth effect at long wavelengths is a characteristic of azimuth recording, and it is necessary to ensure that the spectrum of the signal to be recorded has a small low-frequency content. The signal will need to pass

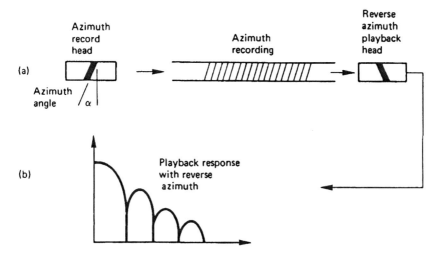

Figure 10.9 In azimuth recording (a), the head gap is titled. If the track is played with the same head, playback is normal, but the response of the reverse azimuth head is attenuated (b).

through a rotary transformer to reach the heads, and cannot therefore contain a DC component.

In recorders such as DAT[3] there is no separate erase process, and erasure is achieved by overwriting with a new waveform. Overwriting is only successful when there are no long wavelengths in the earlier recording, since these penetrate deeper into the tape, and the short wavelengths in a new recording will not be able to erase them. In this case the ratio between the shortest and longest wavelengths recorded on tape should be limited. Restricting the spectrum of the signal to allow erasure by overwrite also eases the design of the rotary transformer.

Azimuth recording requires a different approach from guard-band recording as shown in Figure 10.10. In the first method, there is no separate erase head, and in order to guarantee that no previous recording can survive a new recording, the recorded tracks can be made rather narrower than the head pole simply by reducing the linear speed of the tape so that it does not advance so far between sweeps of the rotary heads. This is shown in Figure 10.10(b). In DAT, for example, the head pole is 20.4 µm wide, but the tracks it records are only 13.59 µm wide. Alternatively, the record head can be the same width as the track pitch (c) and cannot guarantee complete overwrite. A separate erase head will be necessary. The advantage of this approach is that insert editing does not leave a seriously narrowed track.

As azimuth recording rejects crosstalk, it is advantageous if the replay head is some 50% wider than the tracks. It can be seen from Figure 10.11 that there will be crosstalk from tracks at both sides of the home track, but this crosstalk is attenuated by azimuth effect. The amount by which the head overlaps the adjacent track determines the spectrum of the crosstalk, since it changes the delay in the azimuth comb-filtering effect. More importantly, the signal-to-crosstalk ratio becomes independent of tracking error over a small range, because as the head moves to one side, the loss of crosstalk from one

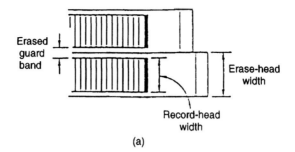

Erased guard band

Erase-head width

Record-head width

(a)

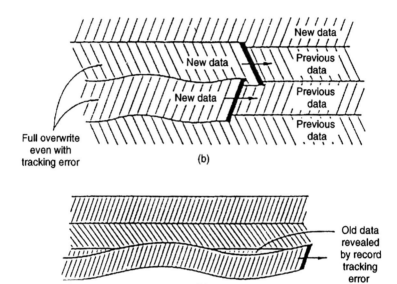

New data

New data

Full overwrite even with tracking error

New data

Previous data

Previous data

Previous data

(b)

Old data revealed by record tracking error

(c)

Figure 10.10 (a) The erase head is wider than the record head in guard-band recording. In (b) if the azimuth record head is wider than the track, full overwrite is obtained even with misalignment. In (c) if the azimuth record head is the same width as the track, misalignment results in failure to overwrite and an erase head becomes necessary.

adjacent track is balanced by the increase of crosstalk from the track on the opposite side. This phenomenon allows for some loss of track straightness and for the residual error which is present in all track-following servo systems.

The hard-disk recorder stores data on concentric tracks which it accesses by moving the head radially. Clearly while the head is moving it cannot transfer data. Using time compression, a hard-disk drive can be made into an audio recorder with the addition of a certain amount of memory.

Figure 10.12 shows the principle. The instantaneous data rate of the disk drive is far in excess of the sampling rate at the converter, and so a large time compression factor can be used. The disk drive can read a block of data from disk, and place it in the timebase corrector in a fraction of the real time it represents in the audio waveform. As the timebase corrector steadily advances

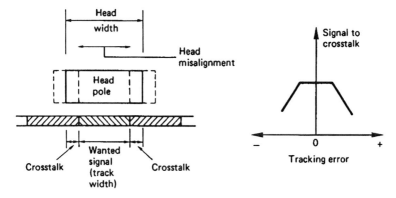

Figure 10.11 When the head pole is wider than the track, the wanted signal is picked up along with crosstalk from the adjacent tracks. If the head is misaligned, the signal-to-crosstalk ratio remains the same until the head fails to register with the whole of the wanted track.

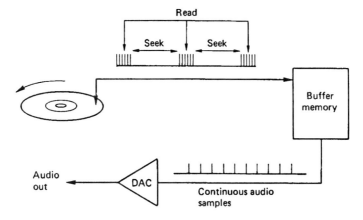

Figure 10.12 In a hard-disk recorder, a large-capacity memory is used as a buffer or timebase corrector between the converters and the disk. The memory allows the converters to run constantly despite the interruptions in disk transfer caused by the head moving between tracks.

through the memory, the disk drive has time to move the heads to another track before the memory runs out of data. When there is sufficient space in the memory for another block, the drive is commanded to read, and fills up the space. Although the data transfer at the medium is highly discontinuous, the buffer memory provides an unbroken stream of samples to the DAC and so continuous audio is obtained.

Recording is performed by using the memory to assemble samples until the contents of one disk block are available. This is then transferred to disk at high data rate. The drive can then reposition the head before the next block is available in memory. An advantage of hard disks is that access to the audio is much quicker than with tape, as all of the data are available within the time taken to move the head.

Disk drives permanently sacrifice storage density in order to offer rapid access. The use of a flying head with a deliberate air gap between it and the medium is necessary because of the high medium speed, but this causes a severe separation loss which restricts the linear density available. The air gap must be accurately maintained, and consequently the head is of low mass and is mounted flexibly.

Figure 10.13 shows that the aerohydrodynamic part of the head is known as the slipper; it is designed to provide lift from the boundary layer which changes rapidly with changes in flying height. It is not initially obvious that the difficulty with disk heads is not making them fly, but making them fly close enough to the disk surface. The boundary layer travelling at the disk surface has the same speed as the disk, but as height increases, it slows down due to drag from the surrounding air. As the lift is a function of relative air speed, the closer the slipper comes to the disk, the greater the lift will be. The slipper is therefore mounted at the end of a rigid cantilever sprung towards the medium. The force with which the head is pressed towards the disk by the spring is equal to the lift at the designed flying height. Because of the spring, the head may rise and fall over small warps in the disk. It would be virtually impossible to manufacture disks flat enough to dispense with this feature. As the slipper negotiates a warp it will pitch and roll in addition to rising and

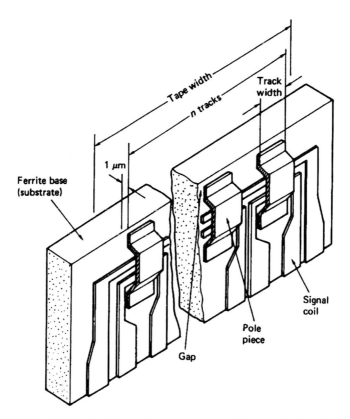

Figure 10.13 The thin-film head shown here can be produced photographically with very small dimensions. Flat structure reduces crosstalk and allows a finer track pitch to be used.

falling, but it must be prevented from yawing, as this would cause an azimuth error. Downthrust is applied to the aerodynamic centre by a spherical thrust button, and the required degrees of freedom are supplied by a thin flexible gimbal.

In a moving-head device it is not practicable to position separate erase, record and playback heads accurately. Erase is by overwriting, and reading and writing are often done by the same head. An exception is where M-R read heads are used. As these cannot write, two separate heads must be closely mounted in the same slipper.

10.5 Channel coding

In most recorders used for storing digital information, the medium carries a track which reproduces a single waveform. Clearly data words representing audio samples contain many bits and so they have to be recorded serially, a bit at a time. Recording data serially is not as simple as connecting the serial output of a shift register to the head. In digital audio, a common sample value is all zeros, as this corresponds to silence. If a shift register is loaded with all zeros and shifted out serially, the output stays at a constant low level, and nothing is recorded on the track. On replay there is nothing to indicate how many zeros were present, or even how fast to move the medium. Clearly serialized raw data cannot be recorded directly, it has to be modulated into a waveform which contains an embedded clock irrespective of the values of the bits in the samples. On replay a circuit called a data separator can lock to the embedded clock and use it to separate strings of identical bits.

The process of modulating serial data to make it self-clocking is called channel coding.[4] Channel coding also shapes the spectrum of the serialized waveform to make it more efficient. With a good channel code, more data can be stored on a given medium. Spectrum shaping is used in DVTRs to produce DC-free signals which will pass through rotary transformers to reach the revolving heads. It is also used in DAT to allow re-recording without erase heads. Channel coding is also needed in digital broadcasting where shaping of the spectrum is an obvious requirement to avoid interference with other services.

The FM code, also known as Manchester code or bi-phase mark code, shown in Figure 10.14 was the first practical self-clocking binary code and it is suitable for both transmission and recording. It is DC free and very easy to encode and decode. It is the code specified for the AES/EBU digital audio interconnect standard described in Chapter 8. In the field of recording it remains in use today only where density is not of prime importance, for example in SMPTE/EBU timecode for professional audio and video recorders and in floppy disks.

In FM there is always a transition at the bit-cell boundary which acts as a clock. For a data 1, there is an additional transition at the bit-cell centre. Figure 10.14 shows that each data bit can be represented by two channel bits. For a data 0, they will be 10, and for a data 1 they will be 11. Since the first bit is always 1, it conveys no information, and is responsible for the density ratio of only one-half. Since there can be two transitions for each data bit, the jitter margin can only be half a bit, and the bandwidth required is high. The high clock content of FM does, however, mean that data recovery is possible

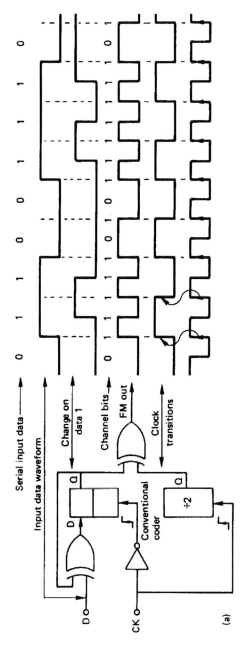

Figure 10.14 The FM waveform and the channel bits which may be used to describe transitions in it.

over a wide range of speeds; hence the use for timecode. The lowest frequency in FM is due to a stream of zeros and is equal to half the bit rate. The highest frequency is due to a stream of ones, and is equal to the bit rate. Thus the fundamentals of FM are within a band of one octave. Effective equalization is generally possible over such a band. FM is not polarity conscious and can be inverted without changing the data.

10.6 Group codes

Further improvements in coding rely on converting patterns of real data to patterns of channel bits with more desirable characteristics using a conversion table known as a codebook. If a data symbol of m bits is considered, it can have 2^m different combinations. As it is intended to discard undesirable patterns to improve the code, it follows that the number of channel bits n must be greater than m. The number of patterns which can be discarded is:

$$2^n - 2^m$$

One name for the principle is group-code recording (GCR), and an important parameter is the code rate, defined as:

$$R = \frac{m}{n}$$

The amount of jitter which can be withstood is proportional to the code rate, and so a figure near to unity is desirable. The choice of patterns which are used in the codebook will be those which give the desired balance between clock content, bandwidth and DC content.

Figure 10.15 shows that the upper spectral limit can be made to be some fraction of the channel bit rate according to the minimum distance between ones in the channel bits. This is known as the minimum transition parameter. It can be obtained by multiplying the number of channel detent periods between transitions by the code rate.

The channel code of DAT will be used here as a good example of a group code. There are many channel codes available, but few of them are suitable for azimuth recording because of the large amount of crosstalk. The crosstalk cancellation of azimuth recording fails at low frequencies, so a suitable channel code must not only be free of DC, but it must suppress low

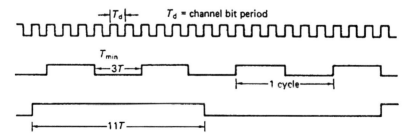

Figure 10.15 A channel code can control its spectrum by placing limits on T_{min} and T_{max} which define upper and lower frequencies. The ratio of T_{max}/T_{min} determines the asymmetry of waveform and predicts DC content and peak shift. Example shown is EFM.

Eight-bit data word	Ten-bit codeword	DSV	Alternative codeword	DSV
00010000	1101010010	0		
00010001	0100010010	2	1100010010	−2
00010010	0101010010	0		
00010011	0101110010	0		
00010100	1101110001	2	0101110001	−2
00010101	1101110011	2	0101110011	−2
00010110	1101110110	2	0101110110	−2
00010111	1101110010	0		

Figure 10.16 Some of the 8/10 codebook for non-zero DSV symbols (two entries) and zero DSV symbols (one entry).

frequencies as well. A further issue is that erasure is by overwriting, and as the heads are optimized for short-wavelength working, best erasure will be when the ratio between the longest and shortest wavelengths in the recording is small.

In Figure 10.16, some examples from the 8/10 group code of DAT are shown.[5] Clearly a channel waveform which spends as much time high as low has no net DC content, and so all 10-bit patterns which meet this criterion of zero disparity can be found. For every bit the channel spends high, the DC content will increase; for every bit the channel spends low, the DC content will decrease. As adjacent channel ones are permitted, the jitter margin will be 0.8 bits. Unfortunately there are not enough DC-free combinations in 10 channel bits to provide the 256 patterns necessary to record eight data bits. A further constraint is that it is desirable to restrict the maximum run length to improve overwrite capability and reduce peak shift. In the 8/10 code of DAT, no more than three channel zeros are permitted between channel ones, which makes the longest wavelength only four times the shortest. There are only 153 10-bit patterns which are within this maximum run length and which have a DSV of zero.

The remaining 103 data combinations are recorded using channel patterns that have non-zero DC content. Two channel patterns are allocated to each of the 103 data patterns. One of these has a DC value of +2, the other has a value of −2. For simplicity, the only difference between them is that the first channel bit is inverted. The choice of which channel bit pattern to use is based on the DC due to the previous code.

For example, if several bytes have been recorded with some of the 153 DC-free patterns, the DC content of the code will be zero. The first data byte is then found which has no zero disparity pattern. If the +2 pattern is used, the code at the end of the pattern will also become +2. When the next pattern of this kind is found, the code having the value of −2 will automatically be selected to return the channel DC content to zero. In this way the code is kept DC free, but the maximum distance between transitions can be shortened. A code of this kind is known as a low-disparity code. Decoding is simpler, because there is a direct relationship between 10-bit codes and 8-bit data.

10.7 Error correction and concealment

In a recording of binary data, a bit is either correct or wrong, with no inter-mediate stage. Small amounts of noise are rejected, but inevitably, infrequent noise impulses cause some individual bits to be in error. Dropouts cause a larger number of bits in one place to be in error. An error of this kind is called a burst error. Whatever the medium and whatever the nature of the mechanism responsible, data are either recovered correctly, or suffer some combination of bit errors and burst errors. In DAT, random errors can be caused by noise, whereas burst errors are due to contamination or scratching of the tape surface.

The audibility of a bit error depends upon which bit of the sample is involved. If the LSB of one sample was in error in a loud passage of music, the effect would be totally masked and no one could detect it. Conversely, if the MSB of one sample was in error in a quiet passage, no one could fail to notice the resulting loud transient. Clearly a means is needed to render errors from the medium inaudible. This is the purpose of error correction.

In binary, a bit has only two states. If it is wrong, it is only necessary to reverse the state and it must be right. Thus the correction process is trivial and perfect. The main difficulty is in identifying the bits which are in error. This is done by coding the data by adding redundant bits. Adding redundancy is not confined to digital technology; airliners have several engines and cars have twin braking systems. Clearly the more failures which have to be handled, the more redundancy is needed. If a four-engined airliner is designed to fly normally with one engine failed, three of the engines have enough power to reach cruise speed, and the fourth one is redundant. The amount of redundancy is equal to the amount of failure which can be handled. In the case of the failure of two engines, the plane can still fly, but it must slow down; this is graceful degradation. Clearly the chances of a two-engine failure on the same flight are remote.

In digital audio, the amount of error which can be corrected is proportional to the amount of redundancy, and within this limit, the samples are returned to exactly their original value. Consequently *corrected* samples are inaudible. If the amount of error exceeds the amount of redundancy, correction is not possible, and, in order to allow graceful degradation, concealment will be used. Concealment is a process where the value of a missing sample is estimated from those nearby. The estimated sample value is not necessarily exactly the same as the original, and so under some circumstances concealment can be audible, especially if it is frequent. However, in a well-designed system, concealments occur with negligible frequency unless there is an actual fault or problem.

Concealment is made possible by rearranging or shuffling the sample sequence prior to recording. This is shown in Figure 10.17 where odd-numbered samples are separated from even-numbered samples prior to recording. The odd and even sets of samples may be recorded in different places, so that an uncorrectable burst error only affects one set. On replay, the samples are recombined into their natural sequence, and the error is now split up so that it results in every other sample being lost. The waveform is now described half as often, but can still be reproduced with some loss of accuracy. This is better than not being reproduced at all even if it is not perfect. Almost

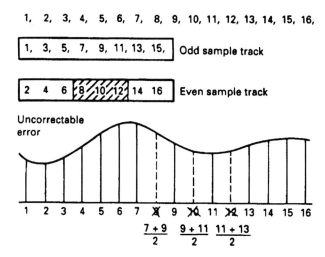

Figure 10.17 In cases where the error correction is inadequate, concealment can be used provided that the samples have been ordered appropriately in the recording. Odd and even samples are recorded in different places as shown here. As a result an uncorrectable error causes incorrect samples to occur singly, between correct samples. In the example shown, sample 8 is incorrect, but samples 7 and 9 are unaffected and an approximation to the value of sample 8 can be had by taking the average value of the two. This interpolated value is substituted for the incorrect value.

all digital recorders use such an odd/even shuffle for concealment. Clearly if any errors are fully correctable, the shuffle is a waste of time; it is only needed if correction is not possible.

In high-density recorders, more data are lost in a given sized dropout. Adding redundancy equal to the size of a dropout to every code is inefficient. Figure 10.18(a) shows that the efficiency of the system can be raised using interleaving. Sequential samples from the ADC are assembled into codes, but these are not recorded in their natural sequence. A number of sequential codes are assembled along rows in a memory. When the memory is full, it is copied

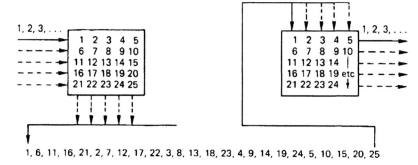

Figure 10.18(a) In interleaving, samples are recorded out of their normal sequence by taking columns from a memory which was filled in rows. On replay the process must be reversed. This puts the samples back in their regular sequence, but breaks up burst errors into many smaller errors which are more efficiently corrected. Interleaving and deinterleaving cause delay.

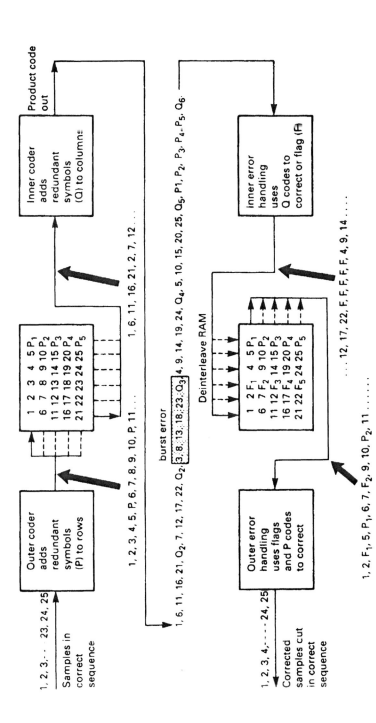

Figure 10.18(b) In addition to the redundancy P on rows, inner redundancy Q is also generated on columns. On replay, the Q code checker will pass on flags F if it finds an error large to handle itself. The flags pass through the deinterleave process and are used by the outer error correction to identify which symbol in the row needs correcting with P redundancy. The concept of crossing two codes in this way is called a product code.

to the medium by reading down columns. On replay, the samples need to be deinterleaved to return them to their natural sequence. This is done by writing samples from tape into a memory in columns, and when it is full, the memory is read in rows. Samples read from the memory are now in their original sequence so there is no effect on the recording. However, if a burst error occurs on the medium, it will damage sequential samples in a vertical direction in the deinterleave memory. When the memory is read, a single large error is broken down into a number of small errors whose size is exactly equal to the correcting power of the codes and the correction is performed with maximum efficiency.

An extension of the process of interleave is where the memory array has not only rows made into codewords, but also columns made into codewords by the addition of vertical redundancy. This is known as a product code.[6] Figure 10.18(b) shows that in a product code the redundancy calculated first and checked last is called the outer code, and the redundancy calculated second and checked first is called the inner code. The inner code is formed along tracks on the medium. Random errors due to noise are corrected by the inner code and do not impair the burst-correcting power of the outer code. Burst errors are declared uncorrectable by the inner code which flags the bad samples on the way into the deinterleave memory. The outer code reads the error flags in order to locate the erroneous data. As it does not have to compute the error locations, the outer code can correct more errors.

The interleave, deinterleave, time compression and timebase correction processes cause delay and this is evident in the time taken before audio emerges after starting a digital machine. Confidence replay takes place later than the distance between record and replay heads would indicate. In DASH format recorders, confidence replay is about one-tenth of a second behind the input. Synchronous recording requires new techniques to overcome the effect of the delays.

The presence of an error-correction system means that the audio quality is independent of the tape/head quality within limits. There is no point in trying to assess the health of a machine by listening to it, as this will not reveal whether the error rate is normal or within a whisker of failure. The only useful procedure is to monitor the frequency with which errors are being corrected, and to compare it with normal figures. Professional digital audio equipment should have an error rate display.

10.8 Introduction to the Reed–Solomon codes

The Reed–Solomon (R–S) codes (Irving Reed and Gustave Solomon) are inherently burst correcting[7] because they work on multi-bit symbols rather than individual bits. The R–S codes are also extremely flexible in use. One code may be used both to detect and correct errors and the number of bursts which are correctable can be chosen at the design stage by the amount of redundancy. A further advantage of the R–S codes is that they can be used in conjunction with a separate error-detection mechanism in which case they perform only the correction by erasure. R–S codes operate at the theoretical limit of correcting efficiency. In other words, no more efficient code can be found and consequently the R–S codes are universally used in modern digital products.

In the simple cyclic codes used on floppy disks, the codeword is created by adding a redundant symbol to the data which is calculated in such a way that the codeword can be divided by a polynomial with no remainder. In the event that a remainder is obtained, there has been an error, but there is not enough information to correct it. In the Reed–Solomon codes, the codeword will divide by a number of polynomials. Clearly if the codeword must divide by, say, two polynomials, it must have two redundant symbols. This is the minimum case of an R–S code. On receiving an R–S coded message there will be two remainders or *syndromes* following the division. In the error-free case, these will both be zero. If either are not zero, there is an error.

The effect of an error is to add an error polynomial to the message polynomial. The number of terms in the error polynomial is the same as the number of errors in the codeword. The codeword divides to zero and the syndromes are a function of the error only. There are two syndromes and two equations. By solving these simultaneous equations it is possible to obtain two unknowns. One of these is the position of the error, known as the *locator*, and the other is the error bit pattern, known as the *corrector*. As the locator is the same size as the code symbol, the length of the codeword is determined by the size of the symbol. A symbol size of eight bits is commonly used because it fits in conveniently with both 16-bit audio samples and byte-oriented computers. An eight-bit syndrome results in a locator of the same wordlength. Eight bits have 2^8 combinations, but one of these is the error-free condition, and so the locator can specify one of only 255 symbols. As each symbol contains eight bits, the codeword will be $255 \times 8 = 2040$ bits long.

As further examples, five-bit symbols could be used to form a codeword 31 symbols long, and three-bit symbols would form a codeword seven symbols long. This latter size is small enough to permit some worked examples, and will be used further here. Figure 10.19 shows that in the seven-symbol codeword, five symbols of three bits each, A–E, are the data, and P and Q are the two redundant symbols. This simple example will locate and correct a single symbol in error. It does not matter, however, how many bits in the symbol are in error.

The two check symbols are solutions to the following equations:

$$A \oplus B \oplus C \oplus D \oplus E \oplus P \oplus Q = 0$$

$$a^7 A \oplus a^6 B \oplus a^5 C \oplus a^4 D \oplus a^3 E \oplus a^2 P \oplus aQ = 0$$

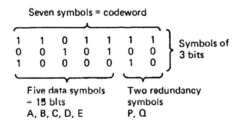

Seven symbols = codeword

Five data symbols Two redundancy
= 15 bits   symbols
A, B, C, D, E   P, Q

Symbols of 3 bits

Figure 10.19 A Reed–Solomon codeword. As the symbols are of three bits, there can only be eight possible syndrome values. One of these is all zeros, the error-free case, and so it is only possible to point to seven errors; hence the codeword length of seven symbols. Two of these are redundant, leaving five data symbols.

where a is a constant. The original data A–E followed by the redundancy P and Q pass through the channel.

The receiver makes two checks on the message to see if it is a codeword. This is done by calculating syndromes using the following expressions, where the (′) implies the received symbol which is not necessarily correct:

$$S_0 = A' \oplus B' \oplus C' \oplus D' \oplus E' \oplus P' \oplus Q'$$

(This is in fact a simple parity check.)

$$S_1 = a^7 A' \oplus a^6 B' \oplus a^5 C' \oplus a^4 D' \oplus a^3 E' \oplus a^2 P' \oplus aQ'$$

If two syndromes of all zeros are not obtained, there has been an error. The information carried in the syndromes will be used to correct the error. For the purpose of illustration, let it be considered that D′ has been corrupted before moving to the general case. D′ can be considered to be the result of adding an error of value E to the original value D such that $D' = D \oplus E$.
As

$$A \oplus B \oplus C \oplus D \oplus E \oplus P \oplus Q = 0$$

then

$$A \oplus B \oplus C \oplus (D \oplus E) \oplus E \oplus P \oplus Q = E = S_0$$

As

$$D' = D \oplus E$$

then

$$D = D' \oplus E = D' \oplus S_0$$

Thus the value of the corrector is known immediately because it is the same as the parity syndrome S_0. The corrected data symbol is obtained simply by adding S_0 to the incorrect symbol.

At this stage, however, the corrupted symbol has not yet been identified, but this is equally straightforward.
As

$$a^7 A \oplus a^6 B \oplus a^5 C \oplus a^4 D \oplus a^3 E \oplus a^2 P \oplus aQ = 0$$

Then:

$$a^7 A \oplus a^6 B \oplus a^5 C \oplus a^4 (D \oplus E) \oplus a^3 E \oplus a^2 P \oplus aQ = a^4 E = S_1$$

Thus the syndrome S_1 is the error bit pattern E, but it has been raised to a power of a which is a function of the position of the error symbol in the block. If the position of the error is in symbol k, then k is the locator value and:

$$S_0 \times a^k = S_1$$

Hence:

$$a^k = \frac{S_1}{S_0}$$

The value of k can be found by multiplying S_0 by various powers of a until the product is the same as S_1. Then the power of a necessary is equal to k. The use of the descending powers of a in the codeword calculation is now clear because the error is then multiplied by a different power of a dependent upon its position, known as the locator, because it gives the position of the error.

10.9 Modulo-n arithmetic

Conventional arithmetic which is in everyday use relates to the real world of counting actual objects, and to obtain correct answers the concepts of borrow and carry are necessary in the calculations.

There is an alternative type of arithmetic which has no borrow or carry which is known as modulo arithmetic. In modulo-n no number can exceed n. If it does, n or whole multiples of n are subtracted until it does not. Thus 25 modulo-16 is 9 and 12 modulo-5 is 2. The count shown in Figure 8.58 is from a four-bit device which overflows when it reaches 1111 because the carry-out is ignored. If a number of clock pulses, m, are applied from the zero state, the state of the counter will be given by m mod. 16. Thus modulo arithmetic is appropriate to systems in which there is a fixed wordlength and this means that the range of values the system can have is restricted by that wordlength. A number range which is restricted in this way is called a finite field.

Modulo-2 is a numbering scheme which is used frequently in digital processes. Figure 10.20 shows that in modulo-2 the conventional addition and subtraction are replaced by the $\oplus$ function such that:

$$A + B \text{ mod. } 2 = A \oplus B$$

When multi-bit values are added mod. 2, each column is computed quite independently of any other. This makes mod. 2 circuitry very fast in operation as it is not necessary to wait for the carries from lower order bits to ripple up to the high-order bits.

Modulo-2 arithmetic is not the same as conventional arithmetic and takes some getting used to. For example, adding something to itself in mod. 2 always gives the answer zero.

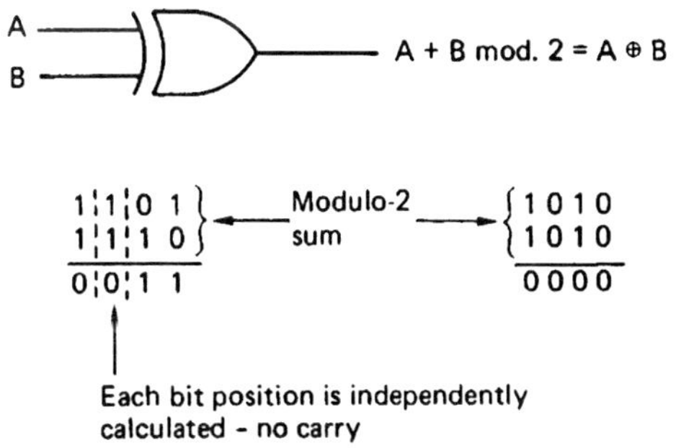

Figure 10.20 In modulo-2 calculations, there can be no carry or borrow operations and conventional addition and subtraction become identical. The XOR gate is a modulo-2 adder.

10.10 The Galois field

Figure 10.21 shows a simple circuit consisting of three D-type latches which are clocked simultaneously. They are connected in series to form a shift register. At (a) a feedback connection has been taken from the output to the input and the result is a ring counter where the bits contained will recirculate endlessly. At (b) one exclusive-OR gate is added so that the output is fed back to more than one stage. The exclusive-OR gate has the property that the output is true when the inputs are different, hence it acts as a modulo-2 adder. The configuration in Figure 10.21 is known as a twisted-ring counter and it has some interesting properties. Whenever the circuit is clocked, the left-hand bit moves to the right-hand latch, the centre bit moves to the left-hand latch and the centre latch becomes the exclusive-OR of the two outer latches. The figure shows that whatever the starting condition of the three bits in the latches, the same state will always be reached again after seven clocks, except if zero is used. The states of the latches form an endless ring of non-sequential numbers called a Galois field after the French mathematical prodigy Evariste Galois who discovered them. The states of the circuit form a maximum length sequence because there are as many states as are permitted by the wordlength. As the states of the sequence have many of the characteristics of random numbers, yet are repeatable, the result can also be called a pseudo-random sequence (prs). As the all-zeros case is disallowed, the length of a maximum length sequence generated by a register of m bits cannot exceed $(2^m - 1)$ states. The Galois field, however, includes the zero term. It is useful to explore the bizarre mathematics of Galois fields which use modulo-2 arithmetic. Familiarity with such manipulations is helpful when studying error correction, particularly

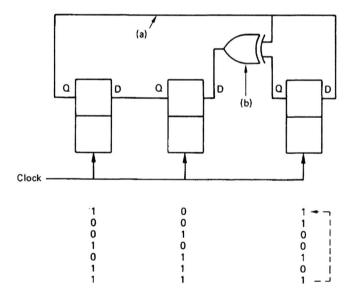

Figure 10.21 The circuit shown is a twisted-ring counter which has an unusual feedback arrangement. Clocking the counter causes it to pass through a series of non-sequential values. See text for details.

the Reed–Solomon codes. They will also be found in processes which require pseudo-random numbers such as digital dither and randomized channel codes.

The circuit of Figure 10.21 can be considered as a counter and the four points shown will then be representing different powers of 2 from the MSB on the left to the LSB on the right. The feedback connection from the MSB to the other stages means that whenever the MSB becomes 1, two other powers are also forced to one so that the code of 1011 is generated.

Each state of the circuit can be described by combinations of powers of x, such as

$$x^2 = 100$$

$$x = 010$$

$$x^2 + x = 110, \text{ etc.}$$

The fact that three bits have the same state because they are connected together is represented by the mod. 2 equation:

$$x^3 + x + 1 = 0$$

Let $x = a$, which is a primitive element. Now:

$$a^3 + a + 1 = 0 \tag{10.1}$$

In modulo-2:

$$a + a = a^2 + a^2 = 0$$

$$a = x = 010$$

$$a^2 = x^2 = 100$$

$$a^3 = a + 1 = 011 \text{ from (10.1)}$$

$$a^4 = a \times a^3 = a(a + 1) = a^2 + a = 110$$

$$a^5 = a^2 + a + 1 = 111$$

$$a^6 = a \times a^5 = a(a^2 + a + 1)$$

$$= a^3 + a^2 + a = a + 1 + a^2 + a$$

$$= a^2 + 1 = 101$$

$$a^7 = a(a^2 + 1) = a^3 + a$$

$$= a + 1 + a = 1 = 001$$

In this way it can be seen that the complete set of elements of the Galois field can be expressed by successive powers of the primitive element. Note that the twisted-ring circuit of Figure 10.21 simply raises a to higher and higher powers as it is clocked; thus the seemingly complex multi bit changes caused by a single clock of the register become simple to calculate using the correct primitive and the appropriate power.

The numbers produced by the twisted-ring counter are not random; they are completely predictable if the equation is known. However, the sequences produced are sufficiently similar to random numbers that in many cases they will be useful. They are thus referred to as pseudo-random sequences. The feedback connection is chosen such that the expression it implements will not factorize. Otherwise a maximum-length sequence could not be generated because the circuit might sequence around one or other of the factors depending on the initial condition. A useful analogy is to compare the operation of a pair of meshed gears. If the gears have a number of teeth which are relatively prime, many revolutions are necessary to make the same pair of teeth touch again. If the number of teeth have a common multiple, far fewer turns are needed.

10.11 R–S calculations

Whilst the expressions above show that the values of P and Q are such that the two syndrome expressions sum to zero, it is not yet clear how P and Q are calculated from the data. Expressions for P and Q can be found by solving the two R–S equations simultaneously. This has been done in Appendix 10.1. The following expressions must be used to calculate P and Q from the data in order to satisfy the codeword equations. These are:

$$P = a^6 A \oplus a B \oplus a^2 C \oplus a^5 D \oplus a^3 E$$

$$Q = a^2 A \oplus a^3 B \oplus a^6 C \oplus a^4 D \oplus a E$$

In both the calculation of the redundancy shown here and the calculation of the corrector and the locator it is necessary to perform numerous multiplications and raising to powers. This appears to present a formidable calculation problem at both the encoder and the decoder. This would be the case if the calculations involved were conventionally executed. However, the calculations can be simplified by using logarithms. Instead of multiplying two numbers, their logarithms are added. In order to find the cube of a number, its logarithm is added three times. Division is performed by subtracting the logarithms. Thus all of the manipulations necessary can be achieved with addition or subtraction, which is straightforward in logic circuits.

The success of this approach depends upon simple implementation of log tables. As was seen in Section 10.10, raising a constant, a, known as the *primitive element*, to successively higher powers in modulo-2 gives rise to a Galois field. Each element of the field represents a different power n of a. It is a fundamental of the R–S codes that all of the symbols used for data, redundancy and syndromes are considered to be elements of a Galois field. The number of bits in the symbol determines the size of the Galois field, and hence the number of symbols in the codeword.

Figure 10.22 shows a Galois field in which the binary values of the elements are shown alongside the power of a they represent. In the R–S codes, symbols are no longer considered simply as binary numbers, but also as equivalent powers of a. In Reed–Solomon coding and decoding, each symbol will be multiplied by some power of a. Thus if the symbol is also known as a power of a it is only necessary to add the two powers. For example, if it is necessary to multiply the data symbol 100 by a^3, the calculation proceeds as follows, referring to Figure 10.22.

$$100 = a^2 \text{ so } 100 \times a^3 = a^{(2+3)} = a^5 = 111$$

Note that the results of a Galois multiplication are quite different from binary multiplication. Because all products must be elements of the field,

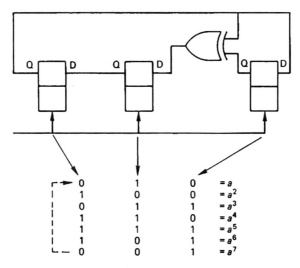

Figure 10.22 The bit patterns of a Galois field expressed as powers of the primitive element a. This diagram can be used as a form of log table in order to multiply binary numbers. Instead of an actual multiplication, the appropriate powers of a are simply added.

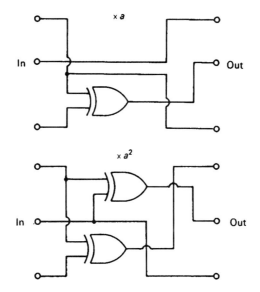

Figure 10.23 Some examples of GF multiplier circuits.

sums of powers which exceed seven wrap around by having seven subtracted. For example:

$$a^5 \times a^6 = a^{11} = a^4 = 110$$

Figure 10.23 shows some examples of circuits which will perform this kind of multiplication. Note that they require a minimum amount of logic.

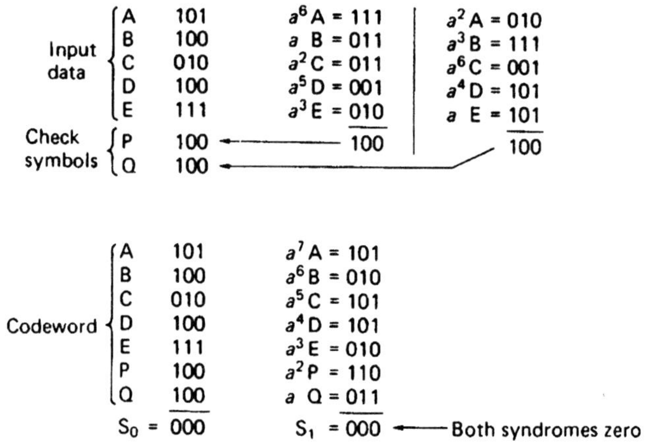

$$
\begin{array}{llll}
& \text{A} & 101 & a^6\text{A} = 111 & a^2\text{A} = 010 \\
\text{Input} & \text{B} & 100 & a\ \text{B} = 011 & a^3\text{B} = 111 \\
\text{data} & \text{C} & 010 & a^2\text{C} = 011 & a^6\text{C} = 001 \\
& \text{D} & 100 & a^5\text{D} = 001 & a^4\text{D} = 101 \\
& \text{E} & 111 & a^3\text{E} = 010 & a\ \text{E} = 101 \\
\end{array}
$$

$$
\begin{array}{ll}
\text{Check} & \text{P} \quad 100 \longleftarrow\!\!\!\longrightarrow 100 \\
\text{symbols} & \text{Q} \quad 100 \longleftarrow\!\!\!\longrightarrow \qquad\qquad 100 \\
\end{array}
$$

$$
\begin{array}{llll}
& \text{A} & 101 & a^7\text{A} = 101 \\
& \text{B} & 100 & a^6\text{B} = 010 \\
& \text{C} & 010 & a^5\text{C} = 101 \\
\text{Codeword} & \text{D} & 100 & a^4\text{D} = 101 \\
& \text{E} & 111 & a^3\text{E} = 010 \\
& \text{P} & 100 & a^2\text{P} = 110 \\
& \text{Q} & 100 & a\ \text{Q} = 011 \\
\end{array}
$$

$$S_0 = \overline{000} \qquad S_1 = \overline{000} \longleftarrow \text{Both syndromes zero}$$

Figure 10.24 Five data symbols A–E are used as terms in the generator polynomials derived in Appendix 10.1 to calculate two redundant symbols P and Q. An example is shown at the top. Below is the result of using the codeword symbols A–Q as terms in the checking polynomials. As there is no error, both syndromes are zero.

Figure 10.24 shows an example of the Reed–Solomon encoding process. The Galois field shown in Figure 10.22 has been used, having the primitive element $a = 010$. At the beginning of the calculation of P, the symbol A is multiplied by a^6. This is done by converting A to a power of a. According to Figure 10.22, $101 = a^6$ and so the product will be $a^{(6+6)} = a^{12} = a^5 = 111$. In the same way, B is multiplied by a, and so on, and the products are added modulo-2. A similar process is used to calculate Q.

Figure 10.25 shows a circuit which can calculate P or Q. The symbols A–E are presented in succession, and the circuit is clocked for each one. On the first clock, a^6A is stored in the left-hand latch. If B is now provided at the input, the second GF multiplier produces aB and this is added to the output of the first latch and when clocked will be stored in the second latch which now contains $a^6\text{A} + a\text{B}$. The process continues in this fashion until the complete expression for P is available in the right-hand latch. The intermediate contents of the right-hand latch are ignored.

The entire codeword now exists, and can be recorded or transmitted. Figure 10.24 also demonstrates that the codeword satisfies the checking equations. The modulo-2 sum of the seven symbols, S_0, is 000 because each column has an even number of ones. The calculation of S_1 requires multiplication by descending powers of a. The modulo-2 sum of the products is again zero. These calculations confirm that the redundancy calculation was properly carried out.

Figure 10.26 gives three examples of error correction based on this codeword. The erroneous symbol is marked with a dash. As there has been an error, the syndromes S_0 and S_1 will not be zero.

Figure 10.27 shows circuits suitable for parallel calculation of the two syndromes at the receiver. The S_0 circuit is a simple parity checker which accumulates the modulo-2 sum of all symbols fed to it. The S_1 circuit is more subtle, because it contains a Galois field (GF) multiplier in a feedback loop,

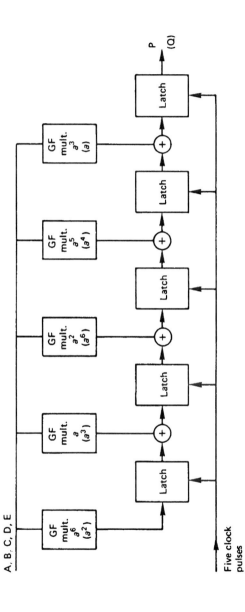

Figure 10.25 If the five data symbols of Figure 10.24 are supplied to this circuit in sequence, after five clocks, one of the check symbols will appear at the output. Terms without brackets will calculate P, bracketed terms calculate Q.

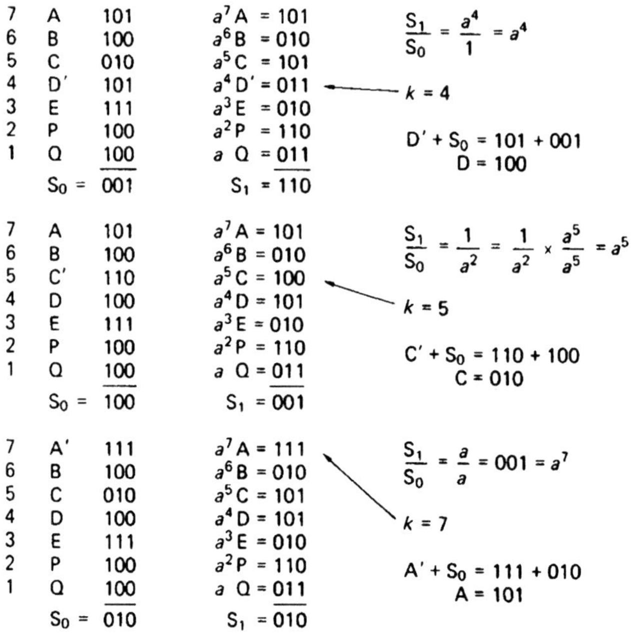

7	A	101	$a^7 A = 101$	$\dfrac{S_1}{S_0} = \dfrac{a^4}{1} = a^4$
6	B	100	$a^6 B = 010$	
5	C	010	$a^5 C = 101$	
4	D'	101	$a^4 D' = 011$ ⟵	$k = 4$
3	E	111	$a^3 E = 010$	
2	P	100	$a^2 P = 110$	$D' + S_0 = 101 + 001$
1	Q	100	$a\, Q = 011$	$D = 100$
	$S_0 = 001$		$S_1 = 110$	

7	A	101	$a^7 A = 101$	$\dfrac{S_1}{S_0} = \dfrac{1}{a^2} = \dfrac{1}{a^2} \times \dfrac{a^5}{a^5} = a^5$
6	B	100	$a^6 B = 010$	
5	C'	110	$a^5 C = 100$ ⟵	
4	D	100	$a^4 D = 101$	$k = 5$
3	E	111	$a^3 E = 010$	
2	P	100	$a^2 P = 110$	$C' + S_0 = 110 + 100$
1	Q	100	$a\, Q = 011$	$C = 010$
	$S_0 = 100$		$S_1 = 001$	

7	A'	111	$a^7 A = 111$	$\dfrac{S_1}{S_0} = \dfrac{a}{a} = 001 = a^7$
6	B	100	$a^6 B = 010$	
5	C	010	$a^5 C = 101$	
4	D	100	$a^4 D = 101$	
3	E	111	$a^3 E = 010$	$k = 7$
2	P	100	$a^2 P = 110$	$A' + S_0 = 111 + 010$
1	Q	100	$a\, Q = 011$	$A = 101$
	$S_0 = 010$		$S_1 = 010$	

Figure 10.26 Three examples of error location and correction. The number of bits in error in a symbol is irrelevant; if all three were wrong, S_0 would be 111, but correction is still possible.

such that early symbols fed in are raised to higher powers than later symbols because they have been recirculated through the GF multiplier more often. It is possible to compare the operation of these circuits with the example of Figure 10.26 and with subsequent examples to confirm that the same results are obtained.

10.12 Correction by erasure

In the examples of Figure 10.26, two redundant symbols P and Q have been used to locate and correct one error symbol. If the positions of errors are known by some separate mechanism (see product codes, Section 10.7) the locator need not be calculated. The simultaneous equations may instead be solved for two correctors. In this case the number of symbols which can be corrected is equal to the number of redundant symbols. In Figure 10.28(a) two errors have taken place, and it is known that they are in symbols C and D. Since S_0 is a simple parity check, it will reflect the modulo-2 sum of the two errors. Hence:

$$S_0 = EC \oplus ED$$

The two errors will have been multiplied by different powers in S_1, such that:

$$S_1 = a^5 EC \oplus a^4 ED$$

These two equations can be solved, as shown in the figure, to find EC and ED, and the correct value of the symbols will be obtained by adding these correctors to the erroneous values. It is, however, easier to set the values of

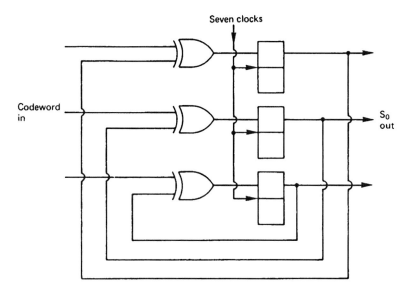

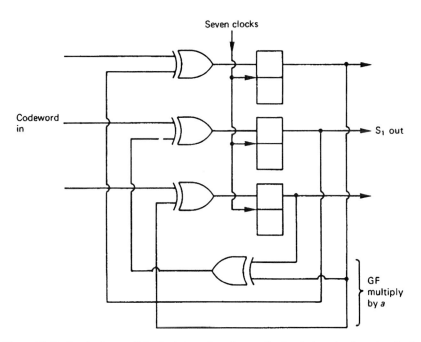

Figure 10.27 Circuits for parallel calculation of syndromes S_0, S_1. S_0 is a simple parity check. S_1 has a GF multiplication by a in the feedback, so that A is multiplied by a^7, B is multiplied by a^6, etc., and all are summed to give S_1.

$$
\begin{array}{llll}
A & 101 & a^7A = & 101 \\
B & 100 & a^6B = & 010 \\
(C \oplus E_C) & 001 & a^5(C \oplus E_C) & 111 \\
(D \oplus E_D) & 010 & a^4(D \oplus E_D) & 111 \\
E & 111 & a^3E = & 010 \\
P & 100 & a^2P = & 110 \\
Q & \underline{100} & a\ Q = & \underline{011} \\
S_1 = & 101 & S_1 = & 000
\end{array}
$$

$$S_0 = E_C \oplus E_D \qquad S_1 = a^5 E_C \oplus a^4 E_D$$

$$S_1 = a^5 E_C \oplus a^4 (S_0 \oplus E_C)$$

$$= a^5 E_C \oplus a^4 S_0 \oplus a^4 E_C$$

$$\therefore E_C = \frac{S_1 \oplus a^4 S_0}{a^5 \oplus a^4} = \frac{000 \oplus 011}{001} = 011$$

$$C = (C \oplus E_C) \oplus E_C = 001 \oplus 011 = 010$$

$$S_1 = a^5 (S_0 \oplus E_D) \oplus a^4 E_D$$

$$= a^5 S_0 \oplus a^5 E_D \oplus a^4 E_D$$

$$\therefore E_D = \frac{S_1 \oplus a^5 S_0}{a^5 \oplus a^4} = \frac{000 \oplus 110}{001} = 110$$

$$D = (D \oplus E_D) + E_D = 010 \oplus 110 = \underline{100}$$

(a)

$$
\begin{array}{lll}
A & 101 & a^7A = 101 \\
B & 100 & a^6B = 010 \quad S_0 = C \oplus D \\
C & \underline{000} & a^5C = 000 \\
D & \underline{000} & a^4D = 000 \quad S_1 = a^5C \oplus a^4D \\
E & 111 & a^3E = 010 \\
P & 100 & a^2P = 110 \\
Q & 100 & a\ Q = 011 \\
S_0 & = 100 & S_1 = 000
\end{array}
$$

$$S_1 = a^5 S_0 \oplus a^5 D \oplus a^4 D = a^5 S_0 \oplus D$$

$$\therefore D = S_1 \oplus a^5 S_0 = 000 \oplus 100 = \underline{100}$$

$$S_1 = a^5 C \oplus a^4 C \oplus a^4 S_0 = C \oplus a^4 S_0$$

$$\therefore C = S_1 \oplus a^4 S_0 = 000 \oplus 010 = \underline{010}$$

(b)

Figure 10.28 If the location of errors is known, then the syndromes are a known function of the two errors as shown in (a). It is, however, much simpler to set the incorrect symbols to zero, i.e. to *erase* them as in (b). Then the syndromes are a function of the wanted symbols and correction is easier.

the symbols in error to zero. In this way the nature of the error is rendered irrelevant and it does not enter the calculation. This setting of symbols to zero gives rise to the term erasure. In this case,

$$S_0 = C \oplus D$$
$$S_1 = a^5 C \oplus a^4 D$$

Erasing the symbols in error makes the errors equal to the correct symbol values and these are found more simply as shown in Figure 10.28(b).

Practical systems will be designed to correct more symbols in error than in the simple examples given here. If it is proposed to correct by erasure an arbitrary number of symbols in error given by t, the codeword must be divisible by t different polynomials. Alternatively if the errors must be located and corrected, $2t$ polynomials will be needed. These will be of the form $(x + a^n)$ where n takes all values up to t or $2t$. a is the primitive element discussed earlier.

Where four symbols are to be corrected by erasure, or two symbols are to be located and corrected, four redundant symbols are necessary, and the codeword polynomial must then be divisible by

$$(x + a^0)(x + a^1)(x + a^2)(x + a^3)$$

Upon receipt of the message, four syndromes must be calculated, and the four correctors or the two error patterns and their positions are determined by solving four simultaneous equations. This generally requires an iterative procedure, and a number of algorithms have been developed for the purpose.[8-10] Modern digital audio formats such as CD and DAT use 8-bit R–S codes and erasure extensively. The primitive polynomial commonly used with GF(256) is

$$x^8 + x^4 + x^3 + x^2 + 1$$

The codeword will be 255 bytes long but will often be shortened by puncturing. The larger Galois fields require less redundancy, but the computational problem increases. LSI chips have been developed specifically for R–S decoding in many high-volume formats.[10-12] As an alternative to dedicated circuitry, it is also possible to perform Reed–Solomon calculations in software using general purpose processors.[13] This may be more economical in small-volume products.

10.13 Introduction to DAT

The rotary-head DAT (digital audio tape) format was the first digital recorder for consumer use to incorporate a dedicated tape deck. By designing for a specific purpose, the tape consumption can be made very much smaller than that of a converted video machine. In fact the DAT format achieved more bits per square inch than any other form of magnetic recorder at the time of its introduction. The origins of DAT are in an experimental machine built by Sony,[3] but the DAT format has grown out of that through a process of standardization involving some 80 companies.

The general appearance of the DAT cassette is shown in Figure 10.29. The overall dimensions are only 73 mm × 54 mm × 10.5 mm which is rather smaller than the Compact Cassette. The design of the cassette incorporates some improvements over its analog ancestor.[14] As shown in Figure 10.30, the apertures through which the heads access the tape are closed by a hinged door, and the hub drive openings are covered by a sliding panel which also locks the door when the cassette is not in the transport. The act of closing the door operates brakes which act on the reel hubs. This results in a cassette which is well sealed against contamination due to handling or storage. The

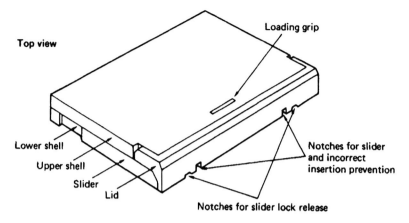

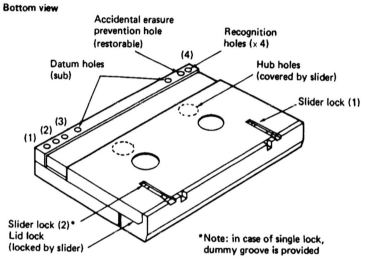

Figure 10.29 Appearance of DAT cassette. Access to the tape is via a hinged lid, and the hub-drive holes are covered by a sliding panel, affording maximum protection to the tape. Further details of the recognition holes are given in Table 10.1. (Courtesy TDK.).

short wavelengths used in digital recording make it more sensitive to spacing loss caused by contamination. As in the Compact Cassette, the tape hubs are flangeless, and the edge guidance of the tape pack is achieved by liner sheets. The flangeless approach allows the hub centres to be closer together for a given length of tape. The cassette has recognition holes in four standard places so that players can automatically determine what type of cassette has been inserted. In addition there is a write-protect (record-lockout) mechanism which is actuated by a small plastic plug sliding between the cassette halves. The end-of-tape (EOT) condition is detected optically and the leader tape is transparent. There is some freedom in the design of the EOT sensor. As can be seen in Figure 10.31, transmitted-light sensing can be used across the corner of the cassette, or reflected-light sensing can be used, because the cassette

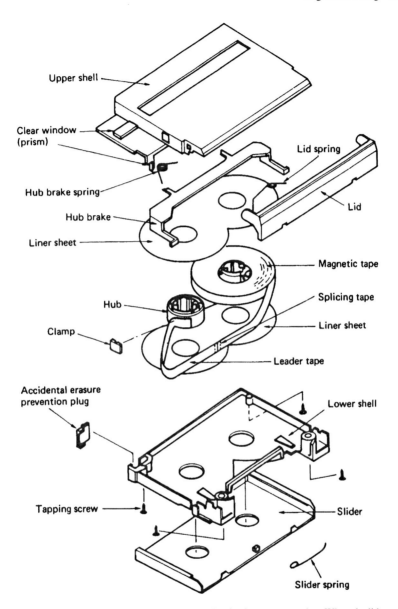

Upper shell

Clear window (prism)

Hub brake spring

Hub brake

Liner sheet

Lid spring

Lid

Magnetic tape

Splicing tape

Hub

Liner sheet

Clamp

Leader tape

Accidental erasure prevention plug

Lower shell

Tapping screw

Slider

Slider spring

Figure 10.30 Exploded view of DAT cassette showing intricate construction. When the lid opens, it pulls the ears on the brake plate, releasing the hubs. Note the EOT/BOT sensor prism moulded into the corners of the clear window. (Courtesy TDK.).

incorporates a prism which reflects light around the back of the tape. Study of Figure 10.31 will reveal that the prisms are moulded integrally with the corners of the transparent insert used for the cassette window. The high coercivity (typically 1480 Oersteds) metal powder tape is 3.81 mm wide, the same width as Compact Cassette tape. The standard overall thickness is 13 μm.

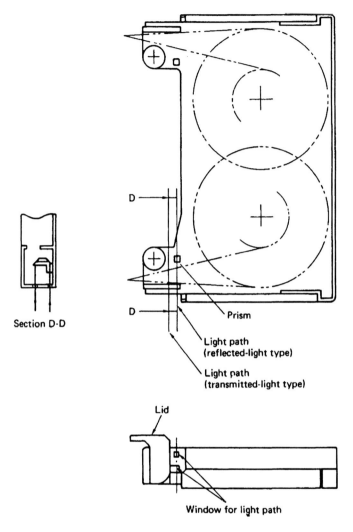

Section D-D

Prism

Light path
(reflected-light type)

Light path
(transmitted-light type)

Lid

Window for light path

Figure 10.31 Tape sensing can be either by transmission across the corner of the cassette, or by reflection through an integral prism. In both cases, the apertures are sealed when the lid closes. (Courtesy TDK.).

When the cassette is placed in the transport, the slider is moved back as it engages. This releases the lid lock. Continued movement into the transport pushes the slider right back, revealing the hub openings. The cassette is then lowered onto the hub-drive spindles and tape guides, and the door is fully opened to allow access to the tape.

As Figure 10.32 shows, DAT extends the technique of time compression used to squeeze continuous samples into an intermittent recording which is interrupted by long pauses. During the pauses in recording, it is not actually necessary for the head to be in contact with the tape, and so the angle of

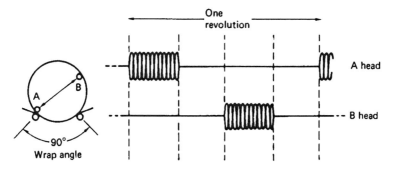

Figure 10.32 The use of time compression reduces the wrap angle necessary, at the expense of raising the frequencies in the channel.

wrap of the tape around the drum can be reduced, which makes threading easier. In DAT the wrap angle is only 90° on the commonest drum size. As the heads are 180° apart, this means that for half the time neither head is in contact with the tape. Figure 10.33 shows that the partial-wrap concept allows the threading mechanism to be very simple indeed. As the cassette is lowered into the transport, the pinch roller and several guide pins pass behind the tape. These then simply move toward the capstan and drum and threading is complete. A further advantage of partial wrap is that the friction between the tape and drum is reduced, allowing power saving in portable applications, and allowing the tape to be shuttled at high speed without the partial unthreading needed by videocassettes. In this way the player can read subcode during shuttle to facilitate rapid track access.

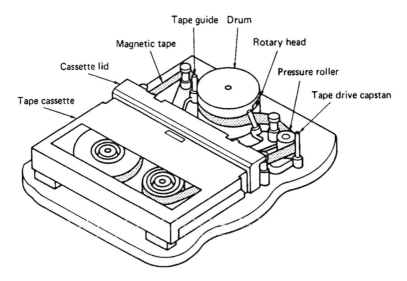

Figure 10.33 The simple mechanism of DAT. The guides and pressure roller move towards the drum and capstan and threading is complete.

The track pattern laid down by the rotary heads is shown in Figure 10.34. The heads rotate at 2000 rev/min in the same direction as tape motion, but because the drum axis is tilted, diagonal tracks 23.5 mm long result, at an angle of just over 6° to the edge. The diameter of the scanner needed is not specified, because it is the track pattern geometry which ensures interchange compatibility. It will be seen from Figure 10.34 that azimuth recording is employed as was described earlier. This requires no spaces or guard bands between the tracks. The chosen azimuth angle of ±20° reduces crosstalk to the same order as the noise, with a loss of only 1 dB due to the apparent reduction in writing speed.

In addition to the diagonal tracks, there are two linear tracks, one at each edge of the tape, where they act as protection for the diagonal tracks against edge damage. Owing to the low linear tape speed the use of these edge tracks is somewhat limited.

Several related modes of operation are available, some of which are mandatory whereas the remainder are optional. These are compared in Table 10.1. The most important modes use a sampling rate of 48 kHz or 44.1 kHz, with 16-bit two's complement uniform quantization. With a linear tape speed of 8.15 mm/s, the standard cassette offers 120 min unbroken playing time. Initially it was proposed that all DAT machines would be able to record and play at 48 kHz, whereas only professional machines would be able to record at 44.1 kHz. For consumer machines, playback only of prerecorded media was proposed at 44.1 kHz, so that the same software could be released on CD or prerecorded DAT tape. Now that a SCMS (serial copying management system) is incorporated in consumer machines, they too can record at 44.1 kHz. For reasons which will be explained later, contact duplicated tapes run at

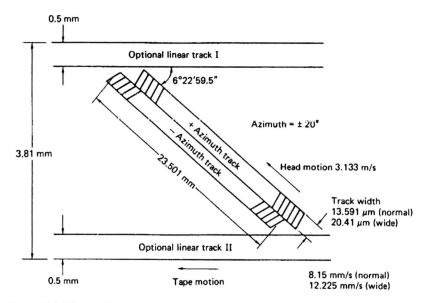

Figure 10.34 The two heads of opposite azimuth angles lay down the above track format. Tape linear speed determines track pitch.

Table 10.1 The significance of the recognition holes on the DAT cassette. Holes 1, 2 and 3 form a coded pattern; whereas hole 4 is independent.

Hole 1	Hole 2	Hole 3	Function
0	0	0	Metal powder tape or equivalent/13 µm thick
0	1	0	MP tape or equivalent/thin tape
0	0	1	1.5 TP/13 µm thick
0	1	1	1.5 TP/thin tape
1	×	×	(Reserved)

Hole 4	
0	Non-prerecorded tape
1	Prerecorded tape

1 = Hole present
0 = Hole blanked off

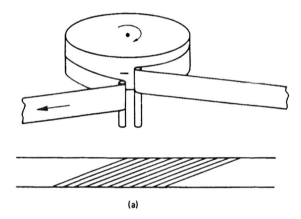

(a)

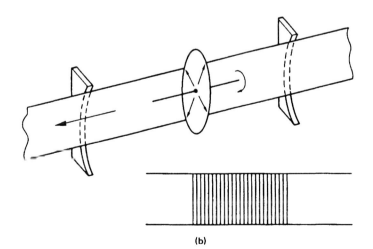

(b)

12.225 mm/s to offer a playing time of 80 min. The above modes are mandatory if a machine is to be considered to meet the format.

Option 1 is identical to 48 kHz mode except that the sampling rate is 32 kHz.

Option 2 is an extra-long-play mode. In order to reduce the data rate, the sampling rate is 32 kHz and the samples change to 12-bit two's complement with non-linear quantizing. Halving the subcode rate allows the overall data rate necessary to be halved. The linear tape speed and the drum speed are both halved to give a playing time of four hours. All of the above modes are stereo, but option 3 uses the sampling parameters of option 2 with four audio channels. This doubles the data rate with respect to option 2, so the standard tape speed of 8.15 mm/s is used.

Figure 10.35 shows a block diagram of a typical DAT recorder. In order to make a recording, an analog signal is fed to an input ADC, or a direct digital input is taken from an AES/EBU interface. The incoming samples are subject to interleaving to reduce the effects of error bursts. Reading the memory at a higher rate than it was written performs the necessary time compression. Additional bytes of redundancy computed from the samples are added to the data stream to permit subsequent error correction. Subcode information such as the content of the AES/EBU channel status message is added, and the parallel byte structure is fed to the channel encoder, which combines a bit clock with the data, and produces a recording signal according to the 8/10 code which is free of DC (see Section 10.6). This signal is fed to the heads via a rotary transformer to make the binary recording, which leaves the tape track with a pattern of transitions between the two magnetic states.

On replay, the transitions on the tape track induce pulses in the head, which are used to recreate the record current waveform. This is fed to the 10/8 decoder which converts it to the original data stream and a separate clock. The subcode data are routed to the subcode output, and the audio samples are fed into a deinterleave memory which, in addition to time-expanding the recording, functions to remove any wow or flutter due to head-to-tape speed variations. Error correction is performed partially before and partially after deinterleave. The corrected output samples can be fed to DACs or to a direct digital output.

In order to keep the rotary heads following the very narrow slant tracks, alignment patterns are recorded as well as the data. The automatic track-following system processes the playback signals from these patterns to control the drum and capstan motors. The subcode and ID information can be used by the control logic to drive the tape to any desired location specified by the user.

10.14 Half-inch and 8 mm rotary formats

A number of manufacturers have developed low-cost digital multitrack recorders for the home studio market. These are based on either VHS or Video-8 rotary-head cassette tape decks and generally offer eight channels of audio. Recording of individual audio channels is possible because the slant tape tracks are divided up into separate blocks for each channel with edit gaps between them. Some models have timecode and include synchronizers so that several machines can be locked together to offer more tracks. These machines represent the future of multitrack recording as their purchase and

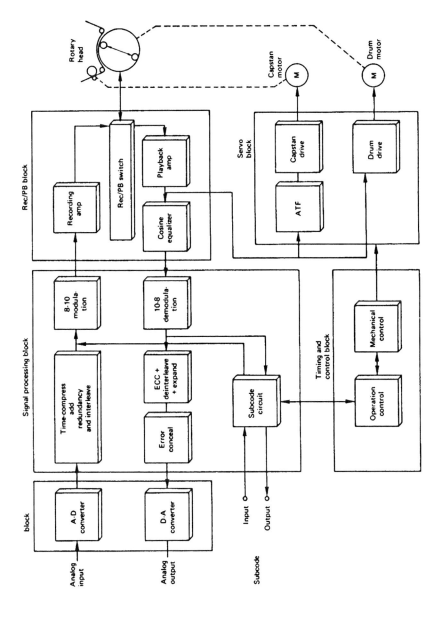

Figure 10.35 Block diagram of DAT.

running costs are considerably lower than that of stationary-head machines. It is only a matter of time before a low-cost 24 track is offered.

10.15 Digital audio disk systems

In order to use disk drives for the storage of audio, a system like the one shown in Figure 10.36 is needed. The control computer determines where

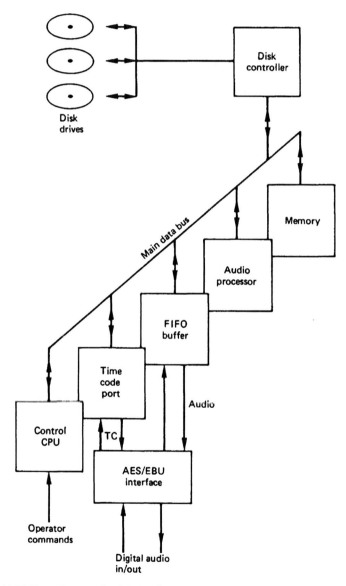

Figure 10.36 The main parts of a digital audio disk system. Memory and FIFO allow continuous audio despite the movement of disk heads between blocks.

and when samples will be stored and retrieved, and sends instructions to the disk controller which causes the drives to read or write, and transfers samples between them and the memory. The instantaneous data rate of a typical drive is roughly 10 times higher than the sampling rate, and this may result in the system data bus becoming choked by the disk transfers so other transactions are locked out. This is avoided by giving the disk controller DMA system a lower priority for bus access so that other devices can use the bus. A rapidly spinning disk cannot wait, and in order to prevent data loss, a silo or FIFO (first in first out) memory is necessary in the disk controller. A silo will be interposed in the disk controller data stream in the fashion shown in Figure 10.37 so that

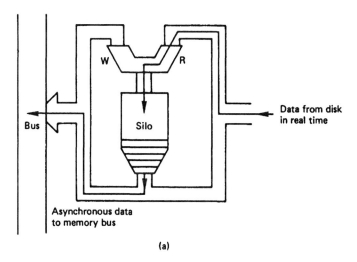

(a)

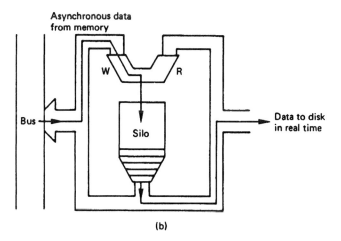

(b)

Figure 10.37 In order to guarantee that the drive can transfer data in real time at regular intervals (determined by disk speed and density) the silo provides buffering to the asynchronous operation of the memory access process. In (a) the silo is configured for a disk read. The same silo is used in (b) for a disk write.

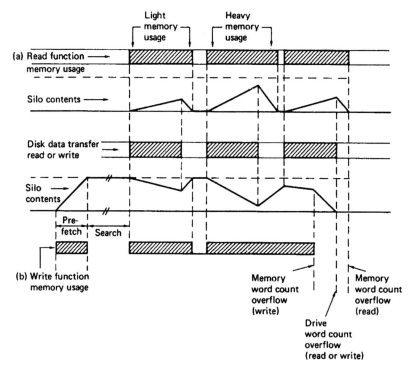

Figure 10.38 The silo contents during read functions (a) appear different from those during write functions (b). In (a) the control logic attempts to keep the silo as empty as possible; in (b) the logic prefills the silo and attempts to keep it full until the memory word count overflows.

it can buffer data both to and from the disk. When reading the disk, the silo starts empty, and if there is bus contention, the silo will start to fill as shown in Figure 10.38(a). Where the bus is free, the disk controller will attempt to empty the silo into the memory. The system can take advantage of the interblock gaps on the disk, containing headers, preambles and redundancy, for in these areas there are no data to transfer, and there is some breathing space to empty the silo before the next block. In practice the silo need not be empty at the start of every block, provided it never becomes full before the end of the transfer. If this happens some data are lost and the function must be aborted. The block containing the silo overflow will generally be re-read on the next revolution. In sophisticated systems, the silo has a kind of dipstick which indicates how full it is, and can interrupt the CPU if the data get too deep. The CPU can then suspend some bus activity to allow the disk controller more time to empty the silo.

When the disk is to be written, a continuous data stream must be provided during each block, as the disk cannot stop. The silo will be prefilled before the disk attempts to write as shown in Figure 10.38(b), and the disk controller attempts to keep it full. In this case all will be well if the silo does not become empty before the end of the transfer.

The disk controller cannot supply samples at a constant rate, because of gaps between blocks, defective blocks and the need to move the heads from one track to another and because of system bus contention. In order to accept a steady audio sample stream for storage, and to return it in the same way on replay, hard disk-based audio recorders must have a quantity of RAM for buffering. Then there is time for the positioner to move whilst the audio output is supplied from the RAM. In replay, the drive controller attempts to keep the RAM as full as possible by issuing a read command as soon as one block space appears in the RAM. This allows the maximum time for a seek to take place before reading must resume. Figure 10.39 shows the action of the RAM during reading. Whilst recording, the drive controller attempts to keep the RAM as empty as possible by issuing write commands as soon as a block of data is present, as in Figure 10.40. In this way the amount of time available to seek is maximized in the presence of a continuous audio sample input.

A hard disk has a discontinuous recording and acts like a RAM in that it must be addressed before data can be retrieved. The rotational speed of the disk is constant and not locked to anything. A vital step in converting a disk drive into an audio recorder is to establish a link between the time through the recording and the location of the data on the disk.

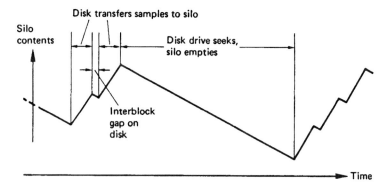

Figure 10.39 During an audio replay sequence, silo is constantly emptied to provide samples, and is refilled in blocks by the drive.

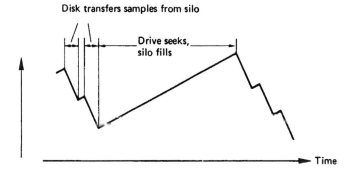

Figure 10.40 During audio recording, the input samples constantly fill the silo, and the drive attempts to keep it empty by reading from it.

When audio samples are fed into a disk-based system, from an AES/EBU interface or from a converter, they will be placed initially in RAM, from which the disk controller will read them by DMA. The continuous-input sample stream will be split up into disk blocks for disk storage. Timecode will be used to assemble a table which contains a conversion from real time in the recording to the physical disk address of the corresponding audio files. In a sound-for-television application, the timecode would be the same as that recorded on the videotape.

The table of disk addresses will also be made into a named disk file and stored in an index which will be in a different area of the disk from the audio files. Several recordings may be fed into the system in this way, and each will have an entry in the index.

If it is desired to playback one or more of the recordings, then it is only necessary to specify the starting timecode and the filename. The system will look up the index file in order to locate the physical address of the first and subsequent sample blocks in the desired recording, and will begin to read them from disk and write them into the RAM. Once the RAM is full, the real-time replay can begin by sending samples from RAM to the output or to local converters. The sampling-rate clock increments the RAM address and the timecode counter. Whenever a new timecode frame is reached, the corresponding disk address can be obtained from the index table, and the disk drive will read a block in order to keep the RAM topped up.

The disk transfers must by definition take varying times to complete because of the rotational latency of the disk. Once all the sectors on a particular cylinder have been read, it will be necessary to seek to the next cylinder, which will cause a further extension of the reading sequence. If a bad block is encountered, the sequence will be interrupted until it has passed. The RAM buffering is sufficient to absorb all of these access time variations. Thus the RAM acts as a delay between the disk transfers and the sound which is heard. A corresponding advance is arranged in timecodes fed to the disk controller. In effect the actual timecode has a constant added to it so that the disk is told to obtain blocks of samples in advance of real time. The disk takes a varying time to obtain the samples, and the RAM then further delays them to the correct timing. Effectively the disk/RAM subsystem is a timecode-controlled memory. One need only put in the time, and out comes the audio corresponding to that time. This is the characteristic of an audio synchronizer. In most audio equipment the synchronizer is extra; the hard disk needs one to work at all, and so every hard disk comes with a free synchronizer. This makes disk-based systems very flexible as they can be made to lock to almost any reference and care little what sampling rate is used or if it varies. They perform well locked to videotape or film via timecode because no matter how the pictures are shuttled or edited, the timecode link always produces the correct sound to go with the pictures. A video edit decision list (EDL) can be used to provide an almost immediate rough edited soundtrack.

A multitrack recording can be stored on a single disk and, for replay, the drive will access the files for each track faster than real time so that they all become present in the memory simultaneously. It is not, however, compulsory to play back the tracks in their original time relationship. For the purpose of synchronization, or other effects, the tracks can be played with any time relationship desired, a feature not possible with multitrack tape drives.

10.16 Audio in digital VTRs

Digital video recorders are capable of exceptional picture quality, but this attribute seems increasingly academic. The real power of digital technology in broadcasting is economic. DVTRs cost less to run than analog equipment. They use less tape and need less maintenance. Cassette formats are suitable for robotic handling, reducing manpower requirements.

A production VTR must offer such features as still frame, slow motion, timecode, editing, pictures in shuttle and so on as well as the further features needed in the digital audio tracks, such as the ability to record the user data in the AES/EBU digital audio interconnect standard. It must be possible to edit the video and each of the audio channels independently, whilst maintaining lip-sync.

An outline of a representative DVTR follows to illustrate some general principles, and to put the audio subsystem in perspective. Figure 10.41 shows that a DVTR generally has converters for both video and audio. Once in digital form, the data are formed into blocks, and the encoding section of the error-correction system supplies redundancy designed to protect the data against errors. These blocks are then converted into some form of channel code which combines the data with clock information so that it is possible to identify how many bits were recorded even if several adjacent bits are identical. The coded data are recorded on tape by a rotary head. Upon replaying the recording of the hypothetical machine of Figure 10.41, the errors caused by various mechanisms will be detected and corrected or concealed using the extra bits appended during the encoding process.

Digital tape tracks are always narrower than analog tracks, since they carry binary data, and a large signal-to-noise ratio is not required. This helps to balance the greater bandwidth required by a digitized video signal, but it does put extra demands on the mechanical accuracy of the transport in order to register the heads with the tracks. The prodigious data rate requires high-frequency circuitry, which causes headaches when complex processing such as error correction is contemplated. Unless a high compression factor is used, DVTRs often use more than one data channel in parallel to reduce the data rate in each. Figure 10.41 shows the distribution of incoming audio and video data between two channels. At any one time there are two heads working on two parallel tape tracks in order to share the data rate.

In the digital domain, samples can be recorded in any structure whatsoever provided they are output in the correct order on replay. This means that a segmented scan is acceptable without performance penalty in a digital video recorder. A segmented scan breaks up each TV picture into a number of different sweeps of the head across the tape. This permits a less than complete wrap of the scanner, which eases the design of the cassette-threading mechanism. The use of very short recorded wavelengths in digital recording makes the effect of spacing loss due to contamination worse, and so a cassette system is an ideal way of keeping the medium clean.

The use of several cassette sizes satisfies the contradicting demands of a portable machine which must be small and a studio machine which must give good playing time. The hubs of DVTR cassettes are at different spacings in the different sizes, so a machine built to play more than one size will need moving reel-motors. The cassette shell contains patterns of holes which can

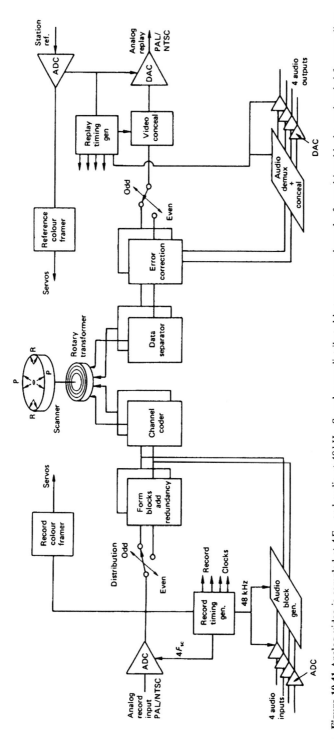

Figure 10.41 Analog video is sampled at $4F_{sc}$ and audio at 48 kHz. Samples are distributed between two channels, formed into blocks, and coded for recording. The audio blocks are time shared with video through a common tape channel consisting of a channel coder, the rotating head assembly and the data separator. On replay, timing is generated by station reference. Errors which cannot be corrected are concealed before conversion.

be read by sensors in the player. One set of these can be used to allow tapes of different thickness to change the tape remaining calculation based on the hub speed automatically. The other set is intended for the user, and one of these causes total record-lockout, or write-protect, when a plug is removed. Another may allow only the audio to be recorded, locking out the recording of the video and control tracks.

The audio samples in a DVTR are binary numbers just like the video samples, and although there is an obvious difference in sampling rate and wordlength, this only affects the relative areas of tape devoted to the audio and video samples. The most important difference between audio and video samples is the tolerance to errors. The acuity of the ear means that uncorrected audio samples must not occur more than once every few hours. There is little redundancy in sound, and concealment of errors is not desirable on a routine basis. In video, the samples are highly redundant, and concealment can be effected using samples from previous or subsequent lines or, with care, from the previous frame. Thus there are major differences between the ways that audio and video samples are handled in a DVTR. One such difference is that the audio samples often have 100% redundancy: every one is recorded using about twice as much space on tape as the same amount of video data.

As digital audio and video are both just data, in DVTRs the audio samples are carried by the same channel as the video samples. This reduces the number of encoders, preamplifiers and data separators needed in the system, whilst increasing the bandwidth requirement by only a few percent even with double recording. Figure 10.42 shows that the audio samples are heavily time-compressed so that the audio and video data have the same bit rate on tape. This increases the amount of circuitry which can be common to both video and audio processing.

In order to permit independent audio and video editing, the tape tracks are given a block structure. Editing will require the heads momentarily to go into record as the appropriate audio block is reached. Accurate synchronization is necessary if the other parts of the recording are to remain uncorrupted. The concept of a head which momentarily records in the centre of a track which it is reading is the normal operating procedure for all computer disk drives.

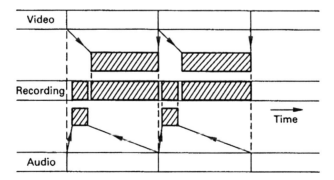

Figure 10.42 Time compression is used to shorten the length of track needed by the video. Heavily time-compressed audio samples can then be recorded on the same track using common circuitry.

There are in fact many parallels between digital helical recorders and disk drives, including the use of the term *sector*. In moving-head disk drives, the sector address is a measure of the angle through which the disk has rotated. This translates to the phase of the scanner in a rotary-head machine. The part of a track which is in one sector is called a block. The word 'sector' is often used instead of 'block' in casual parlance when it is clear that only one head is involved. However, as many DVTRs have two heads in action at any one time, the word 'sector' means the two side-by-side blocks in the segment. As there are four independently recordable audio channels in most production DVTR formats, there are four audio sectors.

There is a requirement for the DVTR to produce pictures in shuttle. In this case, the heads cross tracks randomly, and it is most unlikely that complete video blocks can be recovered. To provide pictures in shuttle, each block is broken down into smaller components called sync blocks in the same way as is done in DAT. These contain their own error checking and an address, which in disk terminology would be called a header, which specifies where in the picture the samples in the sync block belong. In shuttle, if a sync block is read properly, the address can be used to update a frame store. Thus it can be said that a sector is the smallest amount of data which can be written and is that part of a track pair within the same sector address, whereas a sync block is the smallest amount of data which can be read. Clearly there are many sync blocks in a sector.

The same sync block structure continues in the audio because the same read/write circuitry is used for audio and video data. Clearly the address structure must also continue through the audio. In order to prevent audio samples from arriving in the frame store in shuttle, the audio addresses are different from the video addresses. In all formats, the arrangement of the audio blocks is designed to maximize data integrity in the presence of tape defects and head clogs. The allocation of the audio channels to the sectors may change from one segment to the next. If a linear tape scratch damages the data in a given audio channel in one segment, it will damage a different audio channel in the next. Thus the scratch damage is shared between all four audio channels, each of which need correct only one-quarter of the damage. The relationship of the audio channels to the physical tracks may rotate by one track against the direction of tape movement from one audio sector to the next. If a head becomes clogged, the errors will be distributed through all audio channels, instead of causing severe damage in one channel.

Clearly the number of audio sync blocks in a given time is determined by the number of video fields in that time. It is only possible to have a fixed tape structure if the audio sampling rate is locked to video. With 625/50 machines, the sampling rate of 48 kHz results in exactly 960 audio samples in every field.

For use on 525/60, it must be recalled that the 60 Hz is actually 59.94 Hz. As this is slightly slow, it will be found that in 60 fields, exactly 48 048 audio samples will be necessary. Unfortunately 60 will not divide into 48 048 without a remainder. The largest number which will divide 60 and 48 048 is 12; thus in $60/12 = 5$ fields there will be $48\,048/12 = 4004$ samples. Over a five-field sequence the product blocks contain 801, 801, 801, 801 and 800 samples respectively, adding up to 4004 samples.

In order to comply with the AES/EBU digital audio interconnect, wordlengths between 16 and 20 bits can be supported by most DVTRs, but it is necessary to record a code in the sync block to specify the wordlength in use. Pre-emphasis may have been used prior to conversion, and this status is also to be conveyed, along with the four channel-use bits. The AES/EBU digital interface (see Chapter 8) uses a block-sync pattern which repeats after 192 sample periods corresponding to 4 ms at 48 kHz. Since the block size is different to that of the DVTR interleave block, there can be any phase relationship between interleave-block boundaries and the AES/EBU block-sync pattern. In order to recreate the same phase relationship between block sync and sample data on replay, it is necessary to record the position of block sync within the interleave block. It is the function of the interface control word in the audio data to convey these parameters. There is no guarantee that the 192-sample block-sync sequence will remain intact after audio editing; most likely there will be an arbitrary jump in block-sync phase. Strictly speaking a DVTR playing back an edited tape would have to ignore the block-sync positions on the tape, and create new block sync at the standard 192-sample spacing. Unfortunately most DVTR formats are not totally transparent to the whole of the AES/EBU data stream.

Appendix 10.1 Calculation of Reed–Solomon generator polynomials

For a Reed–Solomon codeword over $GF(2^3)$, there will be seven 3-bit symbols. For location and correction of one symbol, there must be two redundant symbols P and Q, leaving A–E for data.

The following expressions must be true, where a is the primitive element of $x^3 \oplus x \oplus 1$ and $\oplus$ is XOR throughout.

$$A \oplus B \oplus C \oplus D \oplus E \oplus P \oplus Q = 0 \qquad (1)$$

$$a^7A \oplus a^6B \oplus a^5C \oplus a^4D \oplus a^3E \oplus a^2P \oplus aQ = 0 \qquad (2)$$

Dividing Eqn (2) by a:

$$a^6A \oplus a^5B \oplus a^4C \oplus a^3D \oplus a^2E \oplus aP \oplus Q = 0$$
$$= A \oplus B \oplus C \oplus D \oplus E \oplus P \oplus Q$$

Cancelling Q, and collecting terms:

$$(a^6 \oplus 1)A \oplus (a^5 \oplus 1)B \oplus (a^4 \oplus 1)C \oplus (a^3 \oplus 1)D \oplus (a^2 \oplus 1)E = (a \oplus 1)P$$

Using Figure 10.22 to calculate $(a^n \oplus 1)$, e.g. $a^6 \oplus 1 = 101 \oplus 001 = 100 = a^2$

$$a^2A \oplus a^4B \oplus a^5C \oplus aD \oplus a^6E = a^3P$$

$$a^6A \oplus aB \oplus a^7C \oplus a^5D \oplus a^3E = P \qquad (3)$$

Multiply Eqn (1) by a^2 and equate to (2):

$$a^2A \oplus a^2B \oplus a^2C \oplus a^2D \oplus a^2E \oplus a^2P \oplus a^2Q = 0$$
$$= a^7A \oplus a^6B \oplus a^5C \oplus a^4D \oplus a^3E \oplus a^2P \oplus aQ$$

Cancelling terms a^2P and collecting terms (remember $a^2 \oplus a^2 = 0$):

$$(a^7 \oplus a^2)A \oplus (a^6 \oplus a^2)B \oplus (a^5 \oplus a^2)C \oplus (a^4 \oplus a^2)D \oplus (a^3 \oplus a^2)E$$
$$= (a^2 \oplus a)Q$$

Adding powers according to Figure 10.22, e.g. $a^7 \oplus a^2 = 001 \oplus 100 = 101 = a^6$:

$$a^6A \oplus B \oplus a^3C \oplus aD \oplus a^5E = a^4Q$$
$$a^2A \oplus a^3B \oplus a^6C \oplus a^4D \oplus aE = Q$$

References

1. Watkinson, J.R., *The Art of Data Recording*, Chapter 3, Oxford: Focal Press (1994)
2. Deeley, E.M., Integrating and differentiating channels in digital tape recording. *Radio Electron. Eng.*, **56** 169–173 (1986)
3. Nakajima, H. and Odaka, K., A rotary-head high-density digital audio tape recorder. *IEEE Trans. Consum. Electron.*, **CE-29**, 430–437 (1983)
4. Watkinson, J.R., *The Art of Data Recording*, Chapter 4, Oxford: Focal Press (1994)
5. Fukuda, S., Kojima, Y., Shimpuku, Y. and Odaka, K., 8/10 modulation codes for digital magnetic recording. *IEEE Trans. Magn.*, **MAG-22**, 1194–1196 (1986)
6. Watkinson, J.R., *The Art of Digital Audio*, Chapter 6, Oxford: Focal Press (1994)
7. Reed, I.S. and Solomon, G., Polynomial codes over certain finite fields. *J. Soc. Indust. Appl. Math.*, **8**, 300–304 (1960)
8. Chien, R.T., Cunningham, B.D. and Oldham, I.B., Hybrid methods for finding roots of a polynomial – with application to BCH decoding. *IEEE Trans. Inf. Theory*, **IT-15**, 329–334 (1969)
9. Berlekamp, E.R., *Algebraic Coding Theory*, New York: McGraw-Hill (1967). Reprint edition: Laguna Hills Ca: Aegean Park Press (1983)
10. Sugiyama, Y. et al., An erasures and errors decoding algorithm for Goppa codes. *IEEE Trans. Inf. Theory*, **IT-22** (1976)
11. Onishi, K., Sugiyama, K., Ishida, Y., Kusonoki, Y. and Yamaguchi, T., An LSI for Reed–Solomon encoder/decoder. Presented at 80th Audio Engineering Society Convention (Montreux, 1986), preprint 2316(A-4)
12. Anon., *Digital Audio Tape Deck Operation Manual*, Sony Corporation (1987)
13. van Kommer, R., Reed–Solomon coding and decoding by digital signal processors. Presented at 84th Audio Engineering Society Convention (Paris, 1988), preprint 2587(D-7)
14. Itoh, F., Shiba, H., Hayama, M. and Satoh, T., Magnetic tape and cartridge of R-DAT. *IEEE Trans. Consum. Electron.*, **CE-32**, 442–452 (1986)

Chapter 11

Optical disks in digital audio

Optical disks are particularly important to digital audio, not least because of the success of the Compact Disc. Ten years later the MiniDisc and magneto-optical mastering recorders took optical disk technology a stage further.

11.1 Types of optical disk

There are numerous types of optical disk, which have different characteristics.[1] There are, however, three broad groups, shown in Figure 11.1, which can be usefully compared.

(1) The Compact Disc and the prerecorded MiniDisc (MD) are read-only laser disks, which are designed for mass duplication by stamping. They cannot be recorded.

(2) Some laser disks can be recorded, but once a recording has been made, it cannot be changed or erased. These are usually referred to as write-once-read-many (WORM) disks. Recordable CDs work on this principle.

(3) Erasable optical disks have essentially the same characteristic as magnetic disks, in that new and different recordings can be made in the same track indefinitely. Recordable Minidick is in this category. Sometimes a separate erase process is necessary before rewriting.

The Compact Disc, generally abbreviated to CD, is a consumer digital audio recording which is intended for mass replication. Philips' approach was to invent an optical medium which would have the same characteristics as the vinyl disk in that it could be mass replicated by moulding or stamping with no requirement for it to be recordable by the user. The information on it is carried in the shape of flat-topped physical deformities in a layer of plastic. Such relief structures lack contrast and must be read with a technique called phase contrast microscopy which allows an apparent contrast to be obtained using optical interference.

Figure 11.2(a) shows that the information layer of CD and the prerecorded MiniDisc is an optically flat mirror upon which microscopic bumps are raised. A thin coating of aluminium renders the layer reflective. When a small spot of light is focused on the information layer, the presence of the bumps affects the way in which the light is reflected back, and variations in the reflected light are

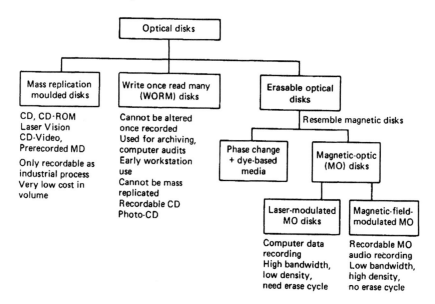

Figure 11.1 The various types of optical disk. See text for details.

detected in order to read the disk. Track dimensions in optical disks are very small. For comparison, some 60 CD/MD tracks can be accommodated in the groove pitch of a vinyl LP. These dimensions demand the utmost cleanliness in manufacture.

Figure 11.2(b) shows that there are two main types of WORM disks. In the first, the disk contains a thin layer of metal; on recording, a powerful laser melts spots on the layer. Surface tension causes a hole to form in the metal, with a thickened rim around the hole. Subsequently a low-power laser can read the disk because the metal reflects light, but the hole passes it through. Computer WORM disks work on this principle. In the second, the layer of metal is extremely thin, and the heat from the laser heats the material below it to the point of decomposition. This causes gassing which raises a blister or bubble in the metal layer. Recordable CDs use this principle as the relief structure can be read like a normal CD. Clearly once such a pattern of holes or blisters has been made, it is permanent.

Re-recordable or erasable optical disks rely on magneto-optics,[2] also known more fully as thermomagneto-optics. Writing in such a device makes use of a thermomagnetic property possessed by all magnetic materials, which is that above a certain temperature, known as the Curie temperature, their coercive force becomes zero. This means that they become magnetically very soft, and take on the flux direction of any externally applied field. On cooling, this field orientation will be frozen in the material, and the coercivity will oppose attempts to change it. Although many materials possess this property, there are relatively few which have a suitably low Curie temperature. Compounds of terbium and gadolinium have been used, and one of the major problems to be overcome is that almost all suitable materials from a magnetic viewpoint corrode very quickly in air.

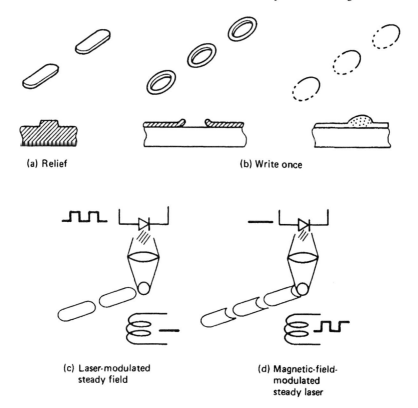

Figure 11.2 (a) The information layer of CD is reflective and uses interference. (b) Write-once disks may burn holes or raise blisters in the information layer. (c) High data rate MO disks modulate the laser and use a constant magnetic field. (d) At low data rates the laser can run continuously and the magnetic field is modulated.

There are two ways in which magneto-optic (MO) disks can be written. Figure 11.2(c) shows the first system, in which the intensity of laser is modulated with the waveform to be recorded. If the disk is considered to be initially magnetized along its axis of rotation with the north pole upwards, it is rotated in a field of the opposite sense, produced by a steady current flowing in a coil which is weaker than the room-temperature coercivity of the medium. The field will therefore have no effect. A laser beam is focused on the medium as it turns, and a pulse from the laser will momentarily heat a very small area of the medium past its Curie temperature, whereby it will take on a reversed flux due to the presence of the field coils. This reversed-flux direction will be retained indefinitely as the medium cools.

Alternatively the waveform to be recorded modulates the magnetic field from the coils as shown in Figure 11.2(d). In this approach, the laser is operating continuously in order to raise the track beneath the beam above the Curie temperature, but the magnetic field recorded is determined by the current in the coil at the instant the track cools. Magnetic field modulation is used in the recordable MiniDisc.

In both of these cases, the storage medium is clearly magnetic, but the writing mechanism is the heat produced by light from a laser; hence the term thermomagneto-optics. The advantage of this writing mechanism is that there is no physical contact between the writing head and the medium. The distance can be several millimetres, some of which is taken up with a protective layer to prevent corrosion. In prototypes, this layer is glass, but commercially available disks use plastics.

The laser beam will supply a relatively high power for writing, since it is supplying heat energy. For reading, the laser power is reduced, such that it cannot heat the medium past the Curie temperature, and it is left on continuously. Readout depends on the so-called Kerr effect, which describes a rotation of the plane of polarization of light due to a magnetic field. The magnetic areas written on the disk will rotate the plane of polarization of incident polarized light to two different planes, and it is possible to detect the change in rotation with a suitable pickup.

11.2 CD and MD contrasted

CD and MD have a great deal in common. Both use a laser of the same wavelength which creates a spot of the same size on the disk. The track pitch and speed are the same and both offer the same playing time. The channel code and error-correction strategy are the same.

CD carries 44.1 kHz 16-bit PCM audio and is intended to be played in a continuous spiral like a vinyl disk. The CD process, from cutting, through pressing and reading, produces no musical degradation whatsoever, since it simply conveys a series of numbers which are exactly those recorded on the master tape. The only part of a CD player which can cause subjective differences in sound quality in normal operation is the DAC, although in the presence of gross errors some players will correct and/or conceal better than others.

MD begins with the same PCM data, but uses a form of compression known as ATRAC (see Chapter 13) having a compression factor of 0.2. After the addition of subcode and housekeeping data MD has an average data rate which is 0.225 that of CD. However, MD has the same recording density and track speed as CD, so the data rate from the disk is greatly in excess of that needed by the audio decoders. The difference is absorbed in RAM as shown in Chapter 10. The RAM in a typical player is capable of buffering about three seconds of audio. When the RAM is full, the disk drive stops transferring data but keeps turning. As the RAM empties into the decoders, the disk drive will top it up in bursts. As the drive need not transfer data for over three-quarters of the time, it can reposition between transfers and so it is capable of editing in the same way as a magnetic hard disk. A further advantage of the RAM buffer is that if the pickup is knocked off track by an external shock the RAM continues to provide data to the audio decoders and provided the pickup can get back to the correct track before the RAM is exhausted there will be no audible effect.

When recording an MO disk, the MiniDisc drive also uses the RAM buffer to allow repositioning so that a continuous recording can be made on a disk which has become chequerboarded through selective erasing. The full total playing time is then always available irrespective of how the disk is divided into different recordings.

The sound quality of MiniDisc is a function of the performance of the converters and of the compression system.

11.3 CD and MD disk construction

Figure 11.3 shows the mechanical specification of CD. Within an overall diameter of 120 mm the program area occupies a 33 mm wide band between the diameters of 50 and 116 mm. Lead-in and lead-out areas increase the width of this band to 35.5 mm. As the track pitch is a constant 1.6 μm, there will be:

$$\frac{35.5 \times 1000}{1.6} = 22\,188$$

tracks crossing a radius of the disk. As the track is a continuous spiral, the track length will be given by the above figure multiplied by the average circumference.

$$\text{Length} = 2 \times \pi \times \frac{58.5 + 23}{2} \times 22\,188 = 5.7\,\text{km}$$

Figure 11.4 shows the mechanical specification of prerecorded MiniDisc. Within an overall diameter of 64 mm the lead-in area begins at a diameter of 29 mm and the program area begins at 32 mm. The track pitch is exactly the same as in CD, but the MiniDisc can be smaller than CD without any sacrifice of playing time because of the use of data reduction. For ease of handling, MiniDisc is permanently enclosed in a shuttered plastic cartridge which is $72 \times 68 \times 5$ mm. The cartridge resembles a smaller version of a $3\frac{1}{2}$-inch floppy disk, but unlike a floppy, it is slotted into the drive with the shutter at the side. An arrow is moulded into the cartridge body to indicate this.

In the prerecorded MiniDisc, it was a requirement that the whole of one side of the cartridge should be available for graphics. Thus the disk is designed to be secured to the spindle from one side only. The centre of the disk is fitted with a ferrous clamping plate and the spindle is magnetic. When the disk is

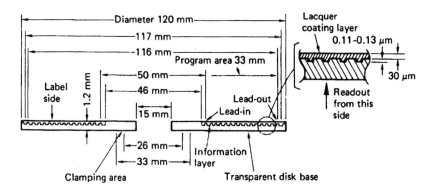

Figure 11.3 Mechanical specification of CD. Between diameters of 46 and 117 mm is a spiral track 5.7 km long.

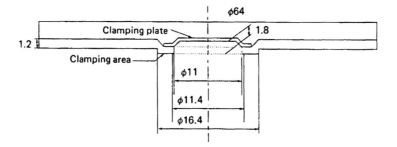

Figure 11.4 The mechanical dimensions of MiniDisc.

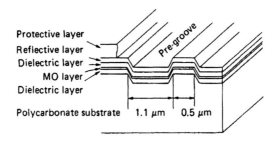

Figure 11.5 The construction of the MO recordable MiniDisc.

lowered into the drive it simply sticks to the spindle. The ferrous disk is only there to provide the clamping force. The disk is still located by the moulded hole in the plastic component. In this way the ferrous component needs no special alignment accuracy when it is fitted in manufacture. The back of the cartridge has a centre opening for the hub and a sliding shutter to allow access by the optical pickup.

The recordable MiniDisc and cartridge has the same dimensions as the prerecorded MiniDisc, but access to both sides of the disk is needed for recording. Thus the recordable MiniDisc has a shutter which opens on both sides of the cartridge, rather like a double-sided floppy disk. The opening on the front allows access by the magnetic head needed for MO recording, leaving a smaller label area.

Figure 11.5 shows the construction of the MO MiniDisc. The 1.1 μm-wide tracks are separated by grooves which can be optically tracked. Once again the track pitch is the same as in CD. The MO layer is sandwiched between protective layers.

11.4 Rejecting surface contamination

A fundamental goal of consumer optical disks is that no special working environment or handling skill is required. The bandwidth required by PCM audio is such that high-density recording is mandatory if reasonable playing time is to be obtained in CD. Although MiniDisc uses compression, it does so in order to make the disk smaller and the recording density is actually the same as for CD.

High-density recording implies short wavelengths. Using a laser focused on the disk from a distance allows short wavelength recordings to be played back without physical contact, whereas conventional magnetic recording requires intimate contact and implies a wear mechanism, the need for periodic cleaning, and susceptibility to contamination.

The information layer of CD and MD is read through the thickness of the disk. Figure 11.6 shows that this approach causes the readout beam to enter and leave the disk surface through the largest possible area. The actual dimensions involved are shown in the figure. Despite the minute spot size of about 1.2 µm diameter, light enters and leaves through a 0.7 mm-diameter circle. As a result, surface debris has to be three orders of magnitude larger than the readout spot before the beam is obscured. This approach has the further advantage in MO drives that the magnetic head, on the opposite side to the laser pickup, is then closer to the magnetic layer in the disk.

When light arrives at an interface with a denser medium, such as the surface of an optical disk, the velocity of propagation is reduced; therefore the wavelength in the medium becomes shorter, causing refraction as was seen in Section 3.21. The ratio of velocity *in vacuo* to velocity in the medium is known as the refractive index of that medium; it determines the relationship between the angles of the incident and refracted wavefronts.

The size of the entry circle in Figure 11.6 is a function of the refractive index of the disk material, the numerical aperture of the objective lens and

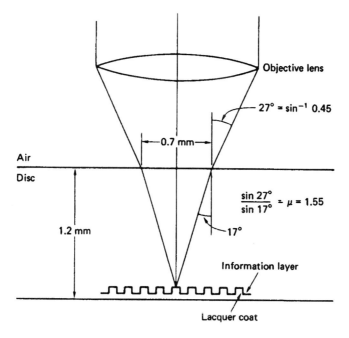

Figure 11.6 The objective lens of a CD pickup has a numerical aperture (NA) of 0.45; thus the outermost rays will be inclined at approximately 27° to the normal. Refraction at the air/disk interface changes this to approximately 17° within the disk. Thus light focused to a spot on the information layer has entered the disk through a 0.7 mm diameter circle, giving good resistance to surface contamination.

the thickness of the disk. MiniDiscs are permanently enclosed in a cartridge, and scratching is unlikely. This is not so for CD, but fortunately the method of readout through the disk thickness tolerates surface scratches very well. In extreme cases of damage, a scratch can often be successfully removed with metal polish. By way of contrast, the label side is actually more vulnerable than the readout side, since the lacquer coating is only 30 µm thick. For this reason, writing on the label side of CD is not recommended.

The base material is in fact a polycarbonate plastic produced by (amongst others) Bayer under the trade name of Makrolon. It has excellent mechanical and optical stability over a wide temperature range, and lends itself to precision moulding and metallization. It is often used for automotive indicator clusters for the same reasons. An alternative material is polymethyl methacrylate (PMMA), one of the first optical plastics, known by such trade names as Perspex and Plexiglas. Polycarbonate is preferred by some manufacturers since it is less hygroscopic than PMMA. The differential change in dimensions of the lacquer coat and the base material can cause warping in a hygroscopic material. Audio disks are too small for this to be a problem, but the larger video disks are actually two disks glued together back to back to prevent this warpage.

11.5 Playing optical disks

A typical laser disk drive resembles a magnetic drive in that it has a spindle drive mechanism to revolve the disk, and a positioner to give radial access across the disk surface. The positioner has to carry a collection of lasers, lenses, prisms, gratings and so on, and cannot be accelerated as fast as a magnetic-drive positioner. A penalty of the very small track pitch possible in laser disks, which gives the enormous storage capacity, is that very accurate track following is needed, and it takes some time to lock onto a track. For this reason tracks on laser disks are usually made as a continuous spiral, rather than the concentric rings of magnetic disks. In this way, a continuous data transfer involves no more than track following once the beginning of the file is located.

In order to record MO disks or replay any optical disk, a source of monochromatic light is required. The light source must have low noise otherwise the variations in intensity due to the noise of the source will mask the variations due to reading the disk. The requirement for a low-noise monochromatic light source is economically met using a semiconductor laser.

The semiconductor laser is a relative of the light-emitting diode (LED). Both operate by raising the energy of electrons to move them from one valence band to another conduction band. Electrons which fall back to the valence band emit a quantum of energy as a photon whose frequency is proportional to the energy difference between the bands. The process is described by Planck's law:

Energy difference $E = h \times f$

where h = Planck's constant

$= 6.6262 \times 10^{-34}$ joules/Hertz

For gallium arsenide, the energy difference is about 1.6 eV, where 1 eV is 1.6×10^{-19} joules.

Using Planck's law, the frequency of emission will be:

$$f = \frac{1.6 \times 1.6 \times 10^{-19}}{6.6262 \times 10^{-34}} \text{Hz}$$

The wavelength will be c/f where

$$c = \text{the velocity of light} = 3 \times 10^8 \text{ m/s}$$

$$\text{Wavelength} = \frac{3 \times 10^8 \times 6.6262 \times 10^{-34}}{2.56 \times 10^{-19}} \text{ m}$$

$$= 780 \text{ nanometres}$$

In the LED, electrons fall back to the valence band randomly, and the light produced is incoherent. In the laser, the ends of the semiconductor are optically flat mirrors, which produce an optically resonant cavity. One photon can bounce to and fro, exciting others in synchronism, to produce coherent light. This is known as Light Amplification by Stimulated Emission of Radiation, mercifully abbreviated to LASER, and can result in a runaway condition, where all available energy is used up in one flash. In injection lasers, an equilibrium is reached between energy input and light output, allowing continuous operation. The equilibrium is delicate, and such devices are usually fed from a current source. To avoid runaway when temperature change disturbs the equilibrium, a photosensor is often fed back to the current source. Such lasers have a finite life, and become steadily less efficient.

Some of the light reflected back from the disk re-enters the aperture of the objective lens. The pickup must be capable of separating the reflected light from the incident light. Figure 11.7 shows two systems. At (a) an intensity beam splitter consisting of a semi-silvered mirror is inserted in the optical path and reflects some of the returning light into the photosensor. This is not very efficient, as half of the replay signal is lost by transmission straight on. In the example at (b) separation is by polarization.

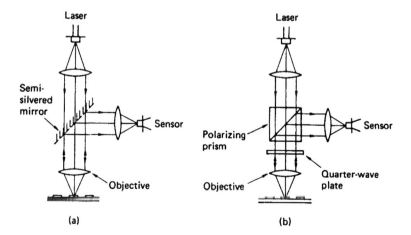

Figure 11.7 (a) Reflected light from the disk is directed to the sensor by a semi-silvered mirror. (b) A combination of polarizing prism and quarter-wave plate separates incident and reflected light.

Rotation of the plane of polarization is a useful method of separating incident and reflected light in a laser pickup. Using a quarter-wave plate, the plane of polarization of light leaving the pickup will have been turned 45°, and on return it will be rotated a further 45°, so that it is now at right angles to the plane of polarization of light from the source. The two can easily be separated by a polarizing prism, which acts as a transparent block to light in one plane, but as a prism to light in the other plane, such that reflected light is directed towards the sensor.

In a CD player, the sensor is concerned only with the intensity of the light falling on it. When playing MO disks, the intensity does not change, but the magnetic recording on the disk rotates the plane of polarization one way or

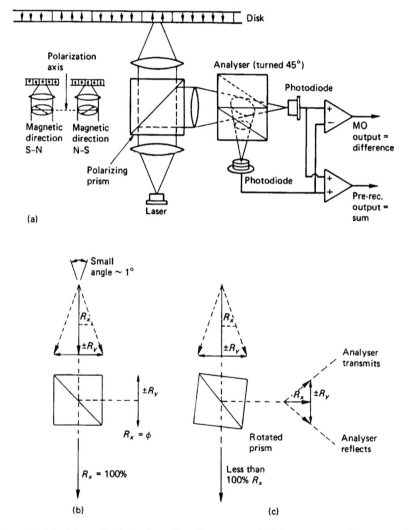

Figure 11.8 A pickup suitable for the replay of magneto-optic disks must respond to very small rotations of the plane of polarization.

the other depending on the direction of the vertical magnetization. MO disks cannot be read with circular polarized light. Light incident on the medium must be plane polarized and so the quarter-wave plate of the CD pickup cannot be used. Figure 11.8(a) shows that a polarizing prism is still required to linearly polarize the light from the laser on its way to the disk. Light returning from the disk has had its plane of polarization rotated by approximately $\pm 1°$. This is an extremely small rotation. Figure 11.8(b) shows that the returning rotated light can be considered to comprise two orthogonal components. R_x is the component which is in the same plane as the illumination and is called the *ordinary* component and R_y in the component due to the Kerr effect rotation and is known as the *magneto-optic* component. A polarizing beam splitter mounted squarely would reflect the magneto-optic component R_y very well because it is at right angles to the transmission plane of the prism, but the ordinary component would pass straight on in the direction of the laser. By rotating the prism slightly a small amount of the ordinary component is also reflected. Figure 11.8(c) shows that when combined with the magneto-optic component, the angle of rotation has increased. Detecting this rotation requires a further polarizing prism or analyser as shown in Figure 11.8(a). The prism is twisted such that the transmission plane is at $45°$ to the planes of R_x and R_y. Thus with an unmagnetized disk, half of the light is transmitted by the prism and half is reflected. If the magnetic field of the disk turns the plane of polarization towards the transmission plane of the prism, more light is transmitted and less is reflected. Conversely if the plane of polarization is rotated away from the transmission plane, less light is transmitted and more is reflected. If two sensors are used, one for transmitted light and one for reflected light, the difference between the two sensor outputs will be a waveform representing the angle of polarization and thus the recording on the disk. This differential analyser eliminates common mode noise in the reflected beam.[3] The output of the two sensors is summed as well as subtracted in a MiniDisc player. When playing MO disks, the difference signal is used. When playing prerecorded disks, the sum signal is used and the effect of the second polarizing prism is disabled.

11.6 Focus and tracking system

The frequency response of the laser pickup and the amount of crosstalk are both a function of the spot size and care must be taken to keep the beam focused on the information layer. Disk warp and thickness irregularities will cause focal-plane movement beyond the depth of focus of the optical system, and a focus-servo system will be needed. The depth of field is related to the numerical aperture, which is defined, and the accuracy of the servo must be sufficient to keep the focal plane within that depth, which is typically $\pm 1\,\mu m$.

The focus servo moves a lens along the optical axis in order to keep the spot in focus. Since dynamic focus changes are largely due to warps, the focus system must have a frequency response in excess of the rotational speed. A focus-error system is necessary to drive the lens. There are a number of ways in which this can be derived, the most common of which will be described here.

In Figure 11.9 a cylindrical lens is installed between the beam splitter and the photosensor. The effect of this lens is that the beam has no focal point on the sensor. In one plane, the cylindrical lens appears parallel-sided, and

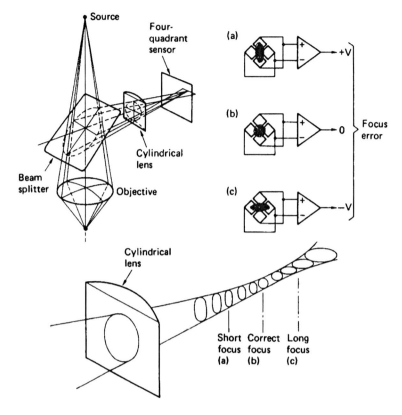

Figure 11.9 The cylindrical lens focus method produces an elliptical spot on the sensor whose aspect ratio is detected by a four-quadrant sensor to produce a focus error.

has negligible effect on the focal length of the main system, whereas in the other plane, the lens shortens the focal length. The image will be an ellipse whose aspect ratio changes as a function of the state of focus. Between the two foci, the image will be circular. The aspect ratio of the ellipse, and hence the focus error, can be found by dividing the sensor into quadrants. When these are connected as shown, the focus-error signal is generated. The data readout signal is the sum of the quadrant outputs.

Figure 11.10 shows the knife-edge method of determining focus. A split sensor is also required. At (a) the focal point is coincident with the knife edge, so it has little effect on the beam. At (b) the focal point is to the right of the knife edge, and rising rays are interrupted, reducing the output of the upper sensor. At (c) the focal point is to the left of the knife edge, and descending rays are interrupted, reducing the output of the lower sensor. The focus error is derived by comparing the outputs of the two halves of the sensor. A drawback of the knife-edge system is that the lateral position of the knife edge is critical, and adjustment is necessary. To overcome this problem, the knife edge can be replaced by a pair of prisms, as shown in (d)–(f). Mechanical tolerances then only affect the sensitivity, without causing a focus offset.

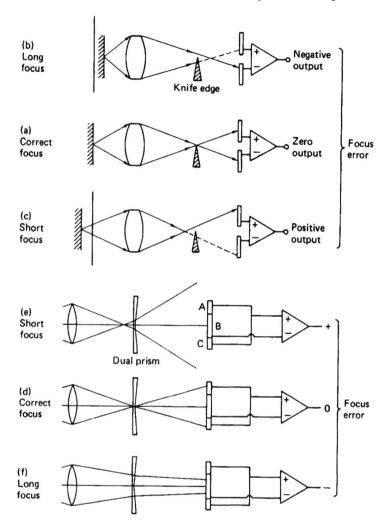

Figure 11.10 (a)–(c) Knife-edge focus method requires only two sensors, but is critically dependent on knife-edge position. (d)–(f) Twin-prism method requires three sensors (A, B, C), where focus error is (A + C) − B. Prism alignment reduces sensitivity without causing focus offset.

The cylindrical lens method is compared with the knife-edge/prism method in Figure 11.11, which shows that the cylindrical lens method has a much smaller capture range. A focus-search mechanism will be required, which moves the focus servo over its entire travel, looking for a zero crossing. At this time the feedback loop will be completed, and the sensor will remain on the linear part of its characteristic. The spiral track of CD and MiniDisc starts at the inside and works outwards. This was deliberately arranged because there is less vertical runout near the hub, and initial focusing will be easier.

The track pitch is only 1.6 μm, and this is much smaller than the accuracy to which the player chuck or the disk centre hole can be made; on a typical

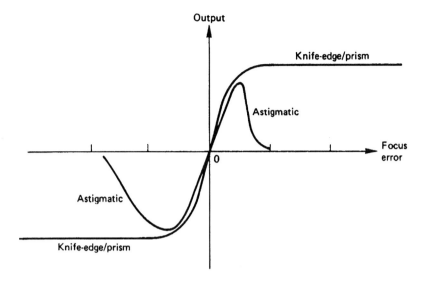

Figure 11.11 Comparison of capture range of knife-edge/prism method and astigmatic (cylindrical lens) system. Knife edge may have a range of 1 mm, whereas astigmatic may only have a range of 40 μm, requiring a focus-search mechanism.

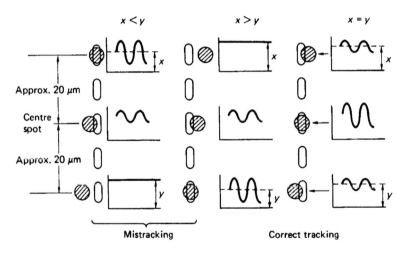

Figure 11.12 Three-spot method of producing tracking error compares average level of side-spot signals. Side spots are produced by a diffraction grating and require their own sensors.

player, runout will swing several tracks past a fixed pickup. A track-following servo is necessary to keep the spot centralized on the track. There are several ways in which a tracking error can be derived.

In the three-spot method, two additional light beams are focused on the disk track, one offset to each side of the track centre line. Figure 11.12 shows that, as one side spot moves away from the track into the mirror area, there is less destructive interference and more reflection. This causes the average amplitude of the side spots to change differentially with tracking error. The laser head contains a diffraction grating which produces the side spots, and two extra photosensors onto which the reflections of the side spots will fall. The side spots feed a differential amplifier, which has a low-pass filter to reject the channel-code information and retain the average brightness difference. Some players use a delay line in one of the side-spot signals whose period is equal

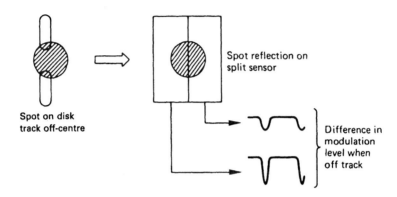

Figure 11.13 Split-sensor method of producing tracking error focuses image of spot onto sensor. One side of spot will have more modulation when off track.

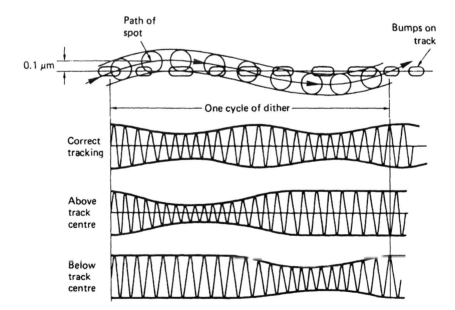

Figure 11.14 Dither applied to readout spot modulates the readout envelope. A tracking error can be derived.

to the time taken for the disk to travel between the side spots. This helps the differential amplifier to cancel the channel code.

The alternative approach to tracking-error detection is to analyse the diffraction pattern of the reflected beam. The effect of an off-centre spot is to rotate the radial diffraction pattern about an axis along the track. Figure 11.13 shows that, if a split sensor is used, one half will see greater modulation than the other when off track. Such a system may be prone to develop an offset due either to drift or to contamination of the optics, although the capture range is large. A further tracking mechanism is often added to obviate the need for periodic adjustment. Figure 11.14 shows a dither-based system in which a sinusoidal drive is fed to the tracking servo, causing a radial oscillation of spot position of about ±50 nm. This results in modulation of the envelope of the readout signal, which can be synchronously detected to obtain the sense of the error. The dither can be produced by vibrating a mirror in the light path, which enables a high frequency to be used, or by oscillating the whole pickup at a lower frequency.

11.7 Typical pickups

It is interesting to compare different designs of laser pickup. Figure 11.15 shows a Philips laser head.[4] The dual-prism focus method is used, which combines the output of two split sensors to produce a focus error. The focus amplifier drives the objective lens which is mounted on a parallel motion formed by two flexural arms. The capture range of the focus system is sufficient to accommodate normal tolerances without assistance. A radial differential tracking signal is extracted from the sensors as shown in the figure. Additionally, a dither frequency of 600 Hz produces envelope modulation which is synchronously rectified to produce a drift-free tracking error. Both errors are combined to drive the tracking system. As only a single spot is used, the pickup is relatively insensitive to angular errors, and a rotary positioner can be used, driven by a moving coil. The assembly is statically balanced to give good resistance to lateral shock.

Figure 11.16 shows a Sony laser head used in consumer players. The cylindrical-lens focus method is used, requiring a four-quadrant sensor. Since this method has a small capture range, a focus-search mechanism is necessary. When a disk is loaded, the objective lens is ramped up and down looking for a zero crossing in the focus error. The three-spot method is used for tracking. The necessary diffraction grating can be seen adjacent to the laser diode. Tracking error is derived from side-spot sensors (E, F). Since the side-spot system is sensitive to angular error, a parallel-tracking laser head traversing a disk radius is essential. A cost-effective linear motion is obtained by using a rack-and-pinion drive for slow, coarse movements, and a laterally moving lens in the light path for fine rapid movements. The same lens will be moved up and down for focus by the so-called two-axis device, which is a dual-moving coil mechanism. In some players this device is not statically balanced, making the unit sensitive to shock, but this was overcome on later heads designed for portable players. Figure 11.17 shows a later Sony design having a prism which reduces the height of the pickup above the disk.

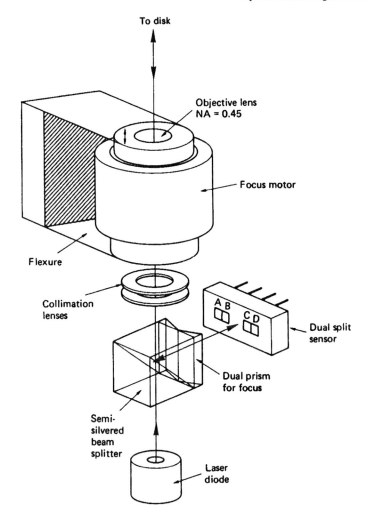

Figure 11.15 Philips laser head showing semi-silvered prism for beam splitting. Focus error is derived from dual-prism method using split sensors. Focus error $(A + D) - (B + C)$ is used to drive focus motor which moves objective lens on parallel action flexure. Radial differential tracking error is derived from split sensor $(A + B) - (C + D)$. Tracking error drives entire pickup on radial arm driven by moving coil. Signal output is $(A + B + C + D)$. System includes 600 Hz dither for tracking. (Courtesy *Philips Technical Review*.).

11.8 CD readout

Figure 11.2 was simplified only to the extent that the light spot was depicted as having a distinct edge of a given diameter. In reality such a neat spot cannot be obtained. It is essential to the commercial success of CD that a useful playing time (75 minutes max.) should be obtained from a recording of reasonable size (12 cm). The size was determined by the European motor industry as being appropriate for car dashboard mounted units. It follows that the smaller the spot of light which can be created, the smaller can be the

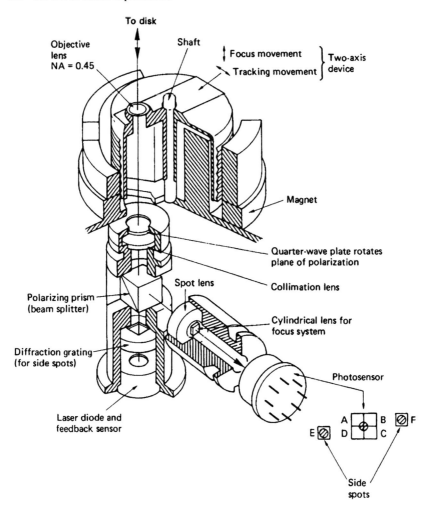

Figure 11.16 Sony laser head showing polarizing prism and quarter-wave plate for beam splitting, and diffraction grating for production of side spots for tracking. The cylindrical lens system is used for focus, with a four-quadrant sensor (A, B, C, D) and two extra sensors, E, F, for the side spots. Tracking error is $E - F$; focus error is $(A + C) - (B + D)$. Signal output is $(A + B + C + D)$. The focus and tracking errors drive the two-axis device. (Courtesy *Sony Broadcast.*).

deformities carrying the information, and so more information per unit area. Development of a successful high-density optical recorder requires an intimate knowledge of the behaviour of light focused into small spots. If it is attempted to focus a uniform beam of light to an infinitely small spot on a surface normal to the optical axis, it will be found that it is not possible. This is probably just as well as an infinitely small spot would have infinite intensity and any matter it fell on would not survive. Instead the result of such an attempt is a distribution of light in the area of the focal point which has no sharply defined boundary. This is called the Airy distribution[5] (sometimes pattern or disk) after

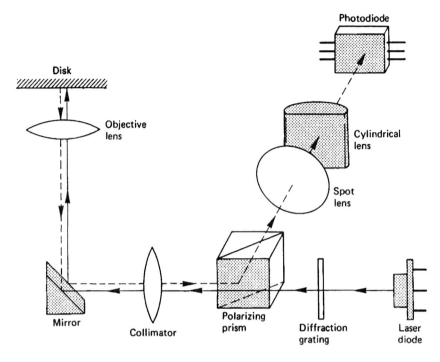

Figure 11.17 For automotive and portable players, the pickup can be made more compact by incorporating a mirror, which allows most of the elements to be parallel to the disk instead of at right angles.

Lord Airy, the then astronomer royal. If a line is considered to pass across the focal plane, through the theoretical focal point, and the intensity of the light is plotted on a graph as a function of the distance along that line, the result is the intensity function shown in Figure 11.18. It will be seen that this contains a central sloping peak surrounded by alternating dark rings and light rings of diminishing intensity. These rings will in theory reach to infinity before their intensity becomes zero. The intensity distribution or function described by Airy is due to diffraction effects across the finite aperture of the objective. For a given wavelength, as the aperture of the objective is increased, so the diameter of the features of the Airy pattern reduces. The Airy pattern vanishes to a singularity of infinite intensity with a lens of infinite aperture which of course cannot be made. The approximation of geometric optics is quite unable to predict the occurrence of the Airy pattern.

An intensity function does not have a diameter, but for practical purposes an effective diameter typically quoted is that at which the intensity has fallen to some convenient fraction of that at the peak. Thus one could state, for example, the half-power diameter.

Since light paths in optical instruments are generally reversible, it is possible to see an interesting corollary which gives a useful insight into the readout principle of CD. Considering light radiating from a phase structure, as in Figure 11.19, the more closely spaced the features of the phase structure,

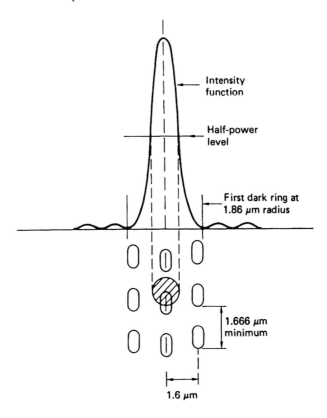

Figure 11.18 The structure of a maximum frequency recording is shown here, related to the intensity function of an objective of 0.45NA with 780 μm light. Note that track spacing puts adjacent tracks in the dark rings, reducing crosstalk. Note also that as the spot has an intensity function it is meaningless to specify the spot diameter without some reference such as an intensity level.

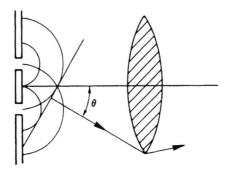

Figure 11.19 Fine detail in an object can only be resolved if the diffracted wavefront due to the higher spatial frequency is collected by the lens. Numerical aperture (NA) = $\sin \theta$, and as θ is the diffraction angle it follows that, for a given wavelength, NA determines resolution.

i.e. the higher the spatial frequency, the more oblique the direction of the wavefronts in the diffraction pattern which results and the larger the aperture of the lens needed to collect the light if the resolution is not to be lost. The corollary of this is that the smaller the Airy distribution it is wished to create, the larger must be the aperture of the lens. Spatial frequency is measured in lines per millimetre and as it increases, the wavefronts of the resultant diffraction pattern become more oblique. In the case of a CD, the smaller the bumps and the spaces between them along the track, the higher the spatial frequency, and the more oblique the diffraction pattern becomes in a plane tangential to the track. With a fixed objective aperture, as the tangential diffraction pattern becomes more oblique, less light passes the aperture and the depth of modulation transmitted by the lens falls. At some spatial frequency, all of the diffracted light falls outside the aperture and the modulation depth transmitted by the lens falls to zero. This is known as the spatial cut-off frequency. Thus a graph of depth of modulation versus spatial frequency can be drawn and which is known as the modulation transfer function (MTF). This is a straight line commencing at unity at zero spatial frequency (no detail) and falling to zero at the cut-off spatial frequency (finest detail). Thus one could describe a lens of finite aperture as a form of spatial low-pass filter. The Airy function is no more than the spatial impulse response of the lens, and the concentric rings of the Airy function are the spatial analogue of the symmetrical ringing in a phase linear electrical filter. The Airy function and the triangular frequency response form a transform pair[6] as shown in Chapter 2.

When an objective lens is used in a conventional microscope, the MTF will allow the resolution to be predicted in lines per millimetre. However, in a scanning microscope the spatial frequency of the detail in the object is multiplied by the scanning velocity to give a temporal frequency measured in Hertz. Thus lines per millimetre multiplied by millimetres per second gives lines per second. Instead of a straight line MTF falling to the spatial cut-off frequency, Figure 11.20 shows that a scanning microscope has a temporal

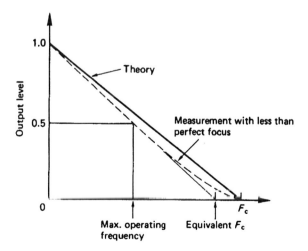

Figure 11.20 Frequency response of laser pickup. Maximum operating frequency is about half of cut-off frequency F_c.

frequency response falling to zero at the optical cut-off frequency which is given by:

$$F_c = \frac{2NA}{\text{wavelength}} \times \text{velocity}$$

The minimum linear velocity of CD is 1.2 m/s, giving a cut-off frequency of:

$$F_c = \frac{2 \times 0.45 \times 1.2}{780 \times 10^{-9}} = 1.38\,\text{MHz}$$

Actual measurements reveal that the optical response is only a little worse than the theory predicts. This characteristic has a large bearing on the type of modulation schemes which can be successfully employed. Clearly, to obtain any noise immunity, the maximum operating frequency must be rather less than the cut-off frequency. The maximum frequency used in CD is 720 kHz, which represents an absolute minimum wavelength of 1.666 µm, or a bump length of 0.833 µm, for the lowest permissible track speed of 1.2 m/s used on the full-length 75 min playing disks. One-hour playing disks have a minimum bump length of 0.972 µm at a track velocity of 1.4 m/s. The maximum frequency is the same in both cases. This maximum frequency should not be confused with the bit rate of CD since this is different owing to the channel code used. Figure 11.18 showed a maximum-frequency recording, and the physical relationship of the intensity function to the track dimensions.

The intensity function can be enlarged if the lens used suffers from optical aberrations. This was studied by Maréchal[7] who established criteria for the accuracy to which the optical surfaces of the lens should be made to allow the ideal Airy distribution to be obtained. CD player lenses must meet the Maréchal criterion. With such a lens, the diameter of the distribution function is determined solely by the combination of numerical aperture (NA) and the wavelength. When the size of the spot is as small as the NA and wavelength allow, the optical system is said to be diffraction-limited. Figure 11.19 showed how NA is defined, and illustrates that the smaller spot needed, the larger must be the NA. Unfortunately the larger the NA the more obliquely to the normal the light arrives at the focal plane and the smaller the depth of focus will be.

11.9 How optical disks are made

The steps used in the production of CDs will be outlined next. Prerecorded MiniDiscs are made in an identical fashion except for detail differences which will be noted. MO disks need to be grooved so that the track-following system will work. The grooved substrate is produced in a similar way to a CD master, except that the laser is on continuously instead of being modulated with a signal to be recorded. As stated, CD is replicated by moulding, and the first step is to produce a suitable mould. This mould must carry deformities of the correct depth for the standard wavelength to be used for reading, and as a practical matter these deformities must have slightly sloping sides so that it is possible to release the CD from the mould.

The major steps in CD manufacture are shown in Figure 11.21. The mastering process commences with an optically flat glass disk about 220 mm

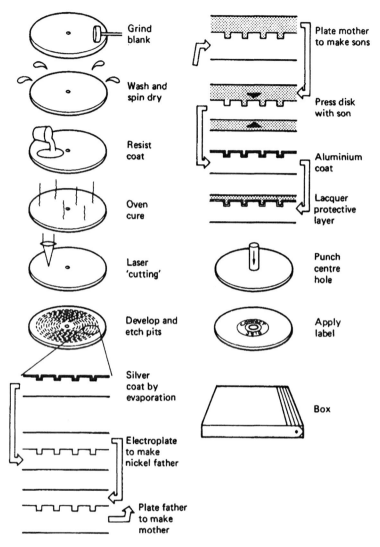

Figure 11.21 The many stages of CD manufacture, most of which require the utmost cleanliness.

in diameter and 6 mm thick. The blank is washed first with an alkaline solution, then with a fluorocarbon solvent, and spun dry prior to polishing to optical flatness. A critical cleaning process is then undertaken using a mixture of deionized water and isopropyl alcohol in the presence of ultrasonic vibration, with a final fluorocarbon wash. The blank must now be inspected for any surface irregularities which would cause data errors. This is done by using a laser beam and monitoring the reflection as the blank rotates. Rejected blanks return to the polishing process, those which pass move on, and an adhesive layer is applied followed by a coating of positive photoresist. This is a chemical substance which softens when exposed to an appropriate intensity of light of a certain wavelength, typically ultraviolet. Upon being thus exposed, the softened resist will be washed away by a developing solution down to the glass to form flat-bottomed pits whose depth is equal to the thickness of the undeveloped resist. During development the master is illuminated with laser light of a wavelength to which it is insensitive. The diffraction pattern changes as the pits are formed. Development is arrested when the appropriate diffraction pattern is obtained.[8] The thickness of the resist layer must be accurately controlled, since it affects the height of the bumps on the finished disk, and an optical scanner is used to check that there are no resist defects which would cause data errors or tracking problems in the end product. Blanks which pass this test are oven-cured, and are ready for cutting. Failed blanks can be stripped of the resist coating and used again.

The cutting process is shown in simplified form in Figure 11.22. A continuously operating helium–cadmium[9] or argon-ion[10] laser is focused on the resist coating as the blank revolves. Focus is achieved by a separate helium–neon laser sharing the same optics. The resist is insensitive to the wavelength of the helium–neon laser. The laser intensity is controlled by a device known as an acousto-optic modulator which is driven by the encoder. When the device is in a relaxed state, light can pass through it, but when the surface is excited by high-frequency vibrations, light is scattered. Information is carried in the lengths of time for which the modulator remains on or remains off. As a result

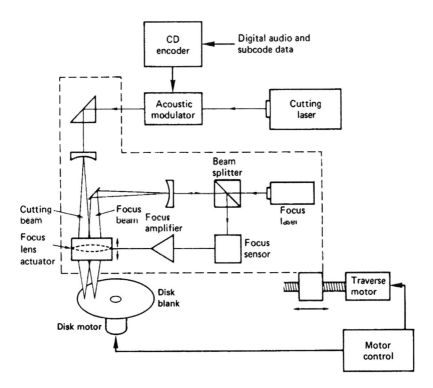

Figure 11.22 CD cutter. The focus subsystem controls this spot size of the main cutting laser on the photosensitive blank. Disk and traverse motors are coordinated to give constant track pitch and velocity. Note that the power of the focus laser is insufficient to expose the photoresist.

the deformities in the resist produced as the disk turns when the modulator allows light to pass are separated by areas unaffected by light when the modulator is shut off. Information is carried solely in the variations of the lengths of these two areas.

The laser makes its way from the inside to the outside as the blank revolves. As the radius of the track increases, the rotational speed is proportionately reduced so that the velocity of the beam over the disk remains constant. This constant linear velocity (CLV) results in rather longer playing time than would be obtained with a constant speed of rotation. Owing to the minute dimensions of the track structure, the cutter has to be constructed to extremely high accuracy. Air bearings are used in the spindle and the laser head, and the whole machine is resiliently supported to prevent vibrations from the building from affecting the track pattern.

As the player is a phase contrast microscope, it must produce an intensity function which straddles the deformities. As a consequence the intensity function which produces the deformities in the photoresist must be smaller in diameter than that in the reader. This is conveniently achieved by using a shorter wavelength of 400–500 nm from a helium–cadmium or argon-ion laser combined with a larger lens NA of 0.9. These are expensive, but only needed for the mastering process.

It is a characteristic of photoresist that its development rate is not linearly proportional to the intensity of light. This non-linearity is known as 'gamma'. As a result there are two intensities of importance when scanning photoresist; the lower sensitivity, or threshold, below which no development takes place, and the upper threshold above which there is full development. As the laser light falling on the resist is an intensity function, it follows that the two thresholds will be reached at different diameters of the function. It can be seen in Figure 11.23 that advantage is taken of this effect to produce tapering

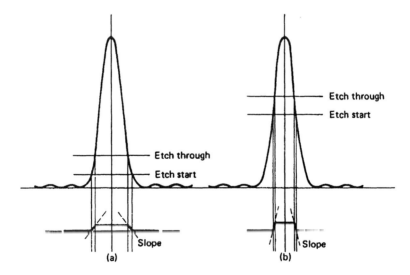

Figure 11.23 The two levels of exposure sensitivity of the resist determine the size and edge slope of the bumps in the CD. (a) Large exposure results in large bump with gentle slope; (b) less exposure results in smaller bump with steeper sloped sides.

sides to the pits formed in the resist. In the centre, the light is intense enough to fully develop the resist right down to the glass. This gives the deformity a flat bottom. At the edge, the intensity falls and as some light is absorbed by the resist, the diameter of the resist which can be developed falls with depth in the resist. By controlling the intensity of the laser, and the development time, the slope of the sides of the pits can be controlled.

The master recording process has produced a phase structure in relatively delicate resist, and this cannot be used for moulding directly. Instead a thin metallic silver layer is sprayed onto the resist to render it electrically conductive so that electroplating can be used to make robust copies of the relief structure.

The electrically conductive resist master is then used as the cathode of an electroplating process where a first layer of metal is laid down over the resist, conforming in every detail to the relief structure thereon. This metal layer can then be separated from the glass and the resist is dissolved away and the silver is recovered leaving a laterally inverted phase structure on the surface of the metal, in which the pits in the photoresist have become bumps in the metal. From this point on, the production of CD is virtually identical to the replication process used for vinyl disks, save only that a good deal more precision and cleanliness is needed.

This first metal layer could itself be used to mould disks, or it could be used as a robust sub-master from which many stampers could be made by pairs of plating steps. The first metal phase structure can itself be used as a cathode in a further electroplating process in which a second metal layer is formed having a mirror image of the first. A third such plating step results in a stamper. The decision to use the master or sub-stampers will be based on the number of disks and the production rate required.

The master is placed in a moulding machine, opposite a flat plate. A suitable quantity of molten plastic is injected between, and the plate and the master are forced together. The flat plate renders one side of the disk smooth, and the bumps in the metal stamper produce pits in the other surface of the disk. The surface containing the pits is next metallized, with any good electrically conductive material, typically aluminium. This metallization is then covered with a lacquer for protection. In the case of CD, the label is printed on the lacquer. In the case of a prerecorded MiniDisc, the ferrous hub needs to be applied prior to fitting the cartridge around the disk.

As CD and prerecorded MDs are simply data disks, they do not need to be mastered in real time. Raising the speed of the mastering process increases the throughput of the expensive equipment. The U-matic-based PCM-1630 CD mastering recorder is incapable of working faster than real time, and pressing plants have been using computer tape streamers in order to supply the cutter with higher data rates. The Sony MO mastering disk drive is designed to operate at up to 2.5 times real time to support high-speed mastering.

11.10 How recordable MiniDiscs are made

Recordable MiniDiscs make the recording as flux patterns in a magnetic layer. However, the disks need to be pre-grooved so that the tracking systems can operate. The grooves have the same pitch as CD and the prerecorded MD, but the tracks are the same width as the laser spot: about $1.1\,\mu m$. The grooves are not a perfect spiral, but have a sinusoidal waviness at a fixed wavelength. Like CD, MD uses constant track linear velocity, not constant speed of rotation. When recording on a blank disk, the recorder needs to know how fast to turn the spindle to get the track speed correct. The wavy grooves will be followed by the tracking servo and the frequency of the tracking error will be proportional to the disk speed. The recorder simply turns the spindle at a speed which makes the grooves wave at the correct frequency. The groove frequency is 75 Hz; the same as the data sector rate. Thus a zero crossing in the groove signal can also be used to indicate where to start recording. The grooves are particularly important when a chequerboarded recording is being replayed. On a CLV disk, every seek to a new track radius results in a different track speed. The wavy grooves allow the track velocity to be corrected as soon as a new track is reached.

The pre-grooves are moulded into the plastics body of the disk when it is made. The mould is made in a similar manner to a prerecorded disk master, except that the laser is not modulated and the spot is larger. The track velocity is held constant by slowing down the resist master as the radius increases, and the waviness is created by injecting 75 Hz into the lens radial positioner. The master is developed and electroplated as normal in order to make stampers. The stampers make pre-grooved disks which are then coated by vacuum deposition with the MO layer, sandwiched between dielectric layers. The MO layer can be made less susceptible to corrosion if it is smooth and homogeneous. Layers which contain voids, asperities or residual gases from the coating process present a larger surface area for attack. The life of an MO disk is affected more by the manufacturing process than by the precise composition of the alloy.

Above the sandwich an optically reflective layer is applied, followed by a protective lacquer layer. The ferrous clamping plate is applied to the centre of the disk, which is then fitted in the cartridge. The recordable cartridge has a double-sided shutter to allow the magnetic head access to the back of the disk.

11.11 Channel code of CD and MD

CD and MiniDisc use the same channel code. This was optimized for the optical readout of CD and prerecorded MiniDisc, but is also used for the recordable version of MiniDisc for simplicity.

The frequency response falling to the optical cut-off frequency is only one of the constraints within which the modulation scheme has to work. There are a number of others. In all players the tracking and focus servos operate by analysing the average amount of light returning to the pickup. If the average amount of light returning to the pickup is affected by the content of the recorded data, then the recording will interfere with the operation of the servos. Debris on the disk surface affects the light intensity and means must be found to prevent this reducing the signal quality excessively. Chapter 10 discussed modulation schemes known as DC-free codes. If such a code is used, the average brightness of the track is constant and independent of the data bits. Figure 11.24(a) shows the replay signal from the pickup being compared with a threshold voltage in order to recover a binary waveform from the analog pickup waveform, a process known as slicing. If the light beam is partially

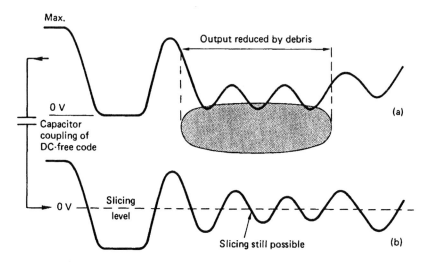

Figure 11.24 A DC-free code allows signal amplitude variations due to debris to be rejected.

obstructed by debris, the pickup signal level falls, and the slicing level is no longer correct and errors occur. If, however, the code is DC free, the waveform from the pickup can be passed through a high-pass filter (e.g. a series capacitor) and Figure 11.24(b) shows that this rejects the falling level and converts it to a reduction in amplitude about the slicing level so that the slicer still works properly. This step cannot be performed unless a DC-free code is used.

As the frequency response on replay falls linearly to the cut-off frequency determined by the aperture of the lens and the wavelength of light used, the shorter bumps and lands produce less modulation than longer ones. Figure 11.25(c) shows that if the recorded waveform is restricted to one which is DC free, as the length of bumps and lands falls with rising density, the replay waveform simply falls in amplitude but the average voltage remains the same and so the slicer still operates correctly. It will be clear that using a DC-free code correct slicing remains possible with much shorter bumps and lands than with direct recording.

CD uses a coding scheme where combinations of the data bits to be recorded are represented by unique waveforms. These waveforms are created by combining various run lengths from $3T$ to $11T$ together to give a channel pattern which is $14T$ long.[11] Within the run-length limits of $3T$ to $11T$, a waveform $14T$ long can have 267 different patterns. This is slightly more than the 256 combinations of eight data bits and so eight bits are represented by a waveform lasting $14T$. Some of these patterns are shown in Figure 11.26. As stated, these patterns are not polarity conscious and they could be inverted without changing the meaning.

Not all of the $14T$ patterns used are DC free, some spend more time in one state than the other. The overall DC content of the recorded waveform is rendered DC free by inserting an extra portion of waveform, known as a packing period, between the $14T$ channel patterns. This packing period is $3T$

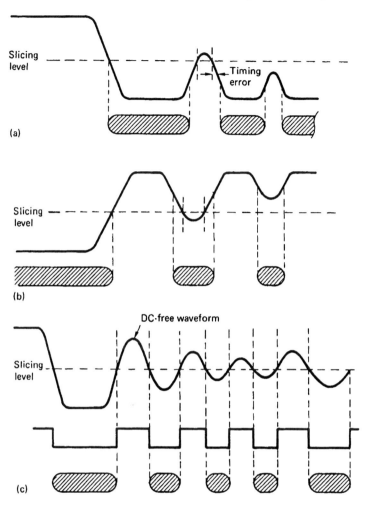

Figure 11.25 If the recorded waveform is not DC free, timing errors occur until slicing becomes impossible. With a DC-free code, jitter-free slicing is possible in the presence of serious amplitude variation.

long and may or may not contain a transition, which if it is present can be in one of three places. The packing period contains no information, but serves to control the DC content of the overall waveform.[12] The packing waveform is generated in such a way that in the long term the amount of time the channel signal spends in one state is equal to the time it spends in the other state. A packing period is placed between every pair of channel patterns and so the overall length of time needed to record eight bits is $17T$.

Thus a group of eight data bits is represented by a code of fourteen channel bits, hence the name of eight to fourteen modulation (EFM). It is a common misconception that the channel bits of a group code are recorded; in fact they are simply a convenient way of synthesizing a coded waveform having uniform time steps. It should be clear that channel bits cannot be recorded as

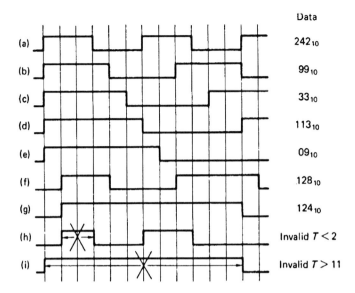

Data

(a) 242_{10}

(b) 99_{10}

(c) 33_{10}

(d) 113_{10}

(e) 09_{10}

(f) 128_{10}

(g) 124_{10}

(h)   Invalid $T < 2$

(i)   Invalid $T > 11$

Figure 11.26 (a–g) Part of the codebook for EFM code showing examples of various run lengths from $3T$ to $11T$. (h, i) Invalid patterns which violate the run-length limits.

they have a rate of 4.3 Mbits/s whereas the optical cut-off frequency of CD is only 1.4 MHz.

Another common misconception is that channel bits are data. If channel bits were data, all combinations of 14 bits, or 16 384 different values could be used. In fact only 267 combinations produce waveforms which can be recorded. In a practical CD modulator, the eight-bit data symbols to be recorded are used as the address of a lookup table which outputs a 14-bit channel-bit pattern. As the highest frequency which can be used in CD is 720 kHz, transitions cannot be closer together than $3T$ and so successive 1s in the channel bit stream must have two or more zeros between them. Similarly transitions cannot be further apart than $11T$ or there will be insufficient clock content. Thus there cannot be more than 10 zeros between channel 1s. Whilst the lookup table can be programmed to prevent code violations within the $14T$ pattern, they could occur at the junction of two successive patterns. Thus a further function of the packing period is to prevent violation of the run-length limits. If the previous pattern ends with a transition and the next begins with one, there will be no packing transition and so the $3T$ minimum requirement can be met. If the patterns either side have long run lengths, the sum of the two might exceed $11T$ unless the packing period contained a transition. In fact the minimum run-length limit could be met with $2T$ of packing, but the requirement for DC control dictated $3T$ of packing.

Decoding the stream of channel bits into data requires that the boundaries between successive $17T$ periods are identified. This is the process of de-serialization. On the disk one $17T$ period runs straight into the next; there are no dividing marks. Symbol separation is performed by counting channel-bit periods and dividing them by 17 from a known reference point. The three packing periods are discarded and the remaining $14T$ symbol is decoded to eight data bits. The reference point is provided by the synchronizing pattern which is given that name because its detection synchronizes the deserialization counter to the replay waveform.

Synchronization has to be as reliable as possible because if it is incorrect all of the data will be corrupted up to the next sync pattern. Synchronization is achieved by the detection of a unique waveform periodically recorded on the track with regular spacing. It must be unique in the strict sense in that nothing else can give rise to it, because the detection of a false sync is just as damaging as failure to detect a correct one. In practice CD synchronizes deserialization with a waveform which is unique in that it is different from any of the 256 waveforms which represent data. For reliability, the sync pattern should have the best signal-to-noise ratio possible, and this is obtained by making it one complete cycle of the lowest frequency ($11T$ plus $11T$) which gives it the largest amplitude and also makes it DC free. Upon detection of the $2 \times T_{max}$ waveform, the deserialization counter which divides the channel-bit count by 17 is reset. This occurs on the next system clock, which is the reason for the 0 in the sync pattern after the third 1 and before the merging bits. CD therefore uses forward synchronization and correctly deserialized data are available immediately after the first sync pattern is detected. The sync pattern is longer than the data symbols, and so clearly no data code value can create it, although it would be possible for certain adjacent data symbols to create a false sync pattern by concatenation were it not for the presence of the packing period. It is a further job of the packing period to prevent false sync patterns being generated at the junction of two channel symbols.

Each data block or frame in CD and MD, shown in Figure 11.27, consists of 33 symbols $17T$ each following the preamble, making a total of $588T$ or $136\,\mu s$. Each symbol represents eight data bits. The first symbol in the block is used for subcode, and the remaining 32 bytes represent 24 audio sample bytes and eight bytes of redundancy for the error-correction system. The subcode byte forms part of a subcode block which is built up over 98 successive data frames.

Figure 11.28 shows an overall block diagram of the record modulation scheme used in CD mastering and the corresponding replay system or data

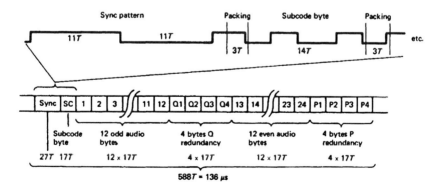

Figure 11.27 One CD data block begins with a unique sync pattern, and one subcode byte, followed by 24 audio bytes and eight redundancy bytes. Note that each byte requires $14T$ in EFM, with $3T$ packing between symbols, making $17T$.

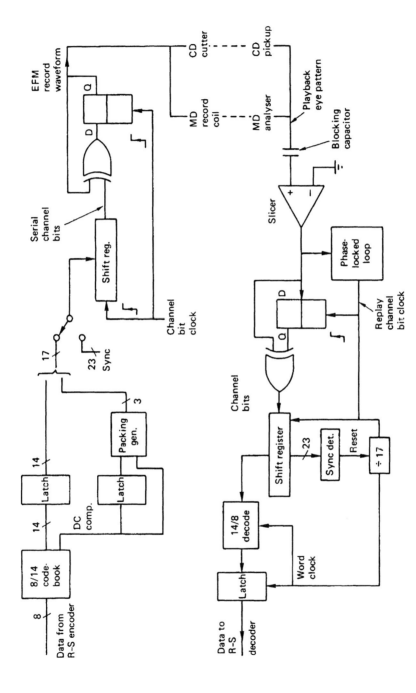

Figure 11.28 Overall block diagram of the EFM encode/decode process. A MiniDisc will contain both. A CD player only has the decoder; the encoding is in the mastering cutter.

separator. The input to the record channel coder consists of 16-bit audio samples which are divided in two to make symbols of eight bits. These symbols are used in the error-correction system which interleaves them and adds redundant symbols. For every 12 audio symbols, there are four symbols of redundancy, but the channel coder is not concerned with the sequence or significance of the symbols and simply records their binary code values.

Symbols are provided to the coder in eight-bit parallel format, with a symbol clock. The symbol clock is obtained by dividing down the 4.3218 MHz T rate clock by a factor of 17. Each symbol is used to address the lookup table which outputs a corresponding 14-channel bit pattern in parallel into a shift register. The T rate clock then shifts the channel bits along the register. The lookup table also outputs data corresponding to the digital sum value (DSV) of the 14-bit symbol to the packing generator. The packing generator determines if action is needed between symbols to control DC content. The packing generator checks for run-length violations and potential false sync patterns. As a result of all the criteria, the packing generator loads three channel bits into the space between the symbols, such that the register then contains 14-bit symbols with three bits of packing between them. At the beginning of each frame, the sync pattern is loaded into the register just before the first symbol is looked up in such a way that the packing bits are correctly calculated between the sync pattern and the first symbol.

A channel bit one indicates that a transition should be generated, and so the serial output of the shift register is fed to the JK bistable along with the T rate clock. The output of the JK bistable is the ideal channel coded waveform containing transitions separated by $3T$ to $11T$. It is a self-clocking, run-length limited waveform. The channel bits and the T rate clock have done their job of changing the state of the JK bistable and do not pass further on. At the output of the JK the sync pattern is simply two $11T$ run lengths in series. At this stage the run-length limited waveform is used to control the acousto-optic modulator in the cutter.

The resist master is developed and used to create stampers. The resulting disks can then be replayed. The track velocity of a given CD is constant, but the rotational speed depends upon the radius. In order to get into lock, the disk must be spun at roughly the right track speed. This is done using the run-length limits of the recording. The pickup is focused and the tracking is enabled. The replay waveform from the pickup is passed through a high-pass filter to remove level variations due to contamination and sliced to return it to a binary waveform. The slicing level is self-adapting as Figure 11.29 shows so that a 50% duty cycle is obtained. The slicer output is then sampled by the unlocked VCO running at approximately T rate. If the disk is running too slowly, the longest run length on the disk will appear as more than $11T$, whereas if the disk is running too fast, the shortest run length will appear as less than $3T$. As a result the disk speed can be brought to approximately the right speed and the VCO will then be able to lock to the clock content of the EFM waveform from the slicer. Once the VCO is locked, it will be possible to sample the replay waveform at the correct T rate. The output of the sampler is then differentiated and the channel bits reappear and are fed into the shift register. The sync pattern detector will then function to reset the deserialization counter which allows the $14T$ symbols to be identified. The $14T$ symbols are then decoded to eight bits in the reverse coding table.

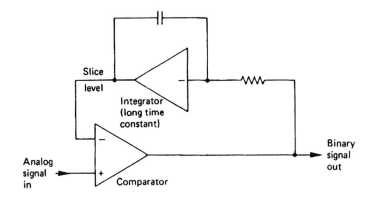

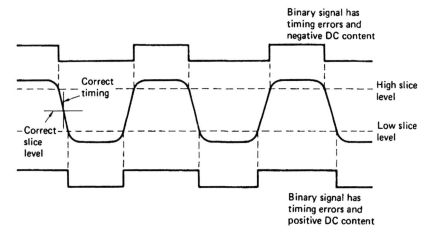

Figure 11.29 Self-slicing a DC-free channel code. Since the channel code signal from the disk is band limited, it has finite rise times, and slicing at the wrong level (as shown here) results in timing errors, which cause the data separator to be less reliable. As the channel code is DC free, the binary signal when correctly sliced should integrate to zero. An incorrect slice level gives the binary output a DC content and, as shown here, this can be fed back to modify the slice level automatically.

Figure 11.30 reveals the timing relationships of the CD format. The sampling rate of 44.1 kHz with 16-bit words in left and right channels results in an audio data rate of 176.4 kb/s (k = 1000 here, not 1024). Since there are 24 audio bytes in a data frame, the frame rate will be:

$$\frac{176.4}{24} \, \text{kHz} = 7.35 \, \text{kHz}$$

If this frame rate is divided by 98, the number of frames in a subcode block, the subcode block or sector rate of 75 Hz results. This frequency can be divided down to provide a running-time display in the player. Note that this is the frequency of the wavy grooves in recordable MDs.

If the frame rate is multiplied by 588, the number of channel bits in a frame, the master clock rate of 4.3218 MHz results. From this the maximum and

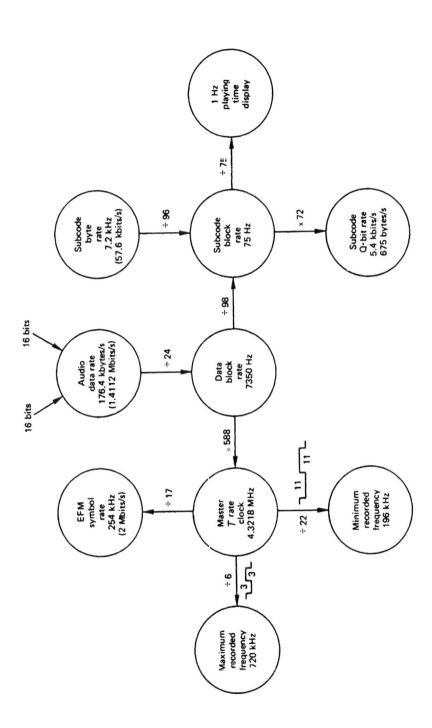

Figure 11.30 CD timing structure.

minimum frequencies in the channel, 720 kHz and 196 kHz, can be obtained using the run-length limits of EFM.

11.12 Error-correction strategy

This section discusses the track structure of CD in detail. The track structure of MiniDisc is based on that of CD and the differences will be noted in the next section.

Each sync block was seen in Figure 11.27 to contain 24 audio bytes, but these are non-contiguous owing to the extensive interleave.[13-15] There are a number of interleaves used in CD, each of which has a specific purpose. The full interleave structure is shown in Figure 11.31. The first stage of interleave is to introduce a delay between odd and even samples. The effect is that uncorrectable errors cause odd samples and even samples to be destroyed at different times, so that interpolation can be used to conceal the errors, with a reduction in audio bandwidth and a risk of aliasing. The odd/even interleave is performed first in the encoder, since concealment is the last function in the decoder. Figure 11.32 shows that an odd/even delay of two

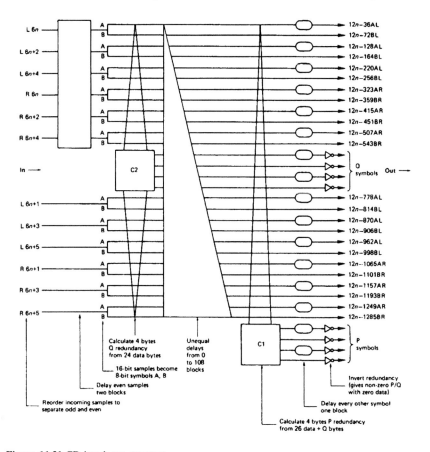

Figure 11.31 CD interleave structure.

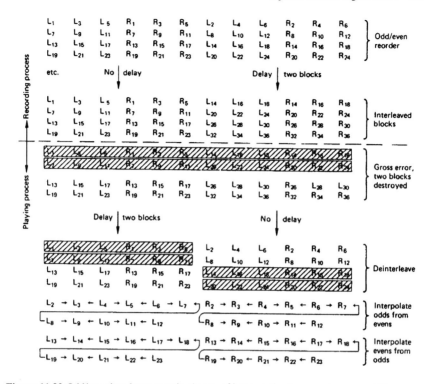

Figure 11.32 Odd/even interleave permits the use of interpolation to conceal uncorrectable errors.

blocks permits interpolation in the case where two uncorrectable blocks leave the error-correction system.

Left and right samples from the same instant form a sample set. As the samples are 16 bits, each sample set consists of four bytes, AL, BL, AR, BR. Six sample sets form a 24-byte parallel word, and the C2 encoder produces four bytes of redundancy Q. By placing the Q symbols in the centre of the block, the odd/even distance is increased, permitting interpolation over the largest possible error burst. The 28 bytes are now subjected to differing delays, which are integer multiples of four blocks. This produces a convolutional interleave, where one C2 codeword is stored in 28 different blocks, spread over a distance of 109 blocks.

At one instant, the C2 encoder will be presented with 28 bytes which have come from 28 different codewords. The C1 encoder produces a further four bytes of redundancy P. Thus the C1 and C2 codewords are produced by crossing an array in two directions. This is known as crossinterleaving.

The final interleave is an odd/even output symbol delay, which causes P codewords to be spread over two blocks on the disk as shown in Figure 11.33. This mechanism prevents small random errors destroying more than one symbol in a P codeword. The choice of eight-bit symbols in EFM assists this strategy. The expressions in Figure 11.31 determine how the interleave is calculated. Figure 11.34 shows an example of the use of these expressions to calculate the contents of a block and to demonstrate the crossinterleave.

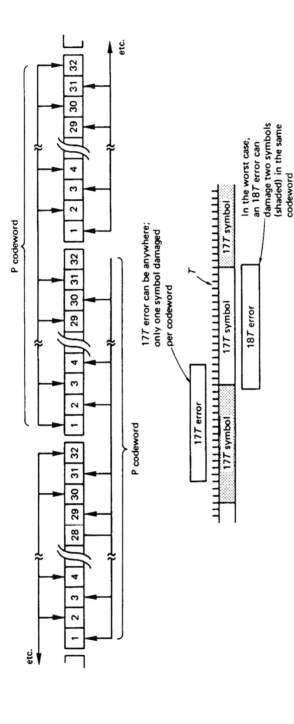

Figure 11.33 The final interleave of the CD format spreads P codewords over two blocks. Thus any small random error can only destroy one symbol in one codeword, even if two adjacent symbols in one block are destroyed. Since the P code is optimized for single-symbol error correction, random errors will always be corrected by the C1 process, maximizing the burst-correcting power of the C2 process after deinterleave.

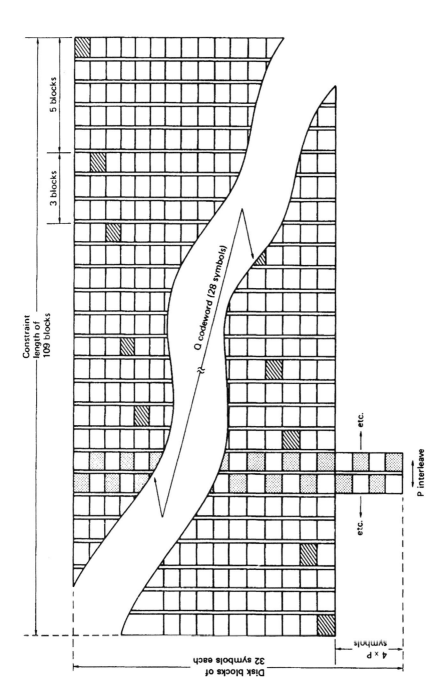

Figure 11.34 Owing to crossinterleave, the 28 symbols from the Q encode process (C2) are spread over 109 blocks, shown hatched. The final interleave of P codewords (as in Figure 11.33) is shown stippled. The result of the latter is that Q codeword has 5, 3, 5, 3 spacing rather than 4, 4.

The calculation of the P and Q redundancy symbols is made using Reed–Solomon polynomial division. The P redundancy symbols are primarily for detecting errors, to act as pointers or error flags for the Q system. The P system can, however, correct single-symbol errors.

11.13 Track layout of MD

MD uses the same channel code and error-correction interleave as CD for simplicity and the sectors are exactly the same size. The interleave of CD is convolutional, which is not a drawback in a continuous recording. However, MD uses random access and the recording is discontinuous. Figure 11.35 shows that the convolutional interleave causes codewords to run between sectors. Re-recording a sector would prevent error correction in the area of the edit. The solution is to use a buffering zone in the area of an edit where the convolution can begin and end. This is the job of the link sectors. Figure 11.36 shows the layout of data on a recordable MD. In each cluster of 36 sectors, 32 are used for encoded audio data. One is used for subcode and the remaining three are link sectors. The cluster is the minimum data quantum which can be recorded and represents just over two seconds of decoded audio. The cluster must be recorded continuously because of the convolutional interleave. Effectively the link sectors form an edit gap which is large enough to absorb both mechanical tolerances and the interleave overrun when a cluster is rewritten. One or more clusters will be assembled in memory before writing to the disk is attempted.

Prerecorded MDs are recorded at one time, and need no link sectors. In order to keep the format consistent between the two types of MiniDisc, three extra subcode sectors are made available. As a result it is not possible to record the entire audio and subcode of a prerecorded MD onto a recordable MD because the link sectors cannot be used to record data.

The ATRAC coder produces what are known as sound groups. Figure 11.36 shows that these contain 212 bytes for each of the two audio channels and are the equivalent of 11.6 milliseconds of real-time audio. Eleven of these sound groups will fit into two standard CD sectors with 20 bytes to spare. The 32

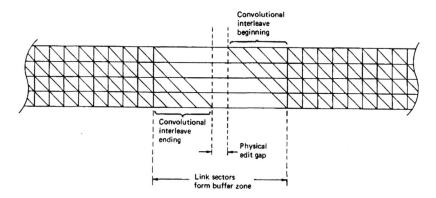

Figure 11.35 The convolutional interleave of CD is retained in MD, but buffer zones are needed to allow the convolutional to finish before a new one begins, otherwise editing is impossible.

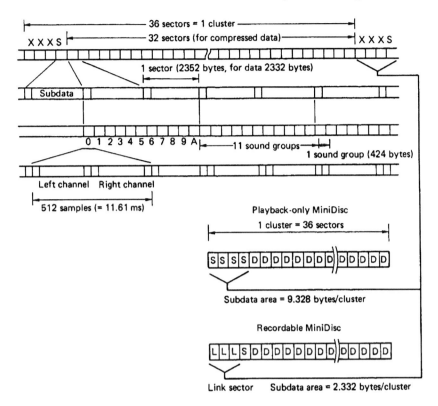

Figure 11.36 Format of MD uses clusters of sectors including link sectors for editing. Prerecorded MDs do not need link sectors, so more subcode capacity is available. The ATRAC coder of MD produces the sound groups shown here.

audio data sectors in a cluster thus contain a total of $16 \times 11 = 176$ sound groups.

11.14 Player structure

The physics of the manufacturing process and the readout mechanism have been described, along with the format on the disk. Here, the details of actual CD and MD players will be explained. One of the design constraints of the CD and MD formats was that the construction of players should be straightforward, since they were to be mass-produced.

Figure 11.38 shows the block diagram of a typical CD player, and illustrates the essential components. The most natural division within the block diagram is into the control/servo system and the data path. The control system provides the interface between the user and the servo mechanisms, and performs the logical interlocking required for safety and the correct sequence of operation.

The servo systems include any power-operated loading drawer and chucking mechanism, the spindle-drive servo, and the focus and tracking servos already described. Power loading is usually implemented on players where the disk

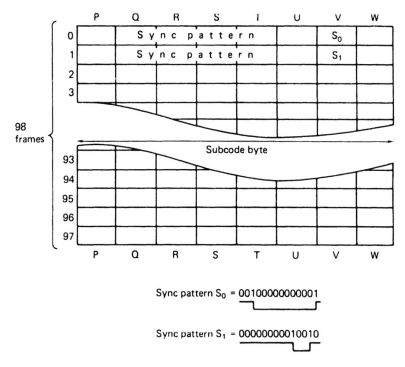

Figure 11.37 Each CD frame contains one subcode byte. After 98 frames, the structure above will repeat. Each subcode byte contains 1 bit from eight 96-bit words following the two synchronizing patterns. These patterns cannot be expressed as a byte, because they are 14-bit EFM patterns additional to those which describe the 256 combination of eight data bits.

is placed in a drawer. Once the drawer has been pulled into the machine, the disk is lowered onto the drive spindle, and clamped at the centre, a process known as chucking. In the simpler top-loading machines, the disk is placed on the spindle by hand, and the clamp is attached to the lid so that it operates as the lid is closed.

The lid or drawer mechanisms have a safety switch which prevents the laser operating if the machine is open. This is to ensure that there can be no conceivable hazard to the user. In actuality there is very little hazard in a CD pickup. This is because the beam is focused a few millimetres away from the objective lens, and beyond the focal point the beam diverges and the intensity falls rapidly. It is almost impossible to position the eye at the focal point when the pickup is mounted in the player, but it would be foolhardy to attempt to disprove this.

The data path consists of the data separator, timebase correction and the deinterleaving and error-correction process followed by the error-concealment mechanism. This results in a sample stream which is fed to the converters. The data separator which converts the readout waveform into data was detailed in the description of the CD channel code. LSI chips have been developed to perform the data-separation function: for example, the Philips SAA 7010 or the Sony CX 7933. The separated output from both of these consists of subcode

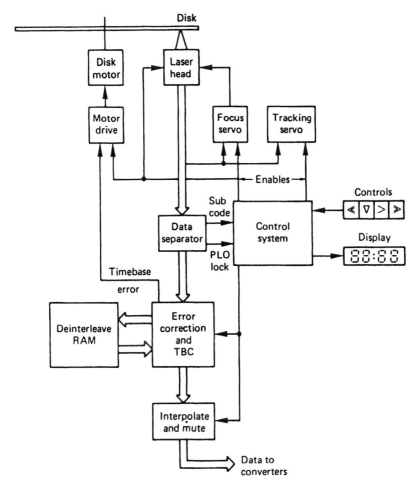

Figure 11.38 Block diagram of CD player showing the data path (broad arrow) and control/servo systems.

bytes, audio samples, redundancy and a clock. The data stream and the clock will contain speed variations due to disk runout and chucking tolerances, and these have to be removed by a timebase corrector.

The timebase corrector is a memory addressed by counters which are arranged to overflow, giving the memory a ring structure as described in Chapter 1. Writing into the memory is done using clocks from the data separator whose frequency rises and falls with runout, whereas reading is done using a crystal-controlled clock, which removes speed variations from the samples, and makes wow and flutter unmeasurable. The timebase-corrector will only function properly if the two addresses are kept apart. This implies that the long-term data rate from the disk must equal the crystal-clock rate. The disk speed must be controlled to ensure that this is always true, and there are two contrasting ways in which it can be done.

The data-separator clock counts samples off the disk. By phase-comparing this clock with the crystal reference, the phase error can be used to drive the spindle motor. This system is used in the Sony CDP-101, where the principle is implemented with a CX-193 chip, which was originally designed for DC turntable motors. The data-separator signal replaces the feedback signal which would originally have come from a toothed wheel on the turntable.

The alternative approach is to analyse the address relationship of the time-base corrector. If the disk is turning too fast, the write address will move towards the read address; if the disk is turning too slowly, the write address moves away from the read address. Subtraction of the two addresses produces an error signal which can be fed to the motor. The TBC RAM in Philips players, which also serves as the deinterleave memory, is a 2 kbyte SSB 2016, and this is controlled by the SAA 7020, which produces the motor-control signal. In these systems, and in all CD players, the speed of the motor is unimportant. The important factor is that the sample rate is correct, and the system will drive the spindle at whatever speed is necessary to achieve the correct rate. As the disk cutter produces constant-bit density along the track by reducing the rate of rotation as the track radius increases, the player will automatically duplicate that speed reduction. The actual linear velocity of the track will be the same as the velocity of the cutter, and although this will be constant for a given disk, it can vary between 1.2 and 1.4 m/s on different disks.

These speed-control systems can only operate when the data separator has phase-locked, and this cannot happen until the disk speed is almost correct. A separate mechanism is necessary to bring the disk up to roughly the right speed. One way of doing this is to make use of the run-length limits of the channel code. Since transitions closer than $3T$ and further apart than $11T$ are not present, it is possible to estimate the disk speed by analysing the run lengths. The period between transitions should be from 694 ns to 2.55 µs. During disk runup the periods between transitions can be measured, and if the longest period found exceeds 2.55 µs, the disk must be turning too slowly, whereas if the shortest period is less than 694 ns, the disk must be turning too fast. Once the data separator locks up, the coarse speed control becomes redundant. The method relies upon the regular occurrence of maximum and minimum run lengths in the channel. Synchronizing patterns have the maximum run length, and occur regularly. The description of the disk format showed that the C1 and C2 redundancy was inverted. This injects some 1s into the channel even when the audio is muted. This is the situation during the lead-in track – the very place that lock must be achieved. The presence of the table of contents in subcode during the lead-in also helps to produce a range of run lengths.

Owing to the use of constant linear velocity, the disk speed will be wrong if the pickup is suddenly made to jump to a different radius using manual search controls. This may force the data separator out of lock, and the player will mute briefly until the correct track speed has been restored, allowing the PLO to lock again. This can be demonstrated with most players, since it follows from the format.

Following data separation and timebase correction, the error-correction and deinterleave processes take place. Because of the crossinterleave system, there are two opportunities for correction, firstly using the C1 redundancy prior to deinterleaving, and secondly using the C2 redundancy after deinterleaving. In Chapter 10 it was shown that interleaving is designed to spread the effects of burst errors among many different codewords, so that the errors in each are reduced. However, the process can be impaired if a small random error, due perhaps to an imperfection in manufacture, occurs close to a burst error caused by surface contamination. The function of the C1 redundancy is to correct single-symbol errors, so that the power of interleaving to handle bursts is undiminished, and to generate error flags for the C2 system when a gross error is encountered.

The EFM coding is a group code which means that a small defect which changes one channel pattern into another will have corrupted up to eight data bits. In the worst case, if the small defect is on the boundary between two channel patterns, two successive bytes could be corrupted. However, the final odd/even interleave on encoding ensures that the two bytes damaged will be in different C1 codewords; thus a random error can never corrupt two bytes in one C1 codeword, and random errors are therefore always correctable by C1. From this it follows that the maximum size of a defect considered random is $17T$ or $3.9\,\mu s$. This corresponds to about a $5\,\mu m$ length of the track. Errors of greater size are, by definition, burst errors.

The deinterleave process is achieved by writing sequentially into a memory and reading out using a sequencer. The RAM can perform the function of the timebase corrector as well. The size of memory necessary follows from the format; the amount of interleave used is a compromise between the resistance to burst errors and the cost of the deinterleave memory. The maximum delay is 108 blocks of 28 bytes, and the minimum delay is negligible. It follows that a memory capacity of $54 \times 28 = 1512$ bytes is necessary. Allowing a little extra for timebase error, odd/even interleave and error flags transmitted from C1 to C2, the convenient capacity of 2048 bytes is reached.

The C2 decoder is designed to locate and correct a single-symbol error, or to correct two symbols whose locations are known. The former case occurs very infrequently, as it implies that the C1 decoder has miscorrected. However, the C1 decoder works before deinterleave, and there is no control over the burst-error size that it sees. There is a small but finite probability that random data in a large burst could produce the same syndrome as a single error in good data. This would cause C1 to miscorrect, and no error flag would accompany the miscorrected symbols. Following deinterleave, the C2 decode could detect and correct the miscorrected symbols as they would now be single-symbol errors in many codewords. The overall miscorrection probability of the system is thus quite minute. Where C1 detects burst errors, error flags will be attached to all symbols in the failing C1 codeword. After deinterleave in the memory, these flags will be used by the C2 decoder to correct up to two corrupt symbols in one C2 codeword. Should more than two flags appear in one C2 codeword, the errors are uncorrectable, and C2 flags the entire codeword bad, and the interpolator will have to be used. The final odd/even sample deinterleave makes interpolation possible because it displaces the odd corrupt samples relative to the even corrupt samples.

If the rate of bad C2 codewords is excessive, the correction system is being overwhelmed, and the output must be muted to prevent unpleasant noise. Unfortunately digital audio cannot be muted by simply switching the sample stream to zero, since this would produce a click. It is necessary to fade down to the mute condition gradually by multiplying sample values by descending coefficients, usually in the form of a half-cycle of a cosine wave. This gradual fadeout requires some advance warning, in order to be able to fade out before the errors arrive. This is achieved by feeding the fader through a delay. The mute status bypasses the delay, and allows the fadeout to begin sufficiently in advance of the error. The final output samples of this system will be either correct, interpolated or muted, and these can then be sent to the converters in the player.

The power of the CD error correction is such that damage to the disk generally results in mistracking before the correction limit is reached. There is thus no point in making it more powerful. CD players vary tremendously in their ability to track imperfect disks and expensive models are not automatically better. It is generally a good idea when selecting a new player to take along some marginal disks to assess tracking performance.

Figure 11.39 contrasts the LSI chip sets used in first-generation Philips and Sony CD players. Figure 11.40 shows the more recent CX-23035 VLSI chip which contains almost all CD functions. This is intended for portable and car players, and replaces the separate LSIs shown.

The control system of a CD player is inevitably microprocessor based, and as such does not differ greatly in hardware terms from any other microprocessor-controlled device. Operator controls will simply interface to processor input ports and the various servo systems will be enabled or overridden by output ports. Software, or more correctly firmware, connects the two. The necessary controls are Play and Eject, with the addition in most players of at least Pause and some buttons which allow rapid skipping through the program material.

Although machines vary in detail, the flowchart of Figure 11.41 shows the logic flow of a simple player, from start being pressed to sound emerging. At the beginning, the emphasis is on bringing the various servos into operation. Towards the end, the disk subcode is read in order to locate the beginning of the first section of the program material.

When track following, the tracking-error feedback loop is closed, but for track crossing, in order to locate a piece of music, the loop is opened, and a microprocessor signal forces the laser head to move. The tracking error becomes an approximate sinusoid as tracks are crossed. The cycles of tracking error can be counted as feedback to determine when the correct number of tracks have been crossed. The 'mirror' signal obtained when the readout spot is half a track away from target is used to brake pickup motion and re-enable the track following feedback.

The control system of a professional player for broadcast use will be more complex because of the requirement for accurate cueing. Professional machines will make extensive use of subcode for rapid access, and in addition are fitted with a hand-operated rotor which simulates turning a vinyl disk by hand. In this mode the disk constantly repeats the same track by performing a single track-jump once every revolution. Turning the rotor moves the jump point to allow a cue point to be located. The machine will commence normal play from the cue point when the start button is depressed or from a switch on the audio fader. An interlock is usually fitted to prevent the rather staccato cueing sound from being broadcast.

Figure 11.42 shows the block diagram of an MD player. There is a great deal of similarity with a conventional CD player in the general arrangement. Focus, tracking and spindle servos are basically the same, as is the EFM

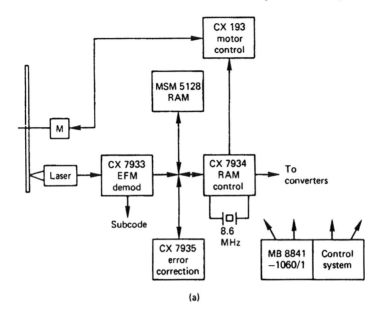

(a)

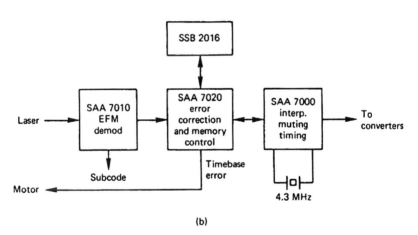

(b)

Figure 11.39 (a) The LSI chip arrangement of the CDP-101, a first-generation Sony consumer CD player. Focus and PLO systems were SSI/discrete. (b) LSI chip arrangement of Philips CD player.

and Reed–Solomon replay circuitry. The main difference is the presence of recording circuitry connected to the magnetic head, the large buffer memory and the data reduction codec. The figure also shows the VLSI chips developed by Sony for MD. Whilst MD machines are capable of accepting 44.1 kHz PCM or analog audio in real time, there is no reason why a twin spindle machine should not be made which can dub at four to five times normal speed.

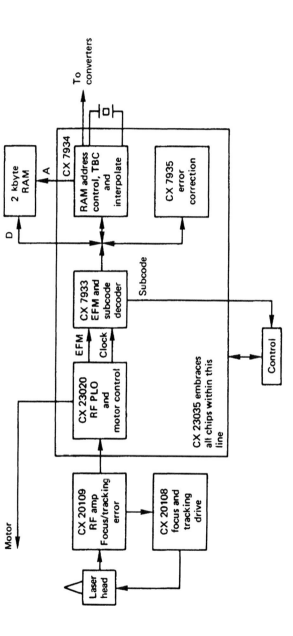

Figure 11.40 Second-generation Sony LSIs put RF, focus and PLO on chips (left). New 80-pin CX 23035 replaces four LSIs, requiring only outboard deinterleave RAM. This chip is intended for car and Discman-type players.

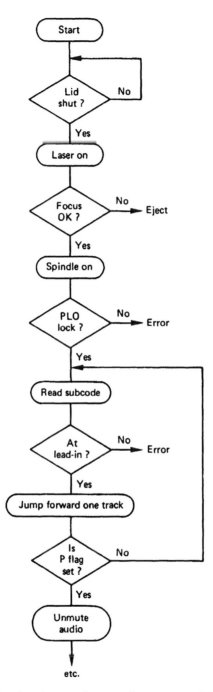

Figure 11.41 Simple flowchart for control system, focuses, starts disk, and reads subcode to locate first item of programme material.

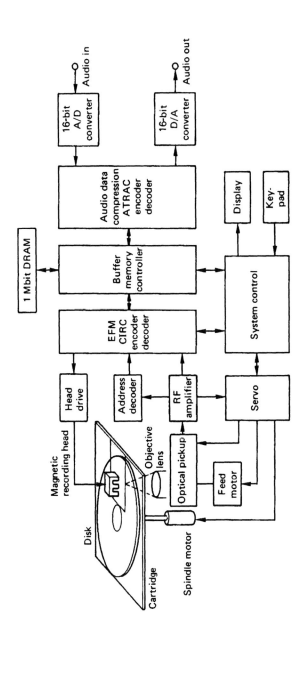

Figure 11.42 MiniDisc block diagram. See text for details.

11.15 Sony mastering disk

Intended as a replacement for the obsolescent U-matic-based PCM adapters used to master CDs and MDs, the Sony mastering disk is a 133 mm diameter random access magneto-optic cartridge disk with a capacity of 1.06 Gbyte of audio data, which is in excess of the playing time of any CD. The disk is similar in concept to the recordable MiniDisc, but larger. It is pre-grooved for track following, and has a groove wobble for timing purposes. The disk drive is not an audio recorder at all, it is a data recorder which connects to a SCSI (small computer systems interface) data bus. As such it behaves exactly the same as a magnetic hard disk, as it transfers at 2.5 times real time and requires memory buffering in order to record real-time audio. The drive is capable of confidence replay in real time. During recording, the disk is actually replayed for monitoring purposes. As with a magnetic disk, the wordlength of the audio is a function of the formatter used with it. 20-bit recording is possible, so that noise shaping can be used when reducing to 16-bit wordlength for CD mastering. As described in Chapter 8, this results in subjectively better quality when the CD is played. Chapter 12 explains how magnetic and magneto-optic data disks can be used for audio editing.

References

1. Bouwhuis, G. et al., *Principles of Optical Disk Systems*, Bristol: Adam Hilger (1985)
2. Mee, C.D. and Danicl, E.D. (eds), *Magnetic Recording Vol. III*, Chapter 6, New York: McGraw-Hill
3. Goldberg, N., A high density magneto-optic memory. *IEEE Trans. Magn.*, **MAG-3**, 605 (1967)
4. Various authors, *Philips Tech. Rev.*, **40**, 149–180 (1982)
5. Airy, G.B., *Trans. Camb. Phil. Soc.* **5**, 283 (1835)
6. Ray, S.F., *Applied Photographic Optics*, Chapter 17, Oxford: Focal Press (1988)
7. Maréchal, A., *Rev. d'Optique*, **26**, 257 (1947)
8. Pasman, J.H.T., Optical diffraction methods for analysis and control of pit geometry on optical disks. *J. Audio Eng. Soc.*, **41**, 19–31 (1993)
9. Verkaik, W., Compact Disc (CD) mastering – an industrial process. In *Digital Audio*, edited by B.A. Blesser, B. Locanthi and T.G. Stockham Jr, New York: Audio Engineering Society, 189–195 (1983)
10. Miyaoka, S., Manufacturing technology of the Compact Disc. In *Digital Audio*, *op. cit.* 196–201
11. Ogawa, H. and Schouhamer Immink, K.A., EFM – the modulation system for the Compact Disc digital audio system. In *Digital Audio*, *op. cit.* 117–124
12. Schouhamer Immink, K.A. and Gross, U., Optimization of low-frequency properties of eight-to-fourteen modulation. *Radio Electron. Eng.*, **53**, 63–66 (1983)
13. Peek, J.B.H., Communications aspects of the Compact Disc digital audio system. *IEEE Commun. Mag.*, **23**, 7–15 (1985)
14. Vries, L.B., et al., The digital Compact Disc – modulation and error correction. *Presented at 67th Audio Engineering Society Convention* (New York, 1980), preprint 1674.
15. Vries, L.B. and Odaka, K., CIRC – the error-correcting code for the Compact Disc digital audio system. In *Digital Audio*, *op. cit.* 178–186

Chapter 12

Audio editing

Initially audio editing was done in the analog domain either by selective copying or by actually cutting and splicing tape. Digital techniques have revolutionized audio editing taking advantage of the freedom to store data in any suitable medium and employing the signal-processing techniques developed in computation. This chapter shows how the edit process is achieved using combinations of analog and digital storage media, processing and control systems.

12.1 Introduction

Editing ranges from a punch-in on a multitrack recorder, or the removal of 'ums and ers' from an interview to the assembly of myriad sound effects and mixing them with timecode locked dialogue in order to create a film soundtrack.

Mastering is a form of editing where various tracks are put together to make an LP, CD, MD or DCC master recording. The duration of each musical piece, the length of any pauses between pieces and the relative levels of the pieces on the album have to be determined at the time of mastering. The master recording will be compiled from source media which may each contain only some of the pieces required on the final album, in any order. The recordings will vary in level, and may contain several retakes of a passage.

The purpose of the mastering editor is to take each piece, and insert sections from retakes to correct errors, and then to assemble the pieces in the correct order, with appropriate pauses between and with the correct relative levels to create the master tape.

Audio editors work in two basic ways, by assembling or by inserting sections of audio waveform to build the finished waveform. Both terms have the same meaning as in the context of video recording. Assembly begins with a blank master tape or file. The beginning of the work is copied from the source, and new material is successively appended to the end of the previous material. Figure 12.1 shows how a master recording is made up by assembly from source recordings. Insert editing begins with an existing recording in which a section is replaced by the edit process. Punch-in on multitrack recorders is a form of insert editing. This technique can be used with analog or digital recordings, or a combination of the two.

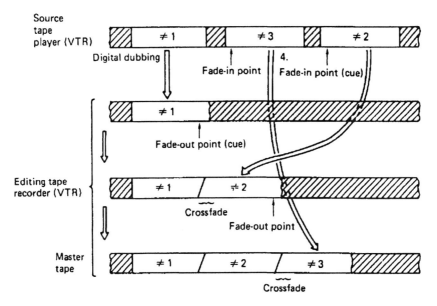

Figure 12.1 The function of an editor is to perform a series of assembles to produce a master tape from source tapes.

In all types of audio editing the goal is the appropriate sequence of sounds at the appropriate times. In analog audio equipment, editing was almost always performed using tape or magnetically striped film. These media have the characteristic that the time through the recording is proportional to the distance along the track. Editing consisted of physically cutting and splicing the medium, in order mechanically to assemble the finished work, or of copying lengths of source medium to the master.

Whilst open-reel digital audio tape formats support splice editing, in all other digital audio editing samples from various sources are brought from the storage media to various pages of RAM. The edit is performed by crossfading between sample streams retrieved from RAM and subsequently rewriting on the output medium. Thus the nature of the storage medium does not affect the form of the edit in any way except the amount of time needed to execute it.

Tapes only allow serial access to data, whereas disks and RAM allow random access and so can be much faster. Editing using random access storage devices is very powerful as the shuttling of tape reels is avoided. The technique is sometimes called non-linear editing.

12.2 Tape-cut editing

One of the oldest editing techniques, tape-cut editing, originally developed using analog tape, but certain early digital tape formats also allowed it.

Analog tape-cut editing requires a tape transport where the capstan can be disabled whilst everything else acts as if the machine were playing. The operator grasps the tape reels and moves the tape to and fro in order to align the wanted point in the recording with the replay head. Traditionally a chinagraph

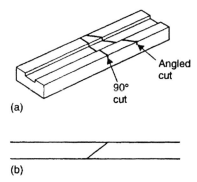

Figure 12.2 (a) A tape-splicing block has a large groove to hold the tape and smaller transverse grooves to guide the razor blade. (b) Angled splice results in short crossfade.

pencil mark would be made at that point. The tape would then be pulled from the heads and set into a splicing block. Figure 12.2(a) shows that a splicing block contains a channel which guides the tape and a number of transverse grooves which guide the razor blade or scalpel used to cut the tape.

The splicing block would also support two free ends of tape in the correct position so that self-adhesive splicing tape could be used to join them. Splicing tape is made slightly narrower than the recording tape so it does not snag on tape guides.

Some professional analog recorders have built-in tape cutters and the transport is designed to automatically advance the tape from the replay head to the cutter blade.

Analog tape-splicing blocks usually have a diagonal cut facility so that the joint between the two tapes is not at right angles to the tape edge. Figure 12.2(b) shows that this produces a short crossfade when the splice is played.

Splicing digital tape is only possible if the format has been designed to handle it.[1,2] The use of interleaving is essential in digital formats to handle burst errors; unfortunately it conflicts with the requirements of tape-cut editing. Figure 12.3 shows that a splice in cross-interleaved data destroys codewords for the entire constraint length of the interleave. The longer the constraint length, the greater the resistance to burst errors, but the more damage is done by a splice.

In order to handle splices, input audio samples are first sorted into odd and even and these two sample streams are displaced by subjecting them to different delays. The odd/even distance has to be greater than the cross-interleave constraint length. In the DASH format (digital audio stationary head), the odd/even delay is 2448 samples. In the case of a splice, samples are destroyed for the constraint length, but Figure 12.4 shows that this occurs at different times for the odd and even samples. Using interpolation, it is possible to obtain simultaneously the end of the old recording and the beginning of the new one. A digital crossfade is made between the old and new recordings.

The interpolation during concealment and splices causes a momentary reduction in frequency response which may result in aliasing if there is significant audio energy above one-quarter of the sampling rate. This was overcome

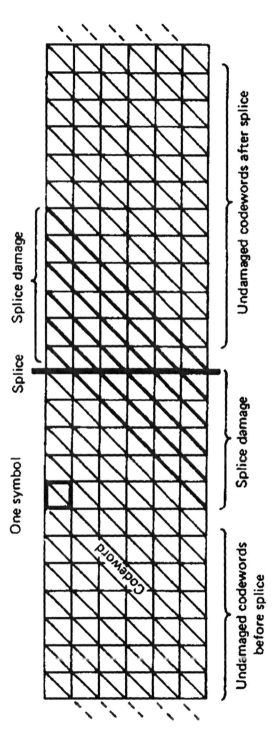

Figure 12.3 Although interleave is a powerful weapon against burst errors, it causes greater data loss when tape is spliced because many codewords are replayed in two unrelated halves.

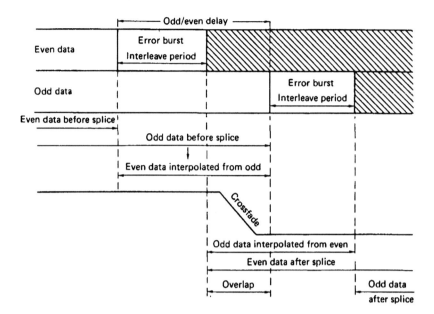

Figure 12.4 Following deinterleave, the effect of a splice is to cause odd and even data to be lost at different times. Interpolation is used to provide the missing samples, and a crossfade is made when both recordings are available in the central overlap.

in twin DASH machines in the following way. All incoming samples will be recorded twice which means twice as many tape tracks or twice the linear speed is necessary. The interleave structure of one of the tracks will be identical to the interleave already described, whereas on the second version of the recording, the odd/even sample shuffle is reversed. When a gross error occurs in twin DASH, it will be seen from Figure 12.5 that the result after deinterleave is that when odd samples are destroyed in one channel, even samples are destroyed in the other. By selecting valid data from both channels, a full bandwidth signal can be obtained and no interpolation is necessary. In the presence of a splice, when odd samples are destroyed in one track, even samples will be destroyed in the other track. Thus at all times, all samples will be available without interpolation, and full bandwidth can be maintained across splices.

When splicing digital tapes it is important to use a right-angle cut. Attempting to use a slant cut simply increases the extent of the data errors. The splice must be assembled with great care as the tracks are much narrower than in analog and the head contact is more critical. After splicing the magnetic surface has to be cleaned to ensure good head contact. It is a good idea to wear gloves when splicing digital tape. Digital splicing was really only introduced to make digital tape recorders seem familiar to those used to analog tradition and its performance was often marginal. In today's computerized world, digital tape-cut editing is an anachronism.

Whether analog or digital, spliced tapes do not archive well. In storage the glue creeps out from under the splicing tape and sticks adjacent turns of tape together. Tape stored under tension may cause splices to slide apart. Tapes containing splices should be copied before archiving.

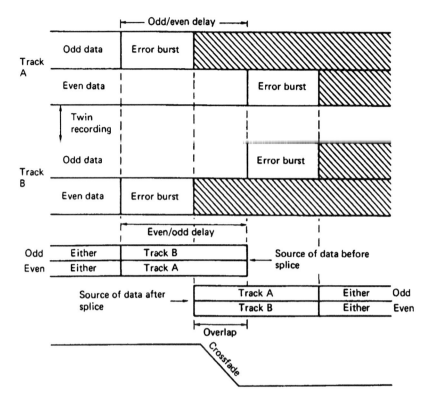

Figure 12.5 In twin DASH, the reversed interleave on the twin recordings means that correct data are always available. This makes twin DASH much more resistant to mishandling.

12.3 Editing on recording media

All digital recording media use error correction which requires an interleave, or re-ordering, of samples to reduce the impact of large errors, and the assembling of many samples into an error-correcting codeword. Codewords are recorded in constant sized blocks on the medium. Audio editing requires the modification of source material in the correct real-time sequence to sample accuracy. This contradicts the interleaved block-based codes of real media.

Editing to sample accuracy simply cannot be performed directly on real media. Even if an individual sample could be located in a block, replacing the samples after it would destroy the codeword structure and render the block uncorrectable. The only solution is to ensure that the medium itself is only edited at block boundaries so that entire error-correction codewords are written down. In order to obtain greater editing accuracy, blocks must be read from the medium and deinterleaved into RAM, modified there and reinterleaved for writing back on the medium, the so-called *read-modify-write* process.

In disks, blocks are often associated into clusters which consist of a fixed number of blocks in order to increase data throughput. When clustering is used, editing on the disk can only take place by rewriting entire clusters.

12.4 The structure of an editor

The audio editor consists of three main areas.[3] First, the various contributory recordings must enter the processing stage at the right time with respect to the master recording. This will be achieved using a combination of timecode, transport synchronization and RAM timebase correction. The synchronizer will take control of the various transports during an edit so that one section reaches its out-point just as another reaches its in-point.

Secondly, the audio signal path of the editor must take the appropriate action, such as a crossfade, at the edit point. This requires some signal-processing circuitry.

Thirdly, the editing operation must be supervised by a control system which coordinates the operation of the transports and the signal processing to achieve the desired result.

Figure 12.6 shows a simple block diagram of an editor. Each source device, be it disk or tape or some other medium must produce timecode locked to the audio samples. The synchronizer section of the control system uses the timecode to determine the relative timing of sources and sends remote control

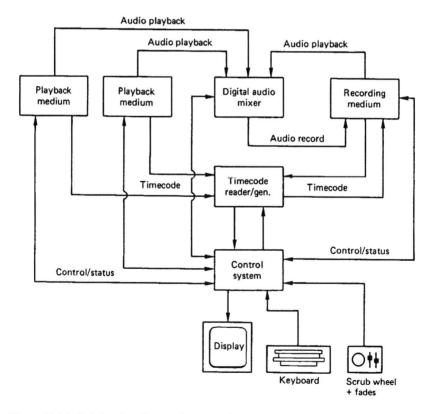

Figure 12.6 A digital audio editor requires an audio path to process the samples, and a timing and synchronizing section to control the time alignment of signals from the various sources. A supervisory control system acts as the interface between the operator and the hardware.

signals to the transport to make the timing correct. The master recorder is also fed with timecode in such a way that it can make a contiguous timecode track when performing assembly edits. The control system also generates a master sampling rate clock to which contributing devices must lock in order to feed samples into the edit process. Any of the playback devices can be replaced by an analog tape deck followed by an ADC which obtains its sampling clock from the system. The audio signal processor takes contributing sources and mixes them as instructed by the control system. The mix is then routed to the recorder.

12.5 Timecode

The timecode used in the PCM-1610/1630 system used for CD mastering is the SMPTE standard originally designed for 525/60 video recorders and is shown in Figure 12.7. SMPTE timecode is also used on US analog audio recorders. European products use EBU timecode which is basically similar to SMPTE except that the field rate is 50 Hz not 60 Hz. Timecode stores hours, minutes, seconds and frames as binary-coded decimal (BCD) numbers, which are serially encoded along with user bits into an FM channel code (see Chapter 10) which is recorded on one of the linear audio tracks of the tape. Disks also use timecode for audio synchronization, but the timecode forms part of the access mechanism so that samples are retrieved by specifying the required timecode.

A problem which arises with the use of video-based timecode is that the accuracy to which the edit must be made in audio is much greater than the frame boundary accuracy needed in video. When the exact edit point is chosen in a digital audio editor, it will be described to great accuracy and is stored as hours, minutes, seconds, frames and the number of the sample within the frame.

12.6 Locating the edit point

Whatever the storage medium, digital audio editors must simulate the 'rock and roll' process of edit-point location in analog tape recorders where the tape reels are moved to and fro by hand. The solution is to transfer the recording in the area of the edit point to RAM in the editor. RAM access can take place at any speed or direction and the precise edit point can then be conveniently found by monitoring audio from the RAM.

Figure 12.8 shows how the area of the edit point is transferred to the memory. The source device, be it disk or tape, is commanded to play, and the operator listens to replay samples via a DAC in the monitoring system. The same samples are continuously written into a memory within the editor. This memory is addressed by a counter which repeatedly overflows to give the memory a ring-like structure rather like that of a timebase corrector, but somewhat larger. When the operator hears the rough area in which the edit is required, he will press a button. This action stops the memory writing, not immediately, but one half of the memory contents later. The effect is then that the memory contains an equal number of samples before and after the rough edit point. Once the recording is in the memory, it can be accessed at leisure,

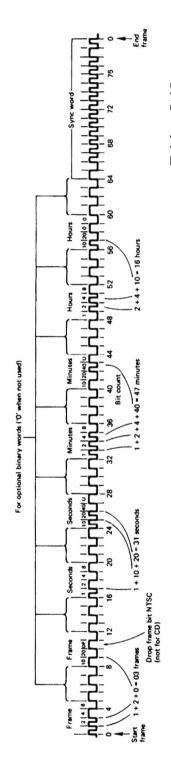

Figure 12.7 In SMPTE standard timecode, the frame number and time are stored as eight BCD symbols. There is also space for 32 user-defined bits. The code repeats every frame. Note the asymmetrical sync word which allows the direction of tape movement to be determined.

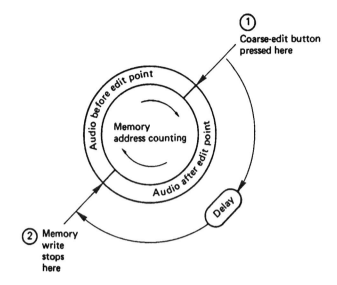

Figure 12.8 The use of a ring memory which overwrites allows storage of samples before and after the coarse edit point.

and the constraints of the source device play no further part in the edit-point location.

There are a number of ways in which the memory can be read. If the memory address is supplied by a counter which is clocked at the appropriate rate, the edit area can be replayed at normal speed, or at some fraction of normal speed repeatedly. In order to simulate the analog method of finding an edit point, the operator is provided with a *scrub wheel* or rotor, and the memory address will change at a rate proportional to the speed with which the rotor is turned, and in the same direction. Thus the sound can be heard forward or backward at any speed, and the effect is exactly that of manually rocking an analog tape past the heads of an ATR.

There are not enough pulses per revolution to create a clock directly from the scrub wheel and the human hand cannot turn the rotor smoothly enough to address the memory directly without flutter. A phase-locked loop is generally employed to damp fluctuations in rotor speed and multiply the frequency. A standard sampling rate must be recreated to feed the monitor DAC and a rate converter, or interpolator, is necessary to restore the sampling rate to normal. These items can be seen in Figure 12.9.

The act of pressing the coarse edit-point button stores the timecode of the source at that point, which is frame-accurate. As the rotor is turned, the memory address is monitored, and used to update the timecode to sample accuracy.

Before assembly can be performed, two edit points must be determined, the out-point at the end of the previously recorded signal, and the in-point at the beginning of the new signal. The editor's microprocessor stores these in an edit decision list (EDL) in order to control the automatic assemble process.

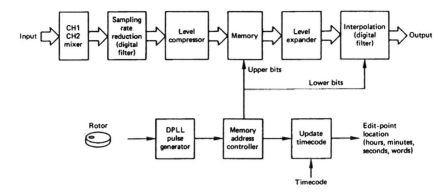

Figure 12.9 In order to simulate the edit location of analog recorders, the samples are read from memory under the control of a hand-operated rotor.

12.7 Non-linear editing

In an editor based on hard disks, using one or other of the above methods, an edit list can be made which contains an in-point, an out-point and an audio filename for each of the segments of audio which need to be assembled to make the final work, along with a crossfade period and a gain parameter. This edit list will also be stored on the disk. When a preview of the edited work is required, the edit list is used to determine what files will be necessary and when, and this information drives the disk controller.

Figure 12.10 shows the events during an edit between two files. The edit list causes the relevant audio blocks from the first file to be transferred from disk to memory, and these will be read by the signal processor to produce the preview output. As the edit point approaches, the disk controller will also place blocks from the incoming file into the memory. It can do this because the rapid data-transfer rate of the drive allows blocks to be transferred to memory much faster than real time, leaving time for the positioner to seek from one file to another. In different areas of the memory there will be simultaneously the end of the outgoing recording and the beginning of the incoming recording. The signal processor will use the fine edit-point parameters to work out the relationship between the actual edit points and the cluster boundaries. The relationship between the cluster on disk and the RAM address to which it was transferred is known, and this allows the memory address to be computed in order to obtain samples with the correct timing. Before the edit point, only samples from the outgoing recording are accessed, but as the crossfade begins, samples from the incoming recording are also accessed, multiplied by the gain parameter and then mixed with samples from the outgoing recording according to the crossfade period required. The output of the signal processor becomes the edited preview material, which can be checked for the required subjective effect. If necessary the in- or out-points can be trimmed, or the crossfade period changed, simply by modifying the edit-list file. The preview can be repeated as often as needed, until the desired effect is obtained. At this stage the edited work does not exist as a file, but is recreated each time by a further execution of the EDL. Thus a lengthy editing session need not fill up the disk.

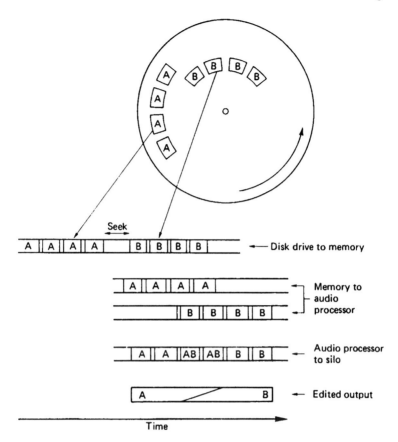

Figure 12.10 In order to edit together two audio files, they are brought to memory sequentially. The audio processor accesses file pages from both together, and performs a crossfade between them. The silo produces the final output at constant steady-sampling rate.

It is important to realize that at no time during the edit process were the original audio files modified in any way. The editing was done solely by reading the audio files. The power of this approach is that if an edit list is created wrongly, the original recording is not damaged, and the problem can be put right simply by correcting the edit list. The advantage of a disk-based system for such work is that location of edit points, previews and reviews are all performed almost instantaneously, because of the random access of the disk. This can reduce the time taken to edit a program to a quarter of that needed with a tape machine.[4]

During an edit, the disk drive has to provide audio files from two different places on the disk simultaneously, and so it has to work much harder than for a simple playback. If there are many close-spaced edits, the drive may be hard-pressed to keep ahead of real time, especially if there are long cross-fades, because during a crossfade the source data rate is twice as great as during replay. A large buffer memory helps this situation because the drive can fill the memory with files before the edit actually begins, and thus the instantaneous sample rate can be met by the memory's emptying during disk-intensive periods. In practice crossfades measured in seconds can be achieved in a disk-based system, a figure which is not matched by tape systems.

Once the editing is finished, it will generally be necessary to transfer the edited material to form a contiguous recording so that the source files can make way for new work. If the source files already exist on tape the disk files can simply be erased. If the disks hold original recordings they will need to be backed up to tape if they will be required again. In large broadcast systems, the edited work can be broadcast directly from the disk file. In smaller systems it will be necessary to output to some removable medium, since the Winchester drives in the editor have fixed media. It is only necessary to connect the AES/EBU output of the signal processor to any type of digital recorder, and then the edit list is executed once more. The edit sequence will be performed again, exactly as it was during the last preview, and the results will be recorded on the external device.

12.8 Editing in DAT

Physical editing on a DAT tape can only take place at the beginning of an interleave block, known as a frame, which is contained in two diagonal tracks. The transport would need to perform a preroll, starting before the edit point, so that the drum and capstan servos would be synchronized to the tape tracks before the edit was reached. Fortunately, the very small drum means that mechanical inertia is minute by the standards of video recorders, and lock-up can be very rapid.

In professional DAT machines designed for editing two sets of heads are fitted in the drum. The standard permits the drum size to be increased and the wrap angle to be reduced provided that the tape tracks are recorded to the same dimensions. In normal recording, the first heads to reach the tape tracks would make the recording, and the second set of heads would be able to replay the recording immediately afterwards for confidence monitoring. For editing, the situation would be reversed. The first heads to meet a given tape track would play back the existing recording, and this would be deinterleaved and corrected, and presented as a sample stream to the record circuitry. The record circuitry would then interleave the samples ready for recording. If the heads are mounted a suitable distance apart in the scanner along the axis of rotation, the time taken for tape to travel from the first set of heads to the second will be equal to the decode/encode delay. If this process goes on for a few blocks, the signal going to the record head will be exactly the same as the pattern already on the tape, so the record head can be switched on at the beginning of an interleave block. Once this has been done, new material can be crossfaded into the sample stream from the advanced replay head, and an edit will be performed.

If insert editing is contemplated, following the above process, it will be necessary to crossfade back to the advanced replay samples before ceasing rerecording at an interleave block boundary. The use of overwrite to produce narrow tracks causes a problem at the end of such an insert. Figure 12.11 shows that this produces a track which is half the width it should be. Normally the error-correction system would take care of the consequences, but if a series of inserts were made at the same point in an attempt to make fine changes to

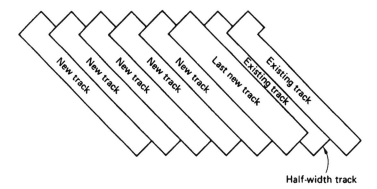

Half-width track

Figure 12.11 When editing a small track-pitch recording, the last track written will be 1.5 times the normal track width, since that is the width of the head. This erases half of the next track of the existing recording.

an edit, the result could be an extremely weak signal for one track duration. One solution is to incorporate an algorithm into the editor so that the points at which the tape begins and ends recording change on every attempt. This does not affect the audible result as this is governed by the times at which the crossfader operates.

12.9 Editing in open-reel digital recorders

On many occasions in studio recording it is necessary to replace a short section of a long recording, because a wrong note was played or something fell over and made a noise. The tape is played back to the musicians before the bad section, and they play along with it. At a musically acceptable point prior to the error, the tape machine passes into record, a process known as punch-in, and the offending section is rerecorded. At another suitable time, the machine ceases recording at the punch-out point, and the musicians can subsequently stop playing.

Once more, a read-modify-write approach is necessary, using a record head positioned *after* the replay head. The mechanism necessary is shown in Figure 12.12. Prior to the punch-in point, the replay-head signal is deinterleaved, and this signal is fed to the record channel. The record channel reinterleaves the samples, and after some time will produce a signal which is identical to what is already on the tape. At a block boundary the record current can be turned on, when the existing recording will be rerecorded. At the punch-in point, the samples fed to the record encoder will be crossfaded to samples from the ADC. The crossfade takes place in the non-interleaved domain. The new recording is made to replace the unsatisfactory section, and at the end, punch-out is commenced by returning the crossfader to the samples from the replay head. After some time, the record head will once more be rerecording what is already on the tape, and at a block boundary the record current can be switched off. The crossfade duration can be chosen according to the nature of the recorded material. It is possible to rehearse the punch-in process and monitor what it would sound like by feeding headphones from the crossfader,

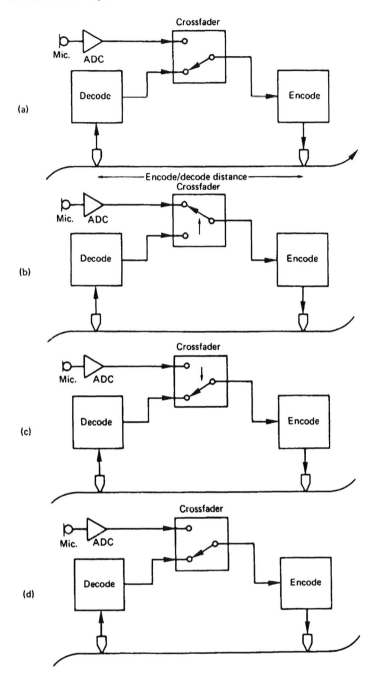

Figure 12.12 The four stages of an insert (punch-in/out) with interleaving: (a) rerecord existing samples for at least one constraint length; (b) crossfade to incoming samples (punch-in point); (c) crossfade to existing replay samples (punch-out point); (d) rerecord existing samples for at least one constraint length. An assemble edit consists of steps (a) and (b) only.

and doing everything described except that the record head is disabled. The punch-in and punch-out points can then be moved to give the best subjective result. The machine can learn the sector addresses at which the punches take place, so the final punch is fully automatic.

Assemble editing, where parts of one or more source tapes are dubbed from one machine to another to produce a continuous recording, is performed in the same way as a punch-in, except that the punch-out never comes. After the new recording from the source machine is faded in, the two machines continue to dub until one of them is stopped. This will be done some time after the next assembly point is reached.

References

1. Watkinson, J.R., Splice-handling mechanisms in DASH format. Presented at 77th Audio Engineering Society Convention (Hamburg, 1985), preprint 2199(A-2)
2. Onishi, K., Ozaki, M., Kawabata, M. and Tanaka, K., Tape-cut editing of stereo PCM tape deck employing stationary head. Presented at 60th Audio Engineering Society Convention (Los Angeles, 1978), preprint 1343(E-6)
3. Ohtsuki, T., Kazami, S., Waiari, M., Tanaka, M. and Doi, T.T., A digital audio editor. *J. Audio Eng. Soc.*, **29**, 358 (1981)
4. Todoroki, S. et al., New PCM editing system and configuration of total professional digital audio system in near future. Presented at 80th Audio Engineering Society Convention (Montreux, 1986), preprint 2319(A-8)

Chapter 13

Audio signal processing

In most practical audio systems some form of signal processing is required, ranging from a simple mixer combining several microphones into one signal, through effects such as artificial reverberation to complex techniques such as digital bit rate reduction or compression. All of these processes are treated here.

13.1 Introduction

Signals from a microphone cannot be used directly without at least some form of level control. In the real world microphone signals may have excessive dynamic range or there could be bass tip-up. Particularly in pop music production the frequency response may need to be altered to obtain some desired characteristic. It is often necessary to mix the output of several microphones into a single signal. In stereo working the spatial location of the sound from each microphone also needs to be controlled by panning. In the case of coincident stereo signals the width may need adjustment. Some recording techniques deliberately use non-reverberant or dry signals and artificial reverberation is added in the production process.

The common processes required for audio production are generally incorporated into a mixing console; an expensive device which once purchased will not be replaced on a whim. Special effects, on the other hand, come and go and these will often be performed by *outboard* equipment which is usually designed to be fitted in a standard RETMA 19-inch rack. Figure 13.1 shows that the mixing console will be fitted with connectors known as *sends* and *returns* to allow signals to be looped through an external device.

Mixing consoles exist in many different configurations depending on the application. Figure 13.2 shows some different types. In multitrack working, the console will be designed to connect many microphones to many tape inputs when recording, and to mix down many replayed tracks to stereo at leisure. The mixing process may well be automated.

For live concerts the requirements are different. Many input signals need to be handled, but a number of different outputs will be required. The signals to drive the main PA speakers are one obvious requirement, but live performers often need on-stage loudspeakers so that they can hear themselves perform. These *foldback* speakers need to be provided with a different mix to the

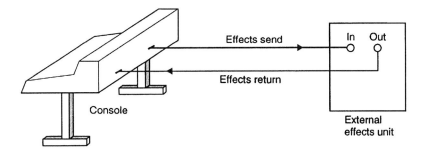

Figure 13.1 Mixing consoles can employ external signal processors using connections called sends and returns.

main output. As the signal levels will be set in real time by the engineer, fader automation is unnecessary. However, on a live performance the console settings for one song may need to be changed completely for the next and there is little time to manipulate the controls. In this case a console having *set-up memories* will be useful because the configurations arrived at in rehearsal can be recalled at will.

Broadcasting has different requirements again. Consider a console which is to be used at a major event which is being broadcast on both radio and television and recorded for an album release. The album recorder needs to be fed with minimally processed signals so that balancing can take place on mix-down. It is unlikely that any commentary will be required. The television mix will be different from the radio mix because the sound has to match shot changes which the radio listener is unaware of. The commentary for radio is different because the scene has to be described. The requirement is for a console which can accept a large number of signals and produce a number of different mixes simultaneously.

At the functional level the engineer does not care whether the signal path of the console is analog or digital. With the adoption of automation and set-up recall the distinction between analog and digital blurs because the control systems are always digital.

As digital audio is just another way of carrying an audio waveform, it follows that any process which can manipulate that waveform in the analog domain can be replicated in the digital domain by processing the sample values. Once the parallels between analog processes and the numerical equivalent are grasped the understanding of digital mixers becomes much easier.

13.2 Level control

Level control is the most fundamental process used in audio. Increasing the level requires *gain* whereas reducing it requires *attenuation*. Figure 13.3(a) shows that a resistor is fitted with a sliding contact or *wiper* so that it acts as a potential divider or *potentiometer* universally abbreviated to *pot*. If the resistor has a uniform cross-section, the output voltage will change linearly with the position of the wiper. This was the original configuration of the potentiometer because it allowed voltage to be measured effectively with a

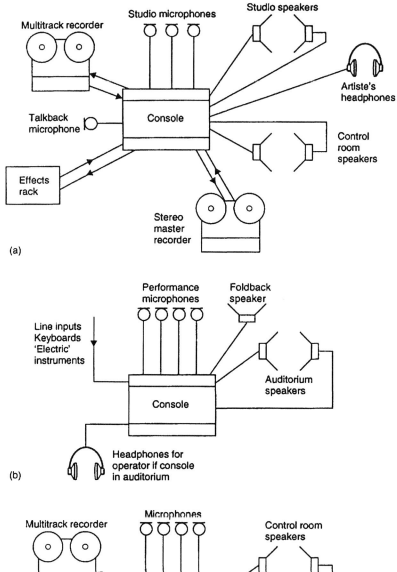

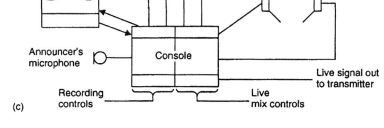

Figure 13.2 Some configurations in which mixing consoles will be used. (a) Multitrack record production. (b) Live PA system. (c) Simultaneous broadcast and recording.

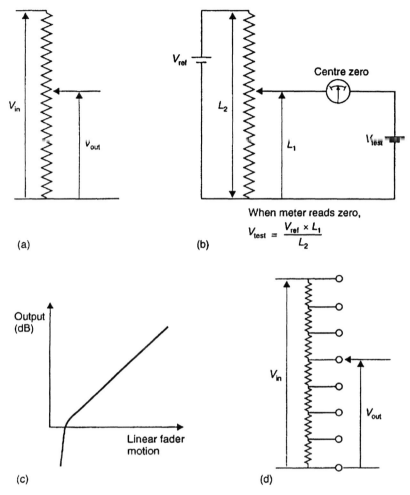

Figure 13.3 (a) A resistor fitted with a sliding contact forms a potential divider. (b) The original potentiometer used a uniform resistor to measure voltage. The slider is moved until the centre zero meter nulls. (c) Audio faders require a logarithmic taper. (d) For precise tracking some faders use switches and discrete precision resistors.

ruler as shown in (b). For audio control the mechanical movement can be rotary, segmental or straight.

The linear potentiometer is of no use for audio level control because human hearing responds logarithmically to level. The solution is to taper the resistive element (c) so that the output level is a logarithmic function of wiper position. Logarithmic taper pots are universally used for audio level control. When a log taper is employed, the control can be marked in dB steps which will be evenly spaced over most of the travel.

When controlling the gain of a stereo signal a dual or *ganged* control is needed. It is important that the two parts of the control offer identical attenuation or *tracking* otherwise the balance will be affected. Instead of a variable control, some esoteric faders use make-before-break switches with a large number of steps (d) so that fixed matched resistors can be used. More recently it has proved possible to make accurately matched variable tracks and these may be fitted with a detent mechanism to give the same feel as a stepped fader.

Where a fixed attenuation is required, perhaps where a close-miked drum signal is overloading an input, a device called a *pad* can be used. For general purpose use these may be built into an XLR barrel which is engraved with the number of dB of attenuation obtained. The stated attenuation will only be obtained when both source and load impedances match the (usually) 600 Ω impedance of the pad. In modern audio systems having low output impedance and high input impedance a different attenuation will be obtained with a 600 Ω

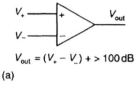

$V_{out} = (V_+ - V_-) + > 100\,dB$

(a)

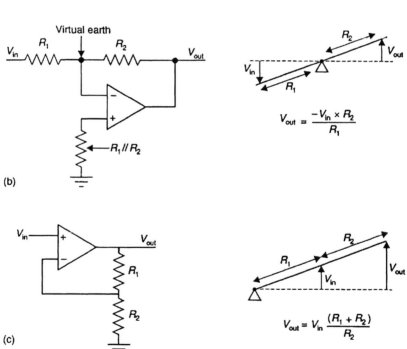

Figure 13.4 (a) Basic operational amplifier has differential inputs and very high gain. (b) Inverting operational amplifier configuration. Gain brings negative input to almost zero volts hence the term 'virtual earth'. (c) Non-inverting configuration uses a potential divider to ground to set gain. (d) Many inputs can be connected to a virtual earth point to perform mixing. (e) Active balance control differentially changes gain with a single control.

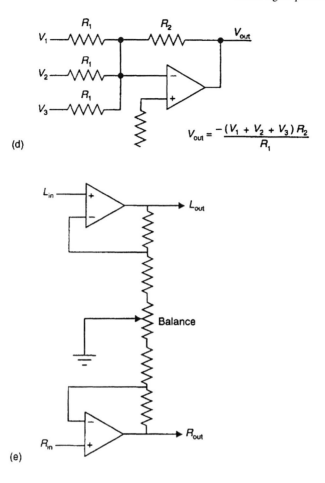

$$V_{out} = \frac{-(V_1 + V_2 + V_3)R_2}{R_1}$$

(d)

(e)

pad. Whilst this might affect precision measurements it is not normally an operational problem. In many mixing consoles a switched pad may be incorporated in the microphone preamp stage.

Gain requires active circuitry and in audio processing the operational amplifier, originally developed for analog computers, is the most common solution. Figure 13.4(a) shows that the operational amplifier has differential inputs and very high gain. The working gain is determined using feedback. The operational amplifier can be arranged as an inverting or non-inverting stage. In the inverting configuration the positive input is held at ground potential and feedback causes the inverting terminal to be driven to the same voltage as ground such that the input and output voltages have a see-saw relationship (b) where the lengths of the see saw arms are proportional to the resistor values. In the non-inverting configuration (c) the feedback has a potential divider and the output will be driven to whatever is needed to make the voltage on the two inputs the same.

As the inverting input has very nearly ground potential it is described as a *virtual earth*. A number of different signals can be summed by connecting each to a virtual earth point through an input resistor as shown in (d). This configuration is the basis of most analog mixing stages. Note that in audio the term mixing means linear addition whereas in radio receivers it means multiplication.

Stereo balance requires a dual control in which the gain of one channel increases as the other falls. A superior solution is to modify the gain of two non-inverting amplifiers differentially with a single control (e).

The panoramic potentiometer or panpot is designed to feed a mono signal into an intensity stereo system in such a way that the location of the virtual sound source can be steered by turning the pot. As Figure 13.5 shows, at both extremes of the pot one output is attenuated to nothing whereas the other reaches full level. In fact at all positions of the panpot the two outputs should sum to the original signal. For example, at the centre position of the pot the two outputs should be equal and 6 dB down. Therefore if the stereo signal is recombined to mono the original level will always be recovered. As there are no phase differences in panpotted intensity stereo the figure of 6 dB at the centre will always be correct. The way the level changes with respect to rotation of the control should be arranged so that the virtual sound source moves linearly. Few console designs actually implement this precisely.

When two non-correlated stereo signals, such as from spaced microphones, are added to create mono, the power is summed to give a 3 dB level increase. Consequently panpots will be found with a central level drop of 3 dB. Some use 4.5 dB as a compromise. These values are quite wrong because the two outputs from a panpot are perfectly correlated and only the figure of 6 dB is correct, the lower values giving a level rise at the centre of the image.

In the digital domain amplification and attenuation are both obtained by multiplying the sample values by a constant known as a *coefficient* as shown

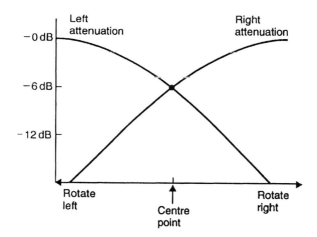

Figure 13.5 A panpot should produce a pair of signals whose levels change differentially but which always add to produce the original input. This means that each output must be 6 dB down at the mid-point.

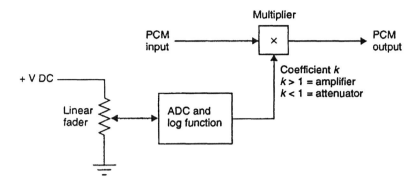

Figure 13.6 In the digital domain a multiplier is needed to amplify or attenuate. Coefficients must have a logarithmic relationship to control setting.

in Figure 13.6. If the coefficient exceeds unity there is gain whereas if it is less than one there is attenuation. As with analog level control the coefficients must have a logarithmic relationship with the gain control. All processes such as mixing, balance and panning are carried out in this way.

13.3 Grouping

Once mixing consoles exceed a certain size it becomes impossible to operate all of the faders individually and some assistance is needed. One solution is to use *audio grouping* shown in Figure 13.7(a) in which some faders produce a sub-mix which can then be controlled by a group fader. Such a sub-mix may be found in multitrack consoles where several channels are assigned to one tape track.

A major step forward in console control was the development of the voltage-controlled amplifier (VCA). Figure 13.7(b) shows that the voltage-controlled amplifier is effectively an analog multiplier; a device whose gain is a function of an external control voltage. These are generally arranged so that 0 V input gives 0 dB gain and gain reduces by 20 dB/volt to +5 V input.

When using VCAs the fader itself is fed with a source of DC power and the wiper produces a control voltage. This can be fed directly to the VCA as in (b), but by adding control circuitry it is easy to arrange flexible *VCA grouping* or *control grouping* (c). When grouping is selected the control voltage to a given VCA is the sum of the control voltage from its own fader and that from the group fader. The number of faders in a group and the group fader which controls them are now selectable, generally by a thumbwheel switch adjacent to each fader. Simply rotating the switch *assigns* the fader to the selected group fader. As many channels as necessary can be assigned to a given group fader.

Note that with control grouping the outputs of the various VCAs are not necessarily mixed to one audio signal. This means that a single group master could control the gain of several independent signals going to, for example, a multitrack recorder.

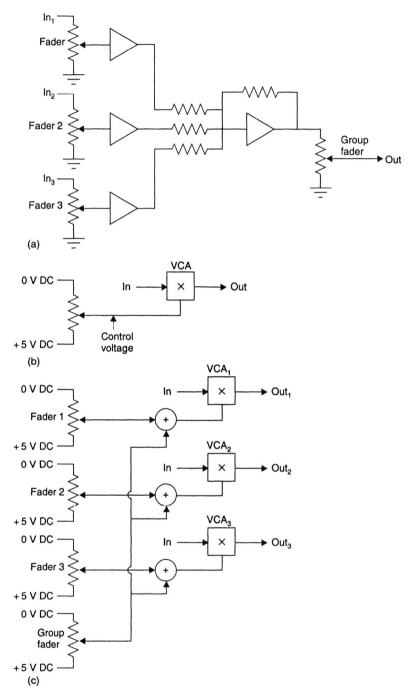

Figure 13.7 (a) In audio grouping several faders are mixed into one group fader. (b) Using a VCA, the fader is supplied with DC and produces a variable control voltage. (c) In VCA grouping the VCA is controlled by the sum of the control voltages from its own fader and the group fader.

13.4 Automation

In an automated system the gain can be programmed to vary automatically as a function of time. In early systems the gain parameters were actually recorded on one track of a multitrack recorder. Updating the automation data required copying to a second track, effectively robbing the recorder of two tracks. With the advent of low-cost computers this technique gave way to computer storage of gain values which were accessed by supplying the computer with timecode from the tape.

Automated consoles may use motorized faders or VCAs. In a motor-fader system shown in Figure 13.8(a) the audio fader has an additional control track which is supplied with DC so that the wiper produces a variable control voltage. The control voltage is digitized and supplied to an input port of the computer. Each fader has unique port addresses which the control program will *poll* or access sequentially at a high enough speed that there appears to be a continuous process.

The output port contains a DAC which reproduces the control voltage. As Figure 13.8(b) shows, the fader control track voltage is compared with the DAC output and the difference drives the fader motor to cancel the error. Thus the fader follows the control codes. Alternatively the control voltage may drive a VCA.

When the automation is 'off', the computer is still active, but it simply passes all input values back to the output unchanged. It is a small step to have the computer substitute level codes so that the gain setting of a single fader is supplied to a number of faders, obtaining a control grouping system which is as flexible as the software allows. In a further step the physical fader controlled by the user can be made quite independent of the fader it controls by using the automation computer as a kind of patchbay. In this way faders can be *assigned* as convenient.

The analog fader can be replaced by an encoder (c) which directly produces a digital coefficient describing its position. This can be connected directly to the computer without an ADC. In a digital console the DAC can also be eliminated (d) as the coefficient can be supplied directly to a digital multiplier. As it is so much easier to make an automated digital console, there is little point in making a digital console without automation and even the most inexpensive semi-professional units incorporate it.

Automation generally functions in three modes: write mode, where the computer learns the operator's fader movements; read mode, where the computer reproduces a previously written mix; and update mode where an earlier mix is modified. In write mode the operator moves the fader and the digitized control voltage is stored in the computer as a function of time. In read mode timecode from the tape is sent to the computer which retrieves the correct data for each timecode frame. The data are output to the fader ports, the DACs reproduce the control voltage and the faders replicate their original gain profile.

In moving fader systems the state of the mix is graphically conveyed by the position of the motorized knobs. If this approach is used, it is impossible to get a gain step when entering update. It is also possible to make the fader

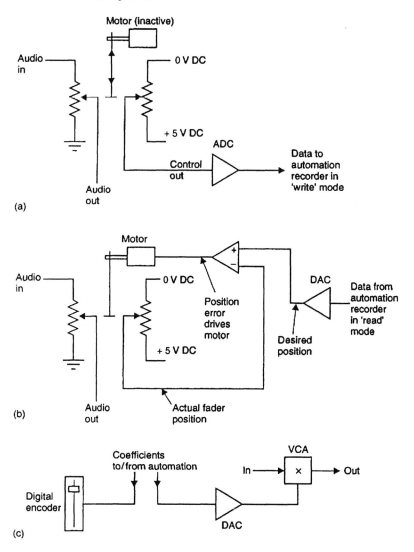

Figure 13.8 (a) A motor fader system has an additional track in the fader to supply level signal in write. (b) In read additional track provides positional feedback to motor servo. (c) The fader track can be replaced by a digital encoder, DAC drives the VCA. In a digital console the output port drives the multiplier directly.

knob touch-sensitive so that if the operator wants to modify the position he simply grasps the knob and moves it accordingly. The act of touching the control will disable the servo motor and switch the computer from read to update so that the new fader position is stored. Upon releasing the fader it returns to its previously programmed position.

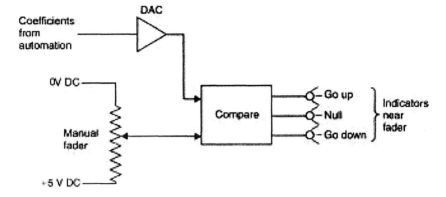

Figure 13.9 In non-motorized automation systems a null indicator is fitted to allow the user to bring the fader to the same position as the recorded gain to allow a step-free edit.

VCA-based automation systems need to indicate the gain being used when under automatic control. This can be done by having an indicator such as a bar LED beside the fader or using the channel level meter. If it is required to modify or update the automated gain profile, it is important that the fader has the same setting as the recorded mix otherwise a gain step will be introduced when the data are edited. This was done in early VCA automation systems by having a comparator between the DAC output and the fader output. The comparator drove a null indicator as in Figure 13.9 so that the operator could position the fader to the correct setting just prior to entering update mode.

An alternative approach was to program the computer so that in update mode the fader would work in a *relative* manner. The fader would be left half-way down the track and on entering update the software would compare the recorded code with the fader code and subtract the difference from the fader data. In this way a gain step would not occur and moving the fader would alter the gain relative to the previous recording rather than absolutely. The null lights can be used to avoid a step when ending an update, but in a relative system returning the fader to its original position would have the same effect.

As many VCAs cannot attenuate the signal indefinitely, in order to silence a channel a cut function is added. This is a push button whose setting can be recalled by the automation system.

13.5 Dynamics

Frequently the dynamic range of a raw microphone signal is considered to be too great and a process called *compression* is used to reduce it. Compression causes audible artifacts which will be discussed later. The related topics of *limiting* and *gating* will also be considered.

Compression has a number of applications. The dynamic range of real sounds can be extremely wide and although high-quality microphones can handle them, reproduction of full dynamic range requires large and powerful loudspeakers which are not always available or practicable. Where more modest equipment is in use, the use of full dynamic range may result in low-level sounds being lost in noise, high-level sounds overloading or both. In pop-recording compression, even analog tape overload may be used as an effect. Tape recorders designed for unskilled users will usually incorporate compressors or automatic level controls so that the recording level control can be dispensed with.

Many people listen to radio stations on transistor radios which are really only suitable for speech. Often these are used in the workplace where ambient noise intrudes. On the AM bands background noise ('static') is extremely high. Consequently it is a practical necessity to compress music signals prior to AM broadcast in order to optimize the transmitter modulation. One of the consequences of compression of this kind is that the programme material appears subjectively louder for given SPL. In a competitive world some broadcasters use large amounts of such compression because it makes transistor radios of limited power seem louder. Generally once one station has taken this step the others feel compelled to follow suit. As a result compression is widely used in FM radio even though the use of FM allows a wider dynamic range.

Compressors work by analysing the signal level and using it to control the gain. The relationship between input and output dynamic range is called the *compression ratio* which is adjustable. Figure 13.10(a) shows that the

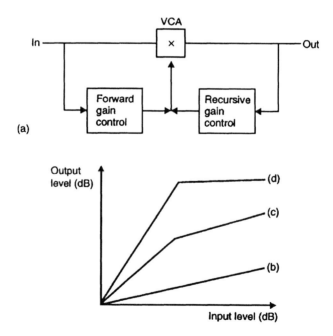

Figure 13.10 (a) Compressors use the input or their own output to reduce dynamic range. (b) Low threshold applies uniform compression. (c) Intermediate threshold puts 'knee' in the characteristic. (d) High threshold produces a limiting action.

compressor can work from the input or the output signal. The instantaneous level signal cannot be used directly or it would distort the audio waveform because rapid gain changes produce sidebands. The gain has to change slowly to prevent this distortion. When the input level is increasing and is in danger of overloading, it is important to reduce the gain promptly. This is known as the *attack* of the compressor and the speed of attack must be a compromise between overload and distortion. Some compressors can change the attack automatically as a function of the slew rate of the input. However, when the level falls after a peak the gain should be restored slowly. This is known as the *release* or recovery of the compressor. Separate controls are provided.

Compressors also have a *threshold* control which only allows compression above a certain level. When manipulated in conjunction with the compression ratio control a variety of transfer functions can be obtained. Figure 13.10 shows some examples. At (b) the threshold is set low so that uniform compression of most of the dynamic range is obtained. At (c) the threshold is raised so that the compression characteristic has a knee. At (d) the knee is set high so that most of the dynamic range is unchanged but peaks are prevented from overloading. In this configuration the action is that of a *limiter*. Limiting is often used prior to recorders to prevent occasional transients entering saturation in analog machines or clipping at the end of the quantizing range in digital machines. The intended dynamic range is entirely uncompressed and provided the limiting only acts on brief transients it will be largely inaudible.

In practice the same unit can be a compressor or a limiter depending upon how the threshold control is set. It is also possible to place an equalizer prior to the compressor control so that the compression is frequency dependent.

The gain variation of compression is frequently audible and the better the quality of the monitoring system the more audible it becomes. In particular, the human voice and classical music sound unnatural if compressed. The timbre of speech and many instruments, particularly piano, changes with level and compression changes that relationship. The dynamics of compression are also audible. Because the signal can only have one gain at any one instant, instruments can amplitude modulate each other. A blow on a tympanum or an organ pedal note has its impact reduced by compression which also depresses the level of the rest of the orchestra. This then slowly recovers as the compressor releases. During release background noises and ambience appear to grow in level.

In multitrack pop music production the compressor will be used on individual tracks. In this case the compression of one instrument cannot affect the level of other sounds in the mix and audibility is reduced.

Of course, when listening on a transistor radio the inherent distortion masks the changes of timbre, the background noise will be below the noise floor and one is just left with the amplitude modulation.

When using compression on stereo or surround sound signals it is important to prevent the gain control process affecting the spatial characteristics. In practice the level of all of the channels must be analysed, but only a single gain control signal is produced which affects each channel equally. Compressor/limiters are often supplied in a two channel configuration. The two channels can work independently for multitrack work, or can be *ganged* for stereo applications so that the gain changes do not affect the balance.

Gating is another application of automatic level control. If a compressor is designed with an inverse characteristic shown in Figure 13.11(a) it will heavily attenuate low-level signals. The result is known as a *noise gate* because the noise floor of the input signal will be removed.

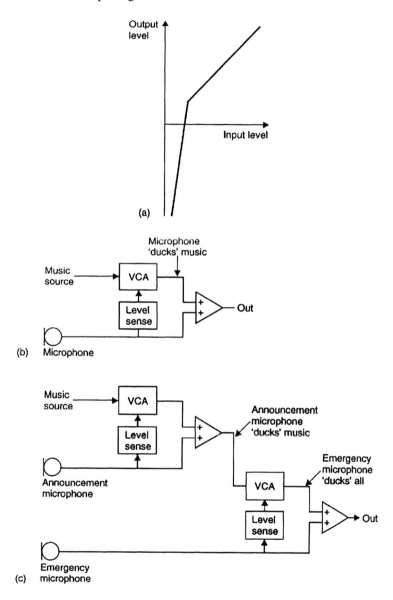

Figure 13.11 (a) Inverting the transfer function of a compressor causes it to attenuate low levels to produce noise gating. (b) Automatic voiceover system in which signal gain is controlled by a different audio signal. (c) Cascaded gating system allows priority for emergency announcements.

Gating can also be used as in (b) where one signal controls the gain of another. This may be used in radio stations or public buildings so that the announcer's voice automatically reduces or *ducks* the level of any background music. Gating may be cascaded as in (c) so that emergency messages will take priority over normal announcements and the background music returns to its normal level when neither type of message occurs.

13.6 Equalization

Equalization or EQ is needed to compensate for shortcomings in a signal or to create some effect, often the tailoring of the timbre of a vocalist. All EQ works by modifying the frequency response. Figure 13.12 shows the common types of EQ and their responses.

At (a) a bass-cut filter rolls off very low frequencies. The slope is generally fixed but the turnover frequency may be adjustable. This filter may be used with microphones to compensate for bass tip-up due to the proximity effect or in vinyl disk players to remove vibrations from the turntable in which case it will be called a *rumble filter*.

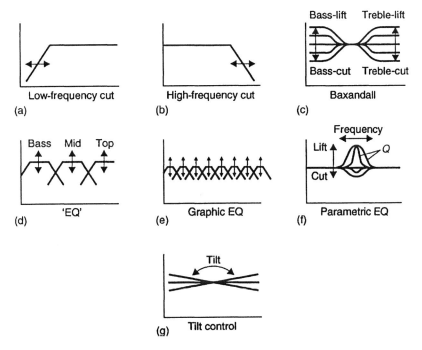

Figure 13.12 Types of EQ. (a) Bass-cut filter for removing vinyl disk rumble or bass tip-up. (b) Top-cut filter removes hiss. (c) Baxandall tone controls fitted to most hi-fi equipment allow separate bass and treble adjustment. (d) Bass-mid-top EQ of simple mixing console. Mid-control is sometimes called 'presence'. (e) Graphic equalizer has many individually controllable bands. (f) Parametric equalizer has variable frequency, level and Q factor. (g) Tilt control slopes whole response.

At (b) a top-cut filter rolls off high frequencies. This may be used in hi-fi systems to reduce the effect of surface noise in vinyl disks, static on AM radio or tape hiss. The slope and turnover point may both be adjustable.

At (c) the bass and treble controls of a hi-fi system affect the level at the extremes of the spectrum. Generally these controls have no effect at 1 kHz. The configuration frequently used is that due to Baxandall, which will be described presently.

At (d) a mixing console typically has bass, mid and top controls. The mid control allows emphasis of vocal frequencies which is not possible with the configuration of Figure 13.12(c). Sometimes two mid controls will be found.

At (e) the graphic equalizer has a large number of fixed frequency controls so that the frequency response can be highly irregular if required.

At (f) the parametric equalizer has a single peak or dip in its response, but the frequency, level and width or Q factor are all independently adjustable.

At (g) the tilt control gently slopes the entire frequency spectrum.

Whilst the frequency response of EQ types has been shown, in most equipment this will be accompanied by a phase response which will also have an audible effect. In practice the listener cannot separate the effects of frequency response and phase but assesses the overall sound. Certain brands of equalizer have achieved cult status because their designers have arrived at a combination which is particularly pleasant. It is often necessary to go to great lengths to ensure that digital equalizers replicate the phase response of their analog predecessors so that they have the same 'sound'.

Analog frequency response modification is invariably done using operational amplifiers in conjunction with reactive components such as capacitors. Whilst inductors could be used, they are large, heavy, prone to hum pickup and are not readily available in a range of values. In any case an inductor can be electronically simulated using other components.

Figure 13.13(a) shows that the impedance of a capacitor falls with frequency. Simple frequency-dependent circuits include the RC network (b). This forms a variable potential divider. At low frequencies the capacitor has a high impedance and is effectively absent. At high frequencies the capacitor has a low impedance and attenuates the signal. The frequency where the capacitor starts to take effect is a function of the time constant RC as shown. Where the impedance of the capacitor is equal to the value of the resistor the level will be 3 dB down. The slope of such a filter will be −6 dB/octave, which is a technical way of saying the output is inversely proportional to frequency.

A high-frequency boost can be obtained by placing such a network in a feed-back loop (c). The slope can be terminated by placing a resistor in series with the capacitor to limit the fall of impedance. This has the effect of producing a shelf response as shown in (d).

Figure 13.14 shows the Baxandall tone control which has reactive components in a bridge so that the frequency and phase response is flat with the controls in a central position. The bridge can be independently imbalanced at bass and treble frequencies.

Where steeper filter slopes are needed, for example in active crossovers, a second-order filter can be used having a 12 dB/octave roll-off.

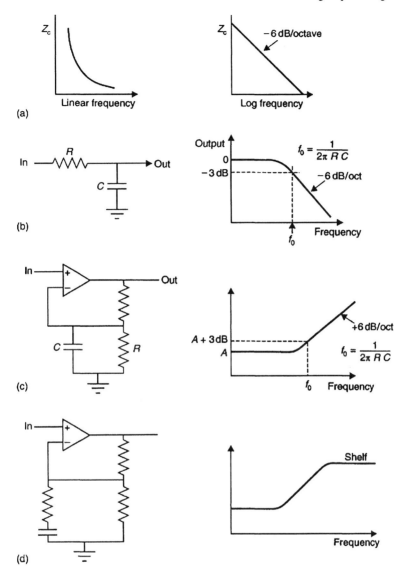

Figure 13.13 (a) Capacitor impedance is inversely proportional to frequency. (b) Simple *RC* network forms 6 dB/octave LPF. (c) Putting a network in the feedback loop produces high-frequency boost. (d) Boost can be terminated by a resistor in series with the capacitor to produce a shelf.

A filter which emphasizes certain frequencies in the middle of the range is called a bandpass filter. Presence controls and parametric equalizers are bandpass filters, as are the individual stages of a graphic equalizer. Figure 13.15 shows the configuration of an analog bandpass filter which essentially combines low- and high-pass functions.

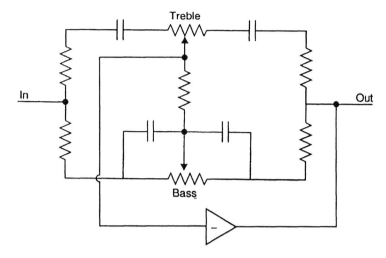

Figure 13.14 Baxandall tone control is nulled at central control position and has no effect on sound.

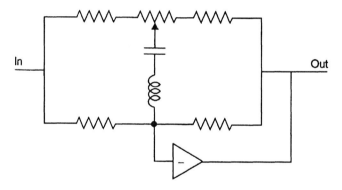

Figure 13.15 Simple analog bandpass filter.

Digital filters were introduced in Chapter 2 and these can be used for equalization. Figure 13.16 illustrates some simple digital filters along with their analog equivalents. Figure 13.17 shows a digital biquadratic bandpass filter which is capable of second-order response.[1]

13.7 Multitrack consoles

Multitrack production proceeds in two distinct phases as shown in Figure 13.18. In the record phase, the console controls the signals from the microphones and sends them to the record inputs of the multitrack machine. It must also be able independently to handle replay signals not only for monitoring the quality of the recording and to create a rough mix during recording but also to play back guide tracks for synchronous recording so that the tape can be built up a track at a time. There must also be provision for overdub or punch-in where a short section of track is replaced synchronously with new material. During recording only the most basic equalization will be used, such as low-frequency cut on close microphones. Generally effects are

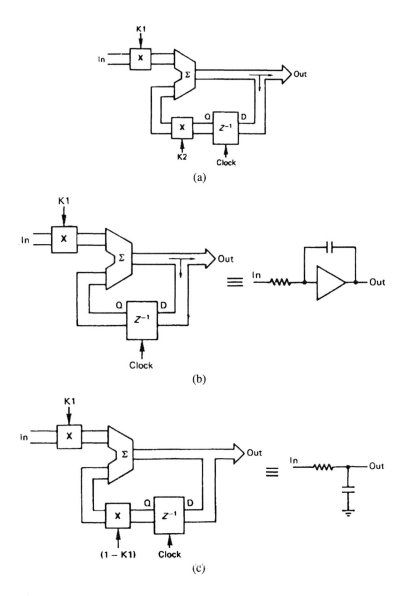

Figure 13.16 (a) First-order lag network IIR filter. Note recursive path through single sample delay latch. (b) The arrangement of (a) becomes an integrator if K2 = 1, since the output is always added to the next sample. (c) When the time coefficients sum to unity, the system behaves as an *RC* lag network.

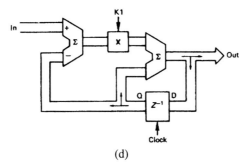

(d)

Figure 13.16 *(continued)* (d) The sample performance as in (c) can be realized with only one multiplier by reconfiguring as shown here..

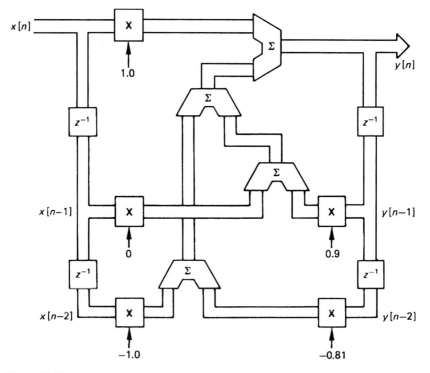

Figure 13.17 Digital biquadratic filter with a bandpass characteristic determined by the value of coefficients.

little used in the record stage, but a limiter may be employed to prevent tape overload.

Once the recording is finished, the console will be used to mix the tape tracks down. The level, EQ and position in the stereo image of each track will be determined and effects can be applied. Special effects may be restricted to one instrument or voice, but reverberation may be applied to the entire

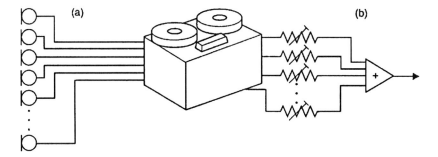

Figure 13.18 Multitrack recording proceeds in two phases. (a) Recording: microphone signals go to tape tracks. (b) Mix-down: multitrack tape is mixed to stereo master tape.

mix. The mix-down process is complex and it is impossible to get everything right immediately. The multitrack recorder with its autolocator will repeatedly playback the tape as the mix is perfected a little at a time, often with the help of a fader automation system.

Two types of console architecture have evolved for working with multitrack tape recorders. The earliest type is called the *split console*, shown in Figure 13.19(a), in which the console is basically constructed in two sections. The left-hand side processes microphone signals and routes them to the record

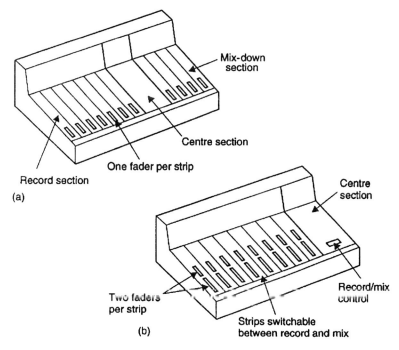

Figure 13.19 (a) Split console puts record and mix-down sections on opposite sides of centre section. (b) Inline console has both functions combined in each channel strip.

inputs of the tape recorder whereas the right-hand side takes tape replay signals and mixes them down to stereo.

The main advantage of a split console is that it is easy to understand the signal flow and this is beneficial ergonomically for live working. It is also advantageous for simultaneous recording and transmission because it can easily accommodate two operators. The split console has a number of drawbacks. When a large number of channels are in use it becomes physically very large because each tape track effectively needs two modules or strips, one for record and one for mix-down. When the monitor speakers are positioned to give a good stereo image when seated at one section of the console, the other section will be off to one side. A further problem is the cost of duplication of processes which will be present in both input and mix-down modules.

Figure 13.19(b) shows the *inline console* which was developed as an alternative to the split console. The inline console has one strip per channel which can handle both record and mix-down functions because it is electrically split into two halves: the *channel* section and the *monitor* section, each of which has its own fader. Generally the channel fader will be physically larger and adjacent to the front edge of the console. The operator of the inline console does not need to move when going from record to mix-down and so enjoys a better stereo perspective. The remainder of this discussion will consider the inline console in detail although many of the concepts mentioned apply to any type of console. The reader will appreciate that it is only possible to consider a generic console and specific products will differ in detail from the description here.

13.8 Console buses

Running across the console at right angles to the channel strips are the signal buses. These act like virtual earth points which will sum signals from any strip which produces them. Most of the summed signals appear in the console centre section.

Figure 13.20 shows the buses in a typical inline console. The tape send bus is actually one bus per tape track which carries signals to be recorded from the strips to the recorder. Each strip carries a set of switches which determine the tape track to which the record output will be routed. As each bus is a virtual earth, if the same track button is pressed on more than one strip the signals from those strips will be mixed as a group onto the same tape track. The group level is set by the channel fader.

There may be up to four stereo buses which carry the main mix-down output. Each strip carries buttons which select the stereo bus to which the output of the strip mix panpot will be routed. Signals switched to the same bus will be mixed together as a group and up to four groups can be created by using all of the buses. In the console centre section each stereo bus will have its own fader controlling the final group mix.

There is also a stereo monitor bus which sums the outputs of the monitor panpots.

There may be up to 16 auxiliary buses which can be configured in mono to mix the outputs of rotary level controls on each strip or configured in pairs where the controls become level and panpot.

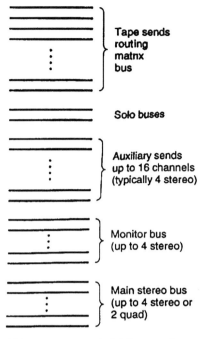

Figure 13.20 The buses which run across a typical inline console.

Finally there will be solo buses which allow an individual strip signal to be selected and monitored, in many cases without disturbing the rest of the mix.

The console centre section handles all of the bus signals except for the tape sends. The auxiliary bus mixes are buffered and provided with minimal tone controls, usually just bass and treble, prior to buffering to feed the auxiliary send sockets.

The stereo buses and the stereo (effects) returns are mixed by the main group faders and the return level controls to produce the stereo mix output. This typically goes to a two-track mastering recorder. The control room monitor speakers are driven by the signals selected in the centre section via the monitor level control which is usually a large rotary control rather than a linear fader. These may be mix, monitor, or solo signals. In addition external sources such as CD players, two-track tapes and DAT machines can be selected so that these can be heard on the control room monitors. A *dim* button will drop the control room monitor level by a significant amount to silence the room.

In a fully featured console the centre section will contain remote controls for the recorder, the tracks to be recorded and the autolocator so that no separate remote is needed.

13.9 The inline strip

The inline strip is a complex item because it has to operate in so many different modes. Paradoxically the more complex the design the easier the console is to operate because the user will find that the control system has automatically

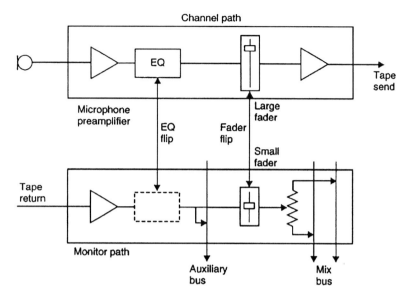

Figure 13.21 Major parts of an inline strip.

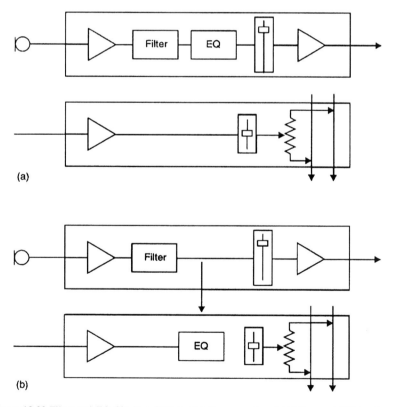

Figure 13.22 Filter and EQ flip is a feature of inline consoles. (a) Filters and EQ in channel path. (b) EQ in monitor path.

produced the desired configuration from a minimum of inputs. It is almost impossible to describe an inline strip with one diagram and the approach here is to consider various layers of functionality individually.

Figure 13.21 shows the inline strip at a conceptual level. The channel path handles microphone signals to be recorded and playback signals to be mixed. The channel path has a large fader which will usually be controlled by the automation system. The monitor path works simultaneously to allow the engineer and the band to listen to various signals.

Auxiliary signals can be sent to the auxiliary buses from the channel path or the monitor path. These auxiliary signals may be tapped off before (pre) or after (post) the faders.

The strip contains band-limiting filters and equalizers. Highly specified consoles may also have dynamics in each strip so no external compressor/limiters are needed. One of the features of the inline approach is that the filters and equalizers can be placed individually in the channel path or the monitor path as required. It is also possible to interchange the function of the large and small faders using a system called *fader flip* which will be detailed later.

Figure 13.22(a) shows the filters and EQ in the channel path, whereas (b) shows the EQ has been moved to the monitor path. Note that the switching maintains signal continuity in both paths.

13.10 Fader flip

The large fader near the front of the console is the one which is most convenient to use and the one which is controlled by the automation system. Consequently it will be used in the channel path for mix-down and recording. However, it is useful during the recording process to create a rough mix which can be used as a reference for the band and the producer. It will also form a basis for the subsequent mix-down process. This rough mix will need to use the large fader so that the automation system can be employed. Eliminating the VCA from the record path may also give a small quality advantage.

Fader flip (also called VCAs to monitor or fader reverse) allows the function of the two faders to be exchanged, i.e. the small fader becomes the channel record level and the large fader becomes the monitor mix fader. As EQ is generally used on mix-down the EQ flip will be employed to put the EQ in the monitor path.

There are several ways of implementing fader flip. Figure 13.23(a) shows the most obvious way is to mimic the switching used in moving the EQ between paths. The faders are literally rewired. This approach allows fader flip to be selected on each strip individually. The alternative approach is to leave the faders connected normally but to turn some or all of the channel path into the monitor path by switching. Figure 13.23(b) shows that this can be done if the two halves of the EQ switching are crossed so that signals flip sections at the EQ stage. The fader outputs are flipped by switching in the centre section which simply makes the stereo buses into monitor buses and vice versa. Figure 13.23(c) shows that a third alternative is to provide both paths with microphone, group and tape returns so the path flip can be done with input switching. The output flip is done with centre section bus switching as in (b). The approaches of (b) and (c) do not allow individual strips to be flipped, but have the advantage that the amount of audio switching is reduced.

When fader flip is used the console control system may automatically flip any auxiliary sends, filter or EQ stages so that as far as the operator is concerned all that happens is the two faders interchange.

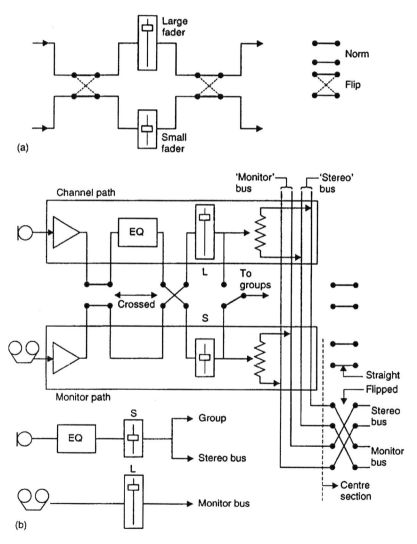

Figure 13.23 Fader flip techniques. (a) Faders are switched directly. (b) Part of channel and monitor paths are interchanged at EQ flip and by exchanging buses.

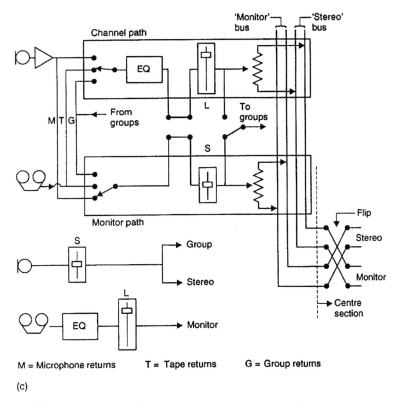

M = Microphone returns   T = Tape returns   G = Group returns

(c)

Figure 13.23 *(continued)* (c) Entire channel and monitor path is flipped by input and bus switching.

13.11 Console operating modes

When recording, microphones are connected to the channel path input. The channel path output is routed to the tape send track selection system and the channel fader controls the recording level. The monitor path may select 'ready group' to monitor the tape send signals or 'ready tape' to monitor the tape returns. Tape return monitoring allows the performance of the tape deck to be verified, although the return signal will have a delay due to the time taken for the tape to travel from the record head to the replay head. Group monitoring can be used to produce a rough mix so that the engineer can hear it via the control room speakers and the band can hear it via headphones.

Once guide tracks are laid down on tape, the remaining tracks are laid using synchronous recording. The recorder switches the tape returns so that the guide track(s) will be played back by the record head to eliminate tape delay. The synchronous replay will be selected on the monitor section with the tape ready button. The band can hear the guide track(s) via the monitor faders and the stereo bus and will play along. The microphones will pick up the sound for a new track and this will be routed to the recorder as well as to the headphone mix via the group input of the monitor channel so that the band hear a mix of the existing guide tracks and their latest efforts.

When sufficient tracks have been laid, the engineer may attempt a rough mix using the automation system. This can take place whilst other tracks are being recorded. Automated mix-down whilst recording will require fader flip to bring the automated fader into the monitor section. The user may decide to flip the EQ into the monitor path.

It happens in practice that something goes wrong with a take, perhaps a wrong note or something falls and makes a noise. Rather than repeat the whole track, it is possible to update part of a track using a process known as overdub or punch-in. Overdub must be performed synchronously with the tape transport playing with the record head. The monitor path must be configured so that the band can hear themselves and the tape. This will be done by selecting both tape ready and group ready so that the new microphone signals can be mixed with the offtape signals. The band will play along with the tape a few times so that the engineer can select subjectively acceptable points at which to begin and end the overdub. This is known as rehearse mode and the engineer can hear what the overdub would sound like by fading between the replay and microphone signals at the in-point, and back again at the out-point. These points can be remembered as timecodes by the automation system.

When the overdub is executed, on the target track the synchronous replay must be terminated because the record head will be needed to record the new material. The loss of the tape return causes a level drop in the band's headphones and this can be disconcerting as it doesn't happen in rehearse mode. The solution is automatically to boost the level of the group signal in the monitor mix during the overdub. Clearly such a precise approach can only be used where the multitrack recorder and the console are closely integrated by a common control system.

When mixing down, the channel path is routed to the stereo buses and the monitor path is redundant unless the tape replay is accompanied by a large number of other signals such as timecode locked sound effects or synthesizers. If these inputs cannot be accommodated on spare strips, it is possible to break into the group input of the monitor path with an external signal. This can be added to the stereo mix using the monitor fader and routing the monitor panpot to a stereo bus. In this way each strip can mix two audio channels.

On occasions it is necessary to make a multitrack recording whilst making a live broadcast. In this case the console can be switched so that the microphone inputs go to both the channel and the monitor paths. The channel path controls the recording and the monitor path controls the broadcast. In broadcast mode it is important that nothing is done to interrupt or change the broadcast output, consequently in this mode the control system may disable certain destructive controls such as certain solo operations and will lock all signal routing into the configuration which existed when the mode was entered.

Replay mode is a simple method of monitoring a tape in which the automation system is disabled.

13.12 Solo

When many different sources are present in a mix it is difficult to concentrate on one alone. If the engineer suspects a subtle defect in a signal due to a faulty microphone or bad positioning it will be necessary to monitor it more closely. The various types of solo function have evolved to help the engineer do just that.

The simplest type of solo to implement is one in which pressing the solo button on a strip simply cuts the output of all other strips. This is called solo-in-place and the result is that only one strip contributes to the mix. The control room monitors reproduce only that signal so it can be assessed. Unfortunately this simple approach is destructive because it affects the main stereo bus output as well so it cannot be used in certain circumstances, although it has the advantage that any effects returns of the soloed signal will still be heard because the centre section effects returns are unaffected by strip cuts.

In order to allow a non-destructive solo function which leaves the stereo mix unchanged whilst giving a solo effect on the control room monitors a number of techniques have been used, all of which require additional buses.

Pre-fade listen (PFL) taps off the channel signal before the fader onto a solo bus, whereas after-fade listen (AFL) taps off after the fader. The monitor bus is cut so that only the PFL or AFL bus signal reaches the monitor speakers. The stereo bus is unaffected and so PFL and AFL solos are non-destructive. Simple PFL and AFL give a mono result giving no indication of the position of the sound in the stereo image and naturally PFL does not reflect the fader level.

If a stereo solo bus is provided, a non-destructive stereo AFL solo is obtained by disabling the monitor bus output to the speakers and routing just the solo buses to the speakers. If a dual ganged panpot is fitted, the PFL signal can be panned in the same way as the mix.

In the up-front solo mode the panpot is routed to the solo bus by the solo button, and the monitor bus mix is dimmed in the monitor speakers so that the solo channel is elevated in the mix.

In some consoles the solo system is further complicated so that if any effects are being used with the soloed strip the relevant effects returns are not cut so that solo excludes everything but the soloed signal and its effects.

13.13 Ancillary functions

In addition to the control room speakers will be necessary in the studio and centre section level and source controls are provided for these.

In some situations the musicians' foldback to headphones will be the stereo bus output, but in overdub an auxiliary bus may be more useful. A small foldback mixer may be provided to control what the band hears.

It is often necessary for the console operator to communicate speech with various parts of the system. This is known as talkback. The console has a microphone, sometimes on a gooseneck, and a talkback control section. Talkback may be routed to the studio monitors or to the foldback mix. For recording take information, the talkback signal can be routed to the stereo mix output. This is known as *slate* and may be accompanied by a low-frequency tone which will become a distinctive beep when the tape is rewound.

It is essential to have some sort of line-up oscillator in a console with means to route it to the mix bus and the tape sends. Whilst 1 kHz for level setting and 10 kHz for bias adjustment are the absolute minimum, some consoles have a more comprehensive tone generator which includes musical pitch references for tuning.

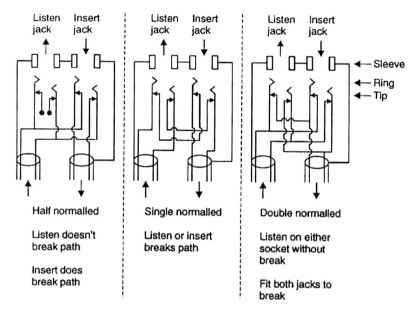

Figure 13.24 A jackfield may be used to extract and insert signals part way through a path. Inserting signals requires the existing input to be disconnected. This is done using switches operated by inserting the jack, a process called normalling.

Most consoles have a jackfield which is an array of sockets. Different input and output points in each strip are brought to the jackfield. Figure 13.24 shows how it can be used. Clearly to extract a signal for external use a jack can be inserted to tap it off from the appropriate point. Signals can also be inserted, and this additional signal may require the existing signal to be disconnected. Switch contacts on the jack sockets operate when the jack is inserted to achieve this in a process called *normalling*.

13.14 Stereo mixers

For classical and television sound work the multitrack approach is either not appropriate or too expensive. Stereo microphones always give greater realism than panned mono, and stereo mixers are designed to handle these. In true stereo recording simple is beautiful and often only a small number of microphones need to be handled. In addition to stereo microphone inputs and stereo line level inputs, provision will be made for a number of spot microphones to be panned into place. Spot microphones can be used when a particular sound source appears indistinct in the stereo microphone signal or to support a close television shot where a tighter acoustic is considered necessary for realism.

Stereo mixers handle pairs of signals in each strip. There is a single fader for stereo level and a balance control instead of a panpot. All of the controls such as level and EQ must track precisely to avoid balance problems and smear. Any dynamics must be ganged so that the gain of the two channels remains identical. Separate level sensing with ganged gain control works better than mono level sensing which tends to compress central sound sources more than those off to the side.

It is extremely useful in practice if a number of switchable sum/difference matrices are installed. This allows direct use of M-S microphones and provision of stereo width control through the use of variable S-gain. An S-phase switch also allows rapid left, right reversal in the case of incorrect wiring or where a fishpole is rotated to change from beneath shot to above shot.

Stereo consoles have different metering requirements. Left, right, M and S mix meters are a minimum, with an audio vectorscope being a useful addition for serious work. It is impractical to have such comprehensive metering on every input and in practice consoles tend to have only one set of such meters with the strips having only left, right meters. It is an advantage if the main meters can be temporarily assigned to specific inputs in what could be described as a metering solo system.

In television work boom operators need foldback from the mixer and it is advantageous if the foldback can be in stereo. Talkback can be mixed in mono into the stereo foldback so that it is clear which sound is which.

13.15 Effects

There is a wide range of effects available which allow almost limitless creativity in the production of popular music and it is only possible to introduce the subject here. Effects modify the sound in some way and in many cases the goal is novelty rather than realism. Many such effects are also used when playing electric guitars and synthesizers and it is difficult to draw the line between the instrument and the production process. In some complex popular music the effects used in the production process are designed to create a recognizable sound. In some cases the sound will be determined so strongly by the production process that the identity of the performers becomes irrelevant.

The most commonly used effect is artificial reverberation. To allow the most freedom in multitrack mixing, the sound on each track is kept isolated from other sounds. This can only be done by recording in a non-reverberant environment and the raw result is extremely dry. Thus artificial reverberation is essential to give life to a panpotted dry mix.

Artificial reverberators are generally designed to operate in stereo. They will accept a dry stereo input and output a stereo image in which the original sounds remain in the same place, but with the addition of reflections and echoes which are designed to appear elsewhere in the image much as they would in real life.

Figure 13.25(a) shows that when listening in a real auditorium, the direct sound arrives first, followed by strong early reflections from around the source, followed later by the decaying reverberation of the room which comes from all around. Artificial reverberators attempt to mimic this process.

Reverberation may be created using a special room, a vibrating plate or spring, or electronically, using delay. A reverberation room (b) consists of a space having solid masonry walls with little absorption. It is preferably an irregular shape to avoid a simple standing wave pattern and generally irregular hard objects are scattered around the room to make the reflections more complex. Even a small room can give the effect of a much larger space. A loudspeaker is placed at some point, and a microphone is suspended at

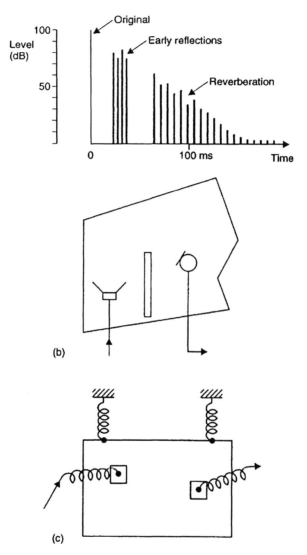

Figure 13.25 (a) Natural reverberation consists of a string of early reflections followed by a decaying tail. (b) Reverberation room has speaker and microphone with irregular dimensions to give complex reverberation. (c) Reverberation plate uses transducers to excite and pick off vibrations in a suspended plate.

another. Obstacles are generally placed so that no direct path is available between the two. The polar diagram of the microphone has a great bearing on the result and it is often overlooked that poor results may indicate the use of a loudspeaker with a poor directivity pattern.

The reverberation plate is an alternative where a dedicated room is impossible. Figure 13.25(c) shows that transducers are used so that the output of an audio power amplifier can launch vibrations in a resiliently supported metal plate. After these have reflected from the plate boundaries they will be picked up by further transducers to form the output.

A simpler but less realistic result is obtained by replacing the plate with a coil spring to create a *spring line delay*. These are popular for enhancing the sound of an electric guitar.

It is possible to produce artificial reverberation using purely electronic means. The basic components are delay, to simulate the additional path travelled by reflected sound, and filtering to simulate the frequency-dependent absorption experienced. As the provision of delay in the digital domain is trivial, the digital reverberator was a natural solution. Most digital reverberators use digital signal processors (DSPs) which are audio specific computers. The action taken is determined purely by the programming and so it is possible to have a reverberator in which the size, shape and texture of the virtual space are all variable. Whilst digital reverberators can be very effective, lack of computation power prevents complete realism which requires extremely complex and detailed modelling.

The variety of effects is too great to treat comprehensively and it is only possible to give an indication here. The *harmonizer* is an effects unit which takes an audio input and produces a more complex output spectrum which is musically related to the input spectrum, resulting in a richer sound. Time variant filtering is commonly used. The first application of this was *phasing* in which a variable frequency notch or comb filter is used. When used with broadband sound sources such as a drum-kit, dynamically changing the filter frequency gives the subjective effect of a pitch change. Originally phasing was performed by making two identical tape recordings and mixing them together on playback. Careful control of the speed of the two capstans would introduce a time delay between the signals, resulting in comb filtering. Nowadays such filtering is entirely electronic.

Dynamically variable filtering can also be used to mimic the formant process of speech to give musical instruments a human characteristic. The simplest of these devices is the *wah-wah*; a sweepable filter often controlled by a foot pedal so it can be operated by a musician. The *vocoder* is a complex device which uses a speech input to modulate the signal from an instrument so that it 'talks'.

Pitch changing is an important requirement where the musical pitch of the input can be altered without otherwise changing the timbre. In television it is often necessary to replay material at the wrong speed. 24 fps film is often televised at 25 fps, resulting in incorrect pitch. A pitch changer can return the pitch to the correct tuning. In public address systems a slight pitch shift prevents the build-up of feedback, allowing a higher gain to be used. Pitch shifters work by multiplying the input signal by a high frequency to produce sidebands. Multiplication of the sideband by a slightly different frequency produces a frequency shift in the output.

13.16 Introduction to audio compression

One of the fundamental concepts of PCM audio is that the signal-to-noise ratio of the channel can be determined by selecting a suitable wordlength. In conjunction with the sampling rate, the resulting data rate is then determined. In many cases the full data rate can be transmitted or recorded as is done in

DASH, DAT, CD and the AES/EBU interface. Professional equipment traditionally has operated with higher sound quality than consumer equipment in order to allow a margin for some inevitable quality loss in production, and so there is little cause to use compression. Where compression has to be used, professional equipment might maintain the quality gap by using a smaller amount or by employing more sophisticated algorithms than consumer devices.

Where there is a practical or economic restriction on channel bandwidth or storage capacity, compression becomes essential. In broadcasting, bandwidth is at a premium as sound radio has to share the spectrum with other services. In NICAM TV sound, the new digital sound carrier has to be squeezed into the existing standardized TV channel spacing. NICAM and DAB both use compression to conserve data rate. Digital recorders based on RAM or ROM, such as samplers and announcement generators have no moving parts and are maintenance free, but suffer from a higher cost per bit than other media. Compression reduces the cost per second of audio by allowing longer recording time in a given memory size.

In the Video-8 system, the PCM audio has to share the tape with an analog video signal. Since Video-8 is a consumer product, compression is once more used to keep tape consumption moderate.

In DCC it was a goal that the cassette would use conventional oxide tape for low cost, and a simple transport mechanism was a requirement. In order to record on low coercivity tape which is to be used in an uncontrolled consumer environment, wavelengths have to be generous, and track widths need to be relatively wide. In order to allow a reasonably slow tape speed and hence good playing time, compression is necessary.

Broadcasters wishing to communicate over long distances are restricted to the data rate conveniently available in the digital telephone network. The BBC is able to pass six compressed high-quality audio channels down the available telephone channels of 2048 kbits/s.[2] Before the introduction of stereo, compression was used to distribute the sound accompanying television broadcasts by putting two PCM samples inside the line-synchronizing pulses, thus obtaining a sampling rate of twice the TV line rate.[3] The modern dual channel sound-in-syncs (DSIS) systems use a similar approach.

There are many different types of compression, each allowing a different compression factor to be obtained before degradations become noticeable. The simplest systems use either non-uniform requantizing or companding on the whole of the audio band. Non-uniform requantizing can be applied to a uniform quantized PCM signal using a lookup table. With this approach, the range of values in which an input sample finds itself determines the factor by which it will be multiplied. For example, a sample value with the most significant bit reset could be multiplied by two to shift the bits up one place. If the two most significant bits are reset, the value could be multiplied by four and so on. Constants are then added to allow the range of the compressed sample value to determine the expansion necessary.

A relative of non-uniform requantizing is floating-point coding. Here the sample value is expressed as a mantissa and a binary exponent which determines how the mantissa needs to be shifted to have its correct absolute value on a PCM scale.

Non-uniform requantizing and floating-point coding work on each sample individually. A floating-point system requires one exponent to be carried with each mantissa. An alternative is near-instantaneous companding, or floating-point block coding, where the magnitude of the largest sample in a block is used to determine the value of a common exponent. Sending one exponent per block requires a lower data rate in than true floating point.[4]

Whilst the above systems do reduce the data rate required, only moderate reductions are possible without subjective quality loss. NICAM has an audio data compression factor of 0.7, Video-8 has a factor of 0.8. Applications such as DCC and DAB require a figure of 0.25. In MiniDisc it is 0.2. Compression factors of this kind can only be realized with much more sophisticated techniques. These differ in principle, but share a dependence on complex digital processing which has only recently become economic due to progress in VLSI. These techniques fall into three basic categories; predictive coding, sub-band coding and transform coding.

Predictive coding uses a knowledge of previous samples to predict the value of the next. It is then only necessary to send the difference between the prediction and the actual value. The receiver contains an identical predictor to which the transmitted difference is added to give the original value.

Sub-band coding splits the audio spectrum into many different frequency bands to exploit the fact that most bands will contain lower level signals than the loudest one.

In spectral coding, a Fourier transform of the waveform is computed periodically. Since the transform of an audio signal generally changes slowly and the ear also responds slowly, a transform need be sent much less often than audio samples. The receiver performs an inverse transform.

Practical compression units will usually use some combination of at least two of these techniques along with non-linear or floating-point requantizing of sub-band samples or transform coefficients.

The sensitivity of the ear to distortion probably deserves more attention than fidelity of dynamic range or frequency response. Except for lossless coding, all audio compression relies to a certain extent on an understanding of the hearing mechanism, particularly the phenomenon of masking, as this determines the audibility of artifacts. Thus compression is a form of perceptual coding.[5]

As was seen in Chapter 3, the basilar membrane in the ear behaves as a kind of spectrum analyser. The part of the basilar membrane which resonates as a result of an applied sound is a function of the frequency. The high frequencies are detected at the end of the membrane nearest to the eardrum and the low frequencies are detected at the opposite end. The ear analyses with frequency bands, known as critical bands, about 100 Hz wide below 500 Hz and from one-sixth to one-third of an octave wide, proportional to frequency, above this.

Thus the membrane has an effective Q factor which is responsible for the phenomenon of auditory masking, defined as the decreased audibility of one sound in the presence of another. The degree of masking depends upon whether the masking tone is a sinusoid, which gives the least masking, or noise.[6] However, harmonic distortion produces widely spaced frequencies, and these are easily detected even in minute quantities by a part of the basilar membrane which is distant from that part which is responding to the fundamental. The masking effect is asymmetrically disposed around the masking frequency.[7] Above the masking frequency, masking is more pronounced, and its extent increases with acoustic level. Below the masking frequency, the extent of masking drops sharply at as much as 90 dB/octave. Clearly very sharp filters are required if noise at frequencies below the masker is to be confined within the masking threshold.

Owing to the resonant nature of the membrane, it cannot start or stop vibrating rapidly. The spectrum sensed changes slowly even if that of the original sound does not. The reduction in information sent to the brain is considerable; masking can take place even when the masking tone begins after and ceases before the masked sound. This is referred to as forward and backward masking.[8]

A detailed model of the masking properties of the ear is necessary to the design of audio compression systems. Since quantizing distortion results in energy moving from one frequency to another, the masking model is essential to estimate how audible the effect of distortion will be. The greater the degree of compression required, the more precise the model must be. If the masking model is inaccurate, then equipment based upon it may produce audible artifacts under some circumstances. Artifacts may also result if the model is not properly implemented. As a result, development of compressors requires careful listening tests with a wide range of source material.[9] The presence of artifacts at a given compression factor indicates only that performance is below expectations; it does not distinguish between the implementation and the model. If the implementation is verified, then a more detailed model must be sought.

Properly conducted listening tests are expensive and time consuming, and alternative methods have been developed which can be used rapidly to evaluate the performance of different techniques. The noise-to-masking ratio (NMR) is one such measurement.[10] Figure 13.26 shows how NMR is measured. Input audio signals are fed simultaneously to a coder and decoder in tandem (known as a codec) and to a compensating delay. At the output of the delay, the coding error is obtained by subtracting the codec output from the original. The original signal is spectrum analysed into critical bands in order to derive the masking threshold of the input audio, and this is compared with the critical band spectrum of the error.

The NMR in each critical band is the ratio between the masking threshold and the quantizing error due to the codec. An average NMR for all bands can

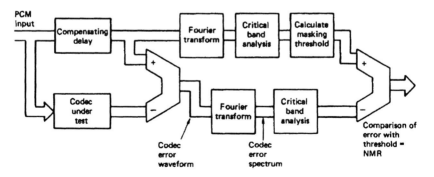

Figure 13.26 The noise-to-masking ratio is derived as shown here.

be computed. A positive NMR in any band indicates that artifacts are potentially audible. Plotting the average NMR against time is a powerful technique, as with an ideal codec the NMR should be stable with different types of program material. NMR excursions can be correlated with the waveform of the audio input to analyse how the extra noise was caused and to redesign the codec to eliminate it.

Practical systems should have a finite NMR in order to give a degree of protection against difficult signals which have not been anticipated and against the use of post codec equalization or several tandem codecs which could change the masking threshold. There is a strong argument that professional devices should have a significantly greater NMR than consumer or program delivery devices.

There are, of course, limits to all technologies. Eventually artifacts will be heard as the amount of compression is increased which no amount of detailed modelling will remove. The ear is only able to perceive a certain proportion of the information in a given sound. This could be called the perceptual entropy,[11] and all additional sound is redundant or irrelevant. Compression works by removing the redundancy, and clearly an ideal system would remove all of it, leaving only the entropy. Once this has been done, the masking capacity of the ear has been reached and the NMR has reached zero over the whole band. Reducing the data rate further must reduce the entropy, because raising noise further at any frequency will render it audible. Thus there is a limit to the degree of compression which can be achieved even with an ideal coder. Systems which go beyond that limit are not appropriate for high-quality music, but are relevant in telephony and communications where intelligibility of speech is the criterion.

Interestingly, the data rate out of a coder is virtually independent of the input sampling rate unless the sampling rate is deliberately reduced. This is because the entropy of the sound is in the waveform, not in the number of samples carrying it.

The compression factor of a coder is only part of the story. All codecs cause delay, and in general the greater the compression the longer the delay. In some applications where the original sound may be heard at the same time as sound which has passed through a codec, a short delay is required.[12] In most applications, the compressed channel will have a constant bit rate, and so a constant compression factor is required. In real programme material, the entropy varies and so the NMR will fluctuate. If greater delay can be accepted, as in a recording application, memory buffering can be used to allow the coder to operate at constant NMR and instantaneously variable data rate. The memory absorbs the instantaneous data-rate differences of the coder and allows a constant rate in the channel. A higher effective degree of compression will then be obtained.

13.17 Non-uniform coding

The digital compression system of Video-8 is illustrated in Figure 13.27. This broadband system reduces samples from 10 bits to 8 bits long and is in addition to an analog compression stage. In order to obtain the most transparent performance, the smallest signals are left unchanged. In a 10-bit system, there are 1024 quantizing intervals, but these are symmetrical about the centre of

Input X	Conversion	Output Y
0–15	Y = X	0–15
16–63	$Y = \dfrac{X}{2} + 8$	16–39
64–319	$Y = \dfrac{X}{4} + 24$	40–103
320–511	$Y = \dfrac{X}{8} + 44$	104–127

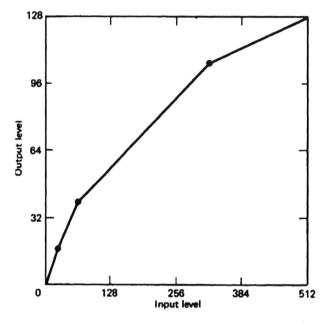

Figure 13.27 The digital compression of PCM samples in 8-mm video. Note that there are only 512 levels in a 10-bit signal, since positive and negative samples can have the same level.

the range; therefore there are only 512 audio levels in the system, an equal number positive and negative. The 10-bit samples are expressed as signed binary, and only the 9-bit magnitude is compressed to 7 bits, following which the sign bit is added on.

The first 16 input levels (0–15) pass unchanged, except that the most significant two bits below the sign bit are removed. The next 48 input levels (16–63) are given a gain of one-half, so that they produce 24 output levels from 16 to 39. The next 256 input levels (64–319) have a gain of one-quarter, so that they produce 64 output levels from 40 to 103. Finally, the remaining 208 levels (320–511) have a gain of one-eighth, so they occupy the remaining output levels from 104 to 127. In this way the coarsening of the effective quantizing intervals is matched by the increasing amplitude of the signal, so that the increased noise is masked. The analog companding further reduces the noise to give remarkably good performance.

13.18 Floating-point coding

In floating-point notation (Figure 13.28), a binary number is represented as a mantissa, which is always a binary fraction with one just to the right of the radix point, and an exponent, which is the power of two the mantissa has to be multiplied by to obtain the fixed-point number. Clearly the signal-to-noise ratio is now defined by the number of bits in the mantissa, and as shown in Figure 13.29, this will vary as a sawtooth function of signal level, as the best value, obtained when the mantissa is near overflow, is replaced by the worst value when the mantissa overflows and the exponent is incremented.

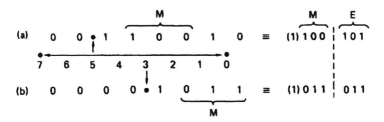

Figure 13.28 In this example of floating-point notation, the radix point can have eight positions determined by the exponent E. The point is placed to the left of the first '1', and the next four bits to the right form the mantissa M. As the MSB of the mantissa is always 1, it need not always be stored.

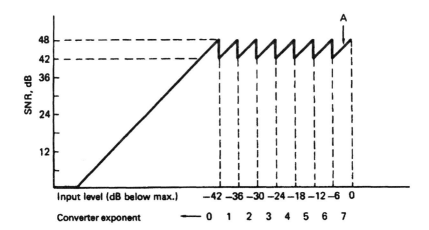

Figure 13.29 In this example of an 8-bit mantissa, 3-bit exponent system, the maximum SNR is 6 dB × 8 = 48 dB with maximum input of 0 dB. As input level falls by 6 dB, the converter noise remains the same, so SNR falls to 42 dB. Further reduction in signal level causes the converter to shift range (point A in the diagram) by increasing the input analog gain by 6 dB. The SNR is restored, and the exponent changes from 7 to 6 in order to cause the same gain change at the receiver. The noise modulation would be audible in this simple system. A longer mantissa word is needed in practice.

Floating-point notation is used within DSP chips as it eases the computational problems involved in handling long wordlengths. For example, when multiplying floating-point numbers, only the mantissae need to be multiplied. The exponents are simply added.

Floating-point coding is at its most useful for compression when several adjacent samples are assembled into a block so that the largest sample value determines a common exponent for the whole block. This technique is known as floating-point block coding.

In NICAM 728, 14-bit samples are converted to 10-bit mantissae in blocks of 32 samples with a common 3-bit exponent. Figure 13.30 shows that the exponent can have five values, which are spaced 6.02 dB or a factor of two in gain apart. When the signal in the block has one or more samples within 6 dB of peak, the lowest gain is used. If the largest sample in the block has a value more than 6 dB below peak, one high-order bit is inactive and the gain is boosted by a factor of two by a 1-bit shift. As a result the system has five transfer functions which are selected to keep the mantissa as large as possible.

The worst case in block coding occurs when there is one large value sample in an otherwise quiet block. The large sample value causes the system to select a low gain and the quiet part is quantized coarsely, resulting in potential distortion. However, distortion products have to be presented to the ear for

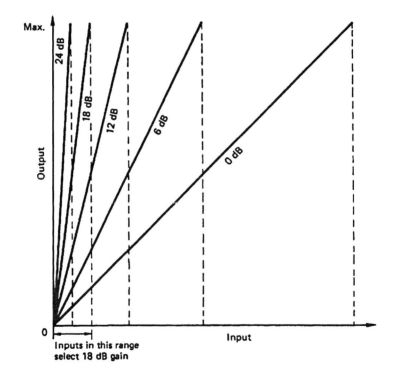

Figure 13.30 The coding scheme of NICAM 728 has five gain ranges. The gain is selected in each block of samples such that the largest coded values are obtained without clipping.

some time before they can be detected as a harmonic structure. With a 1 ms block, the distortion is too brief to be heard.

13.19 Predictive coding

Predictive compression has many principles in common with sigma-delta modulators. As was seen in Chapter 3, a sigma-delta modulator places a high-order filter around a quantizer so that what is quantized is the difference between the filter output and the actual input. The quantizing error serves to drive the filter in such a way that the difference is cancelled. When the difference is small, the filter has anticipated what the next input voltage will be, which leads it to be called by the alternative title of *predictor*. The same process can be modelled entirely in the digital domain as can be seen in Figure 13.31. The input is conventional PCM, and the predictor output is subtracted from the input by a conventional complement and add process. The difference is then requantized, which requires the numerical value to be divided by the new step size. The remainder is discarded. The requantized difference becomes the compressed output.

In order to reduce the size of the difference, the requantizing error must be fed back. The requantized difference must be returned to is original numerical value (except for the requantizing error) by multiplying it by the same step size that was used in the requantizer. This is the function of the inverse quantizer. The inverse-quantized difference is then added to the predictor output. The next input to the predictor is thus the current input plus the requantizing error.

The decoder contains the same predictor and inverse quantizer as the encoder. The inverse quantizer returns the differences to their correct numerical value. The predictor is driven from the system output and will make the same prediction as the encoder. The difference is added to the predictor output to recreate the original value of the sample, which also becomes the next predictor input.

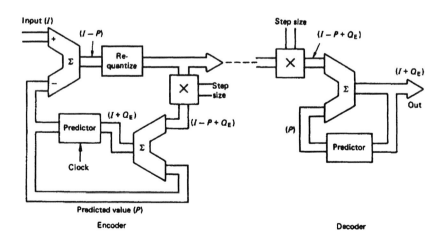

Figure 13.31 A predictive coder resembles the operation of a sigma-delta modulator which has a digital input. See text for details.

Further compression can be obtained by making the requantizing steps of variable size. When the difference signal is large, the step size increases to prevent saturation. When the difference signal is small, the step size reduces to increase resolution. In order to avoid transmitting the step size along with the difference data, it is possible to make the requantizer adaptive. In this case the step size is based on the magnitude of recent requantizer outputs. As the output increases towards the largest code value, the step size increases to avoid clipping. Figure 13.32 shows how an adaptive quantizer can be included in the predictive coder. Clearly the same step size needs to be input both to the inverse quantizer in the coder and to the same device in the decoder. If the step size is obtained from the coder output, the decoder can also recreate it locally so that the same step size is used throughout.

Predictive coders work by finding the redundancy in an audio signal. For example, the fundamental of a note from a musical instrument will be a sine wave which changes in amplitude quite slowly. The predictor will learn the pitch quite quickly and remove it from the data stream which then contains the harmonics only. These are at a lower level than the fundamental and require less data to represent them. When a transient arrives, the predictor will be unprepared, and a large difference will be computed. The adaptive quantizer will react by increasing the step size to prevent clipping.

DPCM coders have the advantage that they work on the signal waveform in the time domain and need a relatively short signal history to operate. Thus they cause a relatively short delay in the coding and decoding stages. A further advantage is that the differential data are actually less sensitive to bit errors than PCM. This is because the difference signals represent a small part of the final signal amplitude.

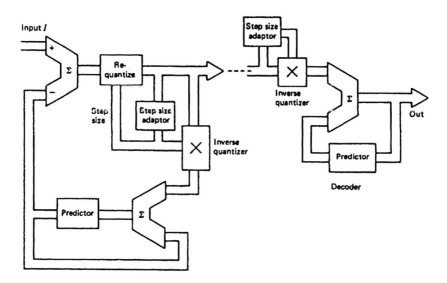

Figure 13.32 A predictive coder incorporating an adaptive quantizer bases the step size on recent outputs. The decoder can recreate the same step size locally so it need not be sent.

13.20 Sub-band coding

Sub-band compression takes advantage of the fact that real sounds do not have uniform spectral energy. The wordlength of PCM audio is based on the dynamic range required and this is generally constant with frequency although any pre-emphasis will affect the situation. When a signal with an uneven spectrum is conveyed by PCM, the whole dynamic range is occupied only by the loudest spectral component, and all of the other components are coded with excessive headroom. In its simplest form, sub-band coding[13] works by splitting the audio signal into a number of frequency bands and companding each band according to its own level. Bands in which there is little energy result in small amplitudes which can be transmitted with short wordlength. Thus each band results in variable length samples, but the sum of all the sample wordlengths is less than that of PCM and so a coding gain can be obtained. Sub-band coding is not restricted to the digital domain; the analog Dolby noise-reduction systems use it extensively.

The number of sub-bands to be used depends upon what other technique is to be combined with the sub-band coding. If it is intended to use compression based on auditory masking, the sub-bands should preferably be narrower than the critical bands of the ear, and therefore a large number will be required, ISO/MPEG and PASC, for example, use 32 sub-bands. Figure 13.33 shows the critical condition where the masking tone is at the top edge of the sub-band. It will be seen that the narrower the sub-band, the higher the requantizing noise that can be masked. The use of an excessive number of sub-bands will, however, raise complexity and the coding delay, as well as risking pre-echo on transients exceeding the temporal masking.

On the other hand, if used in conjunction with predictive coding, relatively few bands are required. The apt-X100 system, for example, uses only four bands as simulations showed that a greater number gave diminishing returns.[14]

The band-splitting process is complex and requires a lot of computation. One band-splitting method which is useful is quadrature mirror filtering (QMF).[15] The QMF is a kind of twin FIR filter which converts a PCM sample stream

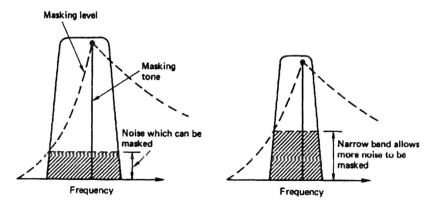

Figure 13.33 In sub-band coding the worst case occurs when the masking tone is at the top edge of the sub-band. The narrower the band, the higher the noise level which can be masked.

into two sample streams of half the input sampling rate, so that the output data rate equals the input data rate. The frequencies in the lower half of the audio spectrum are carried in one sample stream, and the frequencies in the upper half of the spectrum are carried in the other. Whilst the lower frequency output is a PCM band-limited representation of the input waveform, the upper frequency output is not. A moment's thought will reveal that it could not be because the sampling rate is not high enough. In fact the upper half of the input spectrum has been heterodyned down to the same frequency band as the lower half by the clever use of aliasing. The waveform is unrecognizable, but when heterodyned back to its correct place in the spectrum in an inverse step, the correct waveform will result once more.

Figure 13.34 shows the operation of a simple QMF. At (a) the input spectrum of the PCM audio is shown, having an audio baseband extending up to half the sampling rate and the usual lower sideband extending down from there up to the sampling frequency. The input is passed through an FIR low-pass filter which cuts off at one-quarter of the sampling rate to give the spectrum shown at (b). The input also passes in parallel through a second FIR filter which is physically identical, but the coefficients are different. The impulse response of the FIR low-pass filter is multiplied by a cosinusoidal waveform which amplitude-modulates it. The resultant impulse gives the filter a frequency response shown at (c). This is a mirror image of the low-pass filter response.

If certain criteria are met, the overall frequency response of the two filters is flat. The spectra of both (b) and (c) show that both are oversampled by a factor of two because they are half empty. As a result both can be decimated by a factor of two, which is the equivalent of dropping every other sample. In the case of the lower half of the spectrum, nothing remarkable happens. In the case of the upper half of the spectrum, it has been resampled at half the original frequency as shown at (d). The result is that the upper half of the audio spectrum aliases or heterodynes to the lower half.

An inverse QMF will recombine the bands into the original broadband signal. It is a feature of a QMF/inverse QMF pair that any energy near the band edge which appears in both bands due to inadequate selectivity in the filtering reappears at the correct frequency in the inverse filtering process provided that there is uniform quantizing in all of the sub-bands. In practical coders, this criterion is not met, but any residual artifacts are sufficiently small to be masked.

The audio band can be split into as many bands as required by cascading QMFs in a tree. However, each stage can only divide the input spectrum in half. In some coders certain sub-bands will have passed through one splitting stage more than others and will be half their bandwidth.[16] A delay is required in the wider sub-band data for time alignment.

A simple quadrature mirror is computationally intensive because sample values are calculated which are later decimated or discarded, and an alternative is to use polyphase pseudo-QMF filters[17] or wave filters[18] in which the filtering and decimation process is combined. Only wanted sample values are computed. A polyphase QMF operates in a manner not unlike the polyphase operation of an FIR filter used for interpolation in sampling rate conversion (see Chapter 2). In a polyphase filter a set of samples is shifted into position in the transversal register and then these are multiplied by different sets of

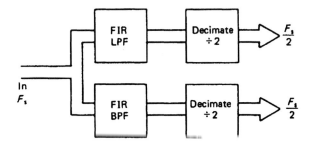

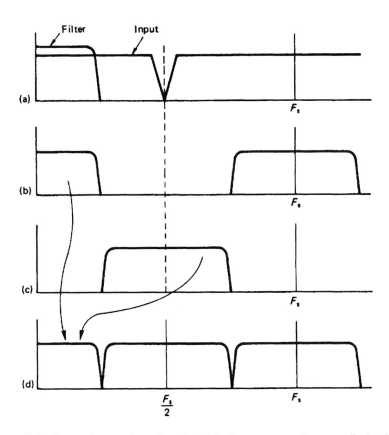

Figure 13.34 The quadrature mirror filter. In (a) the input spectrum has an audio baseband extending up to half the sampling rate. The input is passed through an FIR low-pass filter which cuts off at one-quarter of the sampling rate to give the spectrum shown in (b). The input also passes in parallel through a second FIR filter whose impulse response has been multiplied by a cosinusoidal waveform in order to amplitude-modulate it. The resultant impulse gives the filter a mirror image frequency response shown in (c). The spectra of both (b) and (c) show that both are oversampled by a factor of two because they are half empty. As a result both can be decimated by a factor of two, resulting in (d) in two identical Nyquist-sampled frequency bands of half the original width.

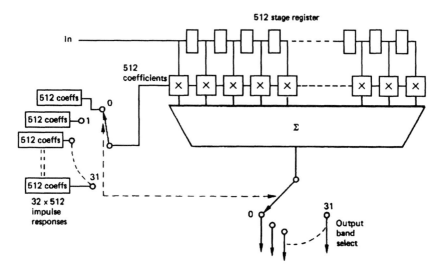

Figure 13.35 In polyphase QMF the same input samples are subject to computation using coefficient sets in many different time-multiplexed phases. The decimation is combined with the filtering so only wanted values are computed.

coefficients and accumulated in each of several phases to give the value of a number of different samples between input samples.

In a polyphase QMF, the same approach is used. Figure 13.35 shows an example of a 32-band polyphase QMF having a 512-sample window. With 32 sub-bands, each band will be decimated to 1/32 of the input sampling rate. Thus only one sample in 32 will be retained after the combined filter/decimate operation. The polyphase QMF only computes the value of the sample which is to be retained in each sub-band. The filter works in 32 different phases with the same samples in the transversal register. In the first phase, the coefficients will describe the impulse response of a low-pass filter, the so-called prototype filter, and the result of 512 multiplications will be accumulated to give a single sample in the first band.

In the second phase the coefficients will be obtained by multiplying the impulse response of the prototype filter by a cosinusoid at the centre frequency of the second band. Once more 512 multiple accumulates will be required to obtain a single sample in the second band. This is repeated for each of the 32 bands, and in each case a different centre frequency is obtained by multiplying the prototype impulse by a different modulating frequency. Following 32 such computations, 32 output samples, one in each band, will have been computed. The transversal register then shifts 32 samples and the process repeats.

The principle of the polyphase QMF is not so different from the techniques used to compute a frequency transform and effectively blurs the distinction between sub-band coding and transform coding.

The QMF technique is restricted to bands of equal width. It might be thought that this is a drawback because the critical bands of the ear are non-uniform. In fact this is only a problem when very large compression factors are required. In all cases it is the masking model of hearing which must have correct critical bands. This model can then be superimposed on bands of any width to determine how much masking and therefore coding gain is possible. Uniform width sub-bands will not be able to obtain as much masking as bands which are matched to critical bands, but for many applications the additional coding gain is not worth the added filter complexity.

13.21 Transform coding

Audio is usually considered to be a time-domain waveform as this is what emerges from a microphone. As has been seen in Chapter 2, Fourier analysis allows any waveform to be represented by a set of harmonically related components of suitable amplitude and phase. In theory it is perfectly possible to decompose an input waveform into its constituent frequencies and phases, and to record or transmit the transform. The transform can then be reversed and the original waveform will be precisely recreated. Although one can think of exceptions, the transform of a typical audio waveform changes relatively slowly. The slow speech of an organ pipe or a violin string, or the slow decay of most musical sounds allow the rate at which the transform is sampled to be reduced, and a coding gain results. A further coding gain will be achieved if the components which will experience masking are quantized more coarsely.

In practice there are some difficulties. The computational task of transforming an entire recording of perhaps one hour's duration as one waveform is staggering. Even if it were feasible, the coding delay would be at least equal to the length of the recording, which is unacceptable for many purposes.

The solution to this difficulty is to cut the waveform into short segments and then to transform each individually. The delay is reduced, as is the computational task, but there is a possibility of artifacts arising because of the truncation of the waveform into rectangular time windows. A solution is to use window functions (see Chapter 2) and to overlap the segments as shown in Figure 13.36. Thus every input sample appears in just two transforms, but with variable weighting depending upon its position along the time axis. Although it appears from the diagram that twice as much data will be generated, in fact certain transforms can eliminate the redundancy.

The DFT (discrete frequency transform) does not produce a continuous spectrum, but instead produces coefficients at discrete frequencies. The frequency resolution (i.e. the number of different frequency coefficients) is equal to the number of samples in the window. If overlapped windows are used, twice as many coefficients are produced as are theoretically necessary. In addition the DFT requires intensive computation, owing to the requirement to use complex arithmetic to render the phase of the components as well as the amplitude. An alternative is to use discrete cosine transforms (DCTs). These are advantageous when used with overlapping windows.

Figure 13.36 Transform coding can only be practically performed on short blocks. These are overlapped using window functions in order to handle continuous waveforms.

In the modified discrete cosine transform (MDCT),[19] windows with 50% overlap are used. Thus twice as many coefficients as necessary are produced. These are sub-sampled by a factor of two to give a critically sampled transform, which results in potential aliasing in the frequency domain. However, by making a slight change to the transform, the alias products in the second half of a given window are equal in size but of opposite polarity to the alias products in the first half of the next window, and so will be cancelled on reconstruction. This is the principle of time-domain aliasing cancellation (TDAC).

The requantizing in the coder raises the quantizing noise in the frequency bin, but it does so over the entire duration of the block. Figure 13.37 shows that if a transient occurs towards the end of a block, the decoder will reproduce the waveform correctly, but the quantizing noise will start at the beginning of the block and may result in a pre-echo where the noise is audible before the transient.

The solution is to use a variable time window according to the transient content of the audio waveform. When musical transients occur, short blocks are necessary and the frequency resolution and hence the coding gain will be low. At other times the blocks become longer and the frequency resolution of the transform rises, allowing a greater coding gain.

The transform of an audio signal is computed in the main signal path in a transform coder, and has sufficient frequency resolution to drive the masking model directly. However, in certain sub-band coders the frequency resolution of the filter bank is good enough to offer a high coding gain, but not good enough to drive the masking model accurately, particularly in respect of the steep slope on the low-frequency side of the masker. In order to overcome this problem, a transform will often be computed for control purposes in a side chain rather than in the main audio path, and so the accuracy in respects other than frequency resolution need not be so high. This approach also permits the use of equal width sub-bands in the main path.

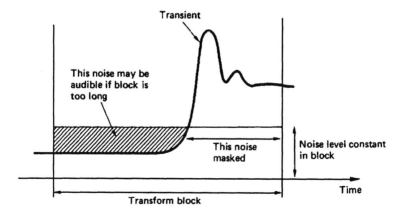

Figure 13.37 If a transient occurs towards the end of a transform block, the quantizing noise will still be present at the beginning of the block and may result in a pre-echo where the noise is audible before the transient.

13.22 A simple sub-band coder

Figure 13.38 shows the block diagram of a simple sub-band coder. At the input, the frequency range is split into sub-bands by a filter bank such as a quadrature mirror filter. The output data rate of the filter bank is no higher than the input rate because each band has been heterodyned to a frequency range from DC upwards. The decomposed sub-band data are then assembled into blocks of fixed size, prior to compression. Whilst all sub-bands may use blocks of the same length, some coders may use blocks which get longer as the sub-band frequency becomes lower. Sub-band blocks are also referred to as frequency bins.

The coding gain is obtained as the waveform in each band passes through a requantizer. The requantization is achieved by multiplying the sample values by a constant and rounding up or down to the required wordlength. For example, if in a given sub-band the waveform is 36 dB down on full scale, there will be at least six bits in each sample which merely replicate the sign bit. Multiplying by 2^6 will bring the high-order bits of the sample into use, allowing bits to be lost at the lower end by rounding to a shorter wordlength.

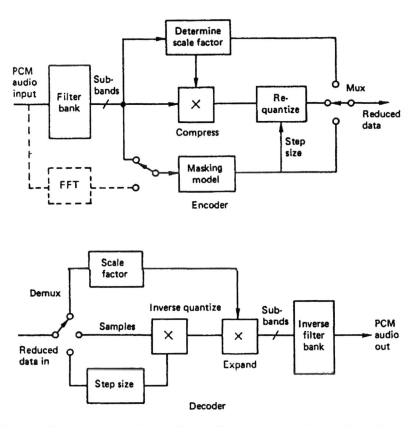

Figure 13.38 A simple sub-band coder. The bit allocation may come from analysis of the sub-band energy, or, for greater reduction, from a spectral analysis in a side chain.

The shorter the wordlength, the greater the coding gain, but the coarser the quantization steps and therefore the level of quantization error.

If a fixed-decompression factor is employed, the size of the coded output block will be fixed. The requantization wordlengths will have to be such that the sum of the bits from each sub-band equals the size of the coded block. Thus some sub-bands can have long wordlength coding if others have short wordlength coding. The process of determining the requantization step size, and hence the wordlength in each sub-band, is known as bit allocation. The bit allocation may be performed by analysing the power in each sub-band, or by a side chain which performs a spectral analysis or transform of the audio. The complexity of the bit allocation depends upon the degree of compression required. The spectral content is compared with an auditory masking model to determine the degree of masking which is taking place in certain bands as a result of higher levels in other bands. Where masking takes place, the signal is quantized more coarsely until the quantizing noise is raised to just below the masking level. The coarse quantization requires shorter wordlengths and allows a coding gain. The bit allocation may be iterative as adjustments are made to obtain the best NMR within the allowable data rate.

The samples of differing wordlength in each bin are then assembled into the output coded block. Unlike a PCM block, which contains samples of fixed wordlength, a coded block contains many different wordlengths and these can vary from one block to the next. In order to deserialize the block into samples of various wordlength and demultiplex the samples into the appropriate frequency bins, the decoder has to be told what bit allocations were used when it was packed, and some synchronizing means is needed to allow the beginning of the block to be identified. Demultiplexing can be done by including the transform of the block which was used to determine the allocation. If the decoder has the same allocation model, it can determine what the coder must have done from the transform, and demultiplex the block accordingly. Once all of the samples are back in their respective frequency bins, the level of each bin is returned to the original value. This is achieved by reversing the gain increase which was applied before the requantizer in the coder. The degree of gain reduction to use in each bin also comes from the transform. The sub-bands can then be recombined into a continuous audio spectrum in the output filter which produces conventional PCM of the original wordlength.

The degree of compression is determined by the bit allocation system. It is not difficult to change the output block size parameter to obtain a different compression. The bit allocator simply iterates until the new block size is filled. Similarly the decoder need only deserialize the larger block correctly into coded samples and then the expansion process is identical except for the fact that expanded words contain less noise. Thus codecs with varying degrees of compression are available which can perform different bandwidth/performance tasks with the same hardware.

13.23 Compression formats

There are currently a large number of compression formats available either as commercially developed products or as international standards under a variety of names. This section sets out the relationships between the major formats.

The ISO (International Standards Organization) and the IEC (International Electrotechnical Commission) recognized that compression would have an important part to play in future digital video products and in 1988 established the ISO/IEC/MPEG (Moving Picture Experts Group) to compare and assess various coding schemes in order to arrive at an international standard. The terms of reference were extended the same year to include audio and the MPEG/Audio group was formed.

As part of the Eureka 147 project, a system known as MUSICAM[20] (Masking pattern adapted Universal Sub-band Integrated Coding And Multiplexing) was developed jointly by CCETT in France, IRT in Germany and Philips in the Netherlands. MUSICAM was designed to be suitable for DAB.

As a parallel development, the ASPEC[21] (Adaptive Spectral Perceptual Entropy Coding) system was developed from a number of earlier systems as a joint proposal by AT&T Bell Labs, Thomson, the Fraunhofer Society and CNET. ASPEC was designed for high degrees of compression to allow audio transmission on ISDN.

These two systems were both fully implemented by July 1990 when comprehensive subjective testing took place at the Swedish Broadcasting Corporation.[9,22,23] As a result of these tests, the MPEG/Audio group combined the attributes of both ASPEC and MUSICAM into a draft standard[24] having three levels of complexity and performance.

As has been seen above, coders can be operated at various compression factors. The performance of a given codec will be improved if the degree of compression is reduced, and so for moderate compression, a simple codec will be more cost-effective. On the other hand, as the degree of compression is increased, the codec performance will be reduced, and it will be necessary to switch to a higher layer, with attendant complexity, to maintain quality. ISO/MPEG coding allows input sampling rates of 32, 44.1 and 48 kHz and supports output bit rates of 32, 48, 56, 64, 96, 112, 128, 192, 256 and 384 kbits/s. The transmission can be mono, dual channel (e.g. bilingual), stereo and joint stereo which is where advantage is taken of redundancy between the two audio channels.

ISO Layer 1 is a simplified version of MUSICAM which is appropriate for the mild compression applications at low cost. It is very similar to PASC. ISO Layer II is identical to MUSICAM and is very likely to be used for DAB. ISO Layer III is a combination of the best features of ASPEC and MUSICAM and is mainly applicable to telecommunications.

13.24 ISO MPEG compression

The simplified MUSICAM system is a sub-band code based on the block diagram of Figure 13.38. A polyphase quadrature mirror filter network divides the audio spectrum into 32 equally spaced bands. Constant size blocks are used, containing 12 samples in each sub-band. The block size was based on the pre-masking phenomenon of Figure 13.37. Like NICAM, the samples in each sub-band block are block compressed according to the peak value in the block. A 6-bit scale factor is used for each sub-band. The sub-bands themselves are used as a spectral analysis of the input in order to determine the bit allocation. The mantissae of the samples in each block are requantized according to the bit allocation. The bit-allocation data are multiplexed with the scale factors and the requantized sub-band samples to form the coded message frames. The bit-allocation codes are necessary to allow the decoder correctly to assemble the variable length samples.

ISO Layer II is identical to MUSICAM. The same filter bank as Layer I is used, and the blocks are the same size. The same block companding scheme is used. In order to give better spectral resolution than the filter bank, a side chain FFT is computed having 1024 points, resulting in an analysis of the audio spectrum eight times better than the sub-bandwidth. The FFT drives the masking model which controls the bit allocation.

Whilst the block-companding scheme is the same as in Layer I, not all of the scale factors are transmitted, because they contain a degree of redundancy on real program material. The difference between scale factors in successive blocks in the same band exceeds 2 dB less than 10 % of the time, and advantage is taken of this characteristic by analysing sets of three successive scale factors. On stationary programmes, only one scale factor out of three is sent. As transient content increases in a given sub-band, two or three scale factors will be sent. A scale factor select code is also sent to allow the decoder to determine what has been sent in each sub-band. This technique effectively halves the scale factor bit rate.[20]

The requantized samples in each sub-band, bit-allocation data, scale factors and scale factor select codes are multiplexed into the output bit stream.

ISO Layer III is the most complex layer of the ISO standard, and is only really necessary when the most severe data rate constraints must be met with high quality. It is a transform code based on the ASPEC system with certain modifications to give a degree of commonality with Layer II. The original ASPEC coder used a direct MDCT on the input samples. In Layer III this was modified to use a hybrid transform incorporating the existing polyphase 32-band QMF of Layers I and II. In Layer 3, the 32 sub-bands from the QMF are each processed by a 12-band MDCT to obtain 384 output coefficients. Two window sizes are used to avoid pre-echo on transients. The window switching is performed by the psychoacoustic model. It has been found that pre-echo is associated with the entropy in the audio rising above the average value.

A highly accurate perceptive model is used to take advantage of the high-frequency resolution available. Non-uniform quantizing is used, along with Huffman coding. This is a technique where the most common code values are allocated the shortest wordlength.

Layer 3 supports additional modes than the four modes of Layers 1 and 2. These are shown in Figure 13.39. M-S coding produces sum and difference signals from the L-R stereo. The S or difference signal will have low entropy when there is correlation between the two input channels. This allows a further saving of bit rate.

Intensity coding is a system where in the upper part of the audio frequency band, for each scale-factor band only one signal is transmitted, along with a code specifying where in the stereo image it belongs. The decoder has the equivalent of a panpot so that it can output the decoded waveform of each band at the appropriate level in each channel. The lower part of the audio band may be sent as L-R or as M-S.

Mono Stereo Dual channel Joint stereo	All levels
M-S Intensity Intensity M-S	Level 3 only

Figure 13.39 Additional audio modes of ISO/MPEG Audio Layer 3.

13.25 Compression artifacts

With the exception of bit-accurate or lossless systems, audio compression always produces a loss of quality. Consequently before using a compression system it is important that stringent listening tests are performed using high-quality loudspeakers, as low-cost speakers will mask the artifacts. The worst degradation occurs in stereo systems where compression tends to eliminate low-level ambience and reverberation leaving the remaining sounds dry.

Cascading compression systems will cause generation loss, particularly when the codecs are not the same. Carrying out an equalization step may well raise noise above the masking threshold. Consequently compression should only be used for final delivery of post-produced material to the consumer. For high-quality production the use of compression should be avoided.

References

1. McNally, G.J., Digital audio: recursive digital filtering for high-quality audio signals. *BBC Res. Dept. Report* RD 1981/10
2. McNally, G.W., Digital audio in broadcasting. *IEEE ASSP Magazine*, 2, 26–44 (1985)
3. Jones, A.H., A PCM sound-in-syncs distribution system. General description. *BBC Res. Dept. Report*, 1969/35
4. Caine, C.R., English, A.R. and O'Clarey, J.W. H., NICAM-3: near-instantaneous companded digital transmission for high-quality sound programmes. *J. IERE*, **50**, 519–530 (1980)
5. Johnston, J.D., Transform coding of audio signals using perceptual noise criteria. *IEEE J. Selected Areas in Comms.*, **JSAC-6**, 314–323 (1988)
6. Ehmer, R.H., Masking of tones vs. noise bands. *J. Audio Eng. Soc.*, **31**, 1253–1256 (1959)
7. Fielder, L.D. and Davidson, G.A., AC-2: a family of low complexity transform based music coders. *Proc. 10th Int. Audio Eng. Soc. Conf.*, 57–70, New York: Audio Eng. Soc. (1991)
8. Carterette, E.C. and Friedman, M.P., *Handbook of Perception*, 305–319, New York: Academic Press (1978)
9. Grewin, C. and Ryden, T., Subjective assessments on low bit-rate audio codecs. *Proc. 10th Int. Audio Eng. Soc. Conf.*, 91–102, New York: Audio Eng. Soc. (1991)
10. Brandenburg, K. and Seitzer, D., Low bit-rate coding of high-quality digital audio: algorithms and evaluation of quality. *Proc. 7th Int. Audio Eng. Soc. Conf.*, 201–209, New York: Audio Eng. Soc. (1989)
11. Johnston, J., Estimation of perceptual entropy using noise masking criteria. *ICASSP*, 2524–2527 (1988)

12. Gilchrist, N.H.C., Delay in broadcasting operations. Presented at 90th Audio Eng. Soc. Conv., preprint 3033 (1991)
13. Crochiere, R.E., Sub-band coding. *Bell System Tech. J.*, **60**, 1633–1653 (1981)
14. Smyth, S.M.F. and McCanny, J.V., 4-bit hi-fi: high-quality music coding for ISDN and broadcasting applications. *Proc. ASSP*, 2532–2535 (1988)
15. Jayant, N.S. and Noll, P., *Digital Coding of Waveforms: Principles and Applications to Speech and Video*, Englewood Cliffs: Prentice-Hall (1984)
16. Theile, G., Stoll, G. and Link, M., Low bit-rate coding of high-quality audio signals: an introduction to the MASCAM system. *EBU Tech. Review*, No. 230, 158–181 (1988)
17. Chu, P.L., Quadrature mirror filter design for an arbitrary number of equal bandwidth channels. *IEEE Trans. ASSP*, **ASSP-33**, 203–218 (1985)
18. Fettweis, A., Wave digital filters: theory and practice. *Proc. IEEE*, **74**, 270–327 (1986)
19. Princen, J.P., Johnson, A. and Bradley, A.B., Sub-band/transform coding using filter bank designs based on time domain aliasing cancellation. *Proc. ICASSP*, 2161–2164 (1987)
20. Wiese, D., MUSICAM: flexible bit rate reduction standard for high-quality audio. Presented at Digital Audio Broadcasting Conference (London, March 1992)
21. Brandenburg, K., ASPEC coding. *Proc. 10th Audio Eng. Soc. Int. Conf.*, 81–90, New York: Audio Eng. Soc. (1991)
22. ISO/IEC JTC1/SC2/WG11 N0030: MPEG/AUDIO test report, Stockholm (1990)
23. ISO/IEC JTC1/SC2/WG11 MPEG 91/010 The SR report on: The MPEG/AUDIO subjective listening test. Stockholm (1991)
24. ISO/IEC JTC1/SC2/WG11 Committee draft 11172

Chapter 14

Sound quality considerations

Sound reproduction can only achieve the highest quality if every step is carried out to an adequate standard. This requires a knowledge of what the standards are and the means to test whether they are being met.

14.1 Introduction

In principle quality can be lost almost anywhere in the audio chain, whether by the correct use of poorly designed equipment or the incorrect use of good equipment. The best results will only be obtained when good equipment is used correctly.

There is only one symptom of quality loss, which is where the reproduced waveform differs from the original. For convenience the differences are often categorized.

Any error in an audio waveform can be considered as an unwanted signal which has been linearly added to the wanted signal. Figure 14.1 shows that there are only two classes of waveform error. The first is where the error is not a function of the audio signal, but results from some uncorrelated process. This is the definition of noise. The second is where the error is a direct function of the audio signal which is the definition of distortion.

Noise can be broken into categories according to its characteristics. Noise due to thermal and electronic effects in components or tape hiss has essentially stationary statistics, and forms a constant background which is subjectively benign in comparison with most other errors. Noise can be periodic or impulsive. Power frequency related hum is often rich in harmonics. Interference due to electrical switching or lightning is generally impulsive. Crosstalk from other signals is also noise.

Distortion is a signal dependent error and has two main categories. Non-linear distortion arises because the transfer function is not straight. The result in analog systems is harmonic distortion where new frequencies which are integer multiples of the signal frequency are added, changing the spectrum of the signal. In digital systems non-linearity can also result in anharmonic distortion because harmonics above half the sampling rate will alias. Non-linear distortions are subjectively the least acceptable as the resulting harmonics are frequently not masked, especially on pure tones.

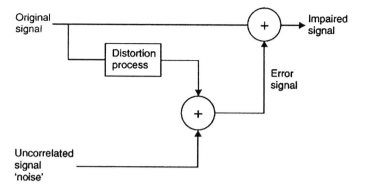

Figure 14.1 All waveform errors can be broken down into two classes: those due to distortions of the signal and those due to an unwanted additional signal.

Linear distortion is a signal dependent error in which different frequencies propagate at different speeds due to lack of phase linearity. In complex signals this has the effect of altering the waveform but without changing the spectrum. As no harmonics are produced, this form of distortion is more benign than non-linear distortion. Whilst a completely linear phase system is ideal, the limited phase accuracy of the ear means that in practice a *minimum phase system* is probably good enough. Minimum phase implies that phase error changes smoothly and continuously over the audio band without any sudden discontinuities. Loudspeakers often fail to achieve minimum phase, with techniques such as reflex tuning and passive crossovers being particularly unsatisfactory.

As an audio waveform is simply a voltage changing with time within a limited spectrum, any error introduced by a given device can in principle be extracted and analysed. If this approach is used during the design process, the performance of any unit can be refined until the error is arbitrarily small. This is an expensive approach when it is only necessary to reduce the error until it is inaudible.

Naturally the determination of what is meant by inaudible has to be done carefully using psychoacoustic experiment. Using such subjective results it is possible to design objective tests which will measure different aspects of the signal to determine if it is acceptable. It is equally valid to use listening tests where the reproduced sound is compared with that of other audio systems or of live performances. In fact the combination of careful listening tests with objective technical measurements is the only way to achieve outstanding results. The reason is that the ear can be thought of as making all tests simultaneously. If any aspect of the system has been overlooked the ear will detect it. Then objective tests can be made to find the deficiency the ear detected and remedy it. Unfortunately this ideal combination of subjective and objective testing is not achieved as often as might be thought. Because of the narrowness of most education systems, people tend to be divided into two camps where audio quality is concerned.

Subjectivists are those who listen at length to both live and recorded music and can detect quite small defects in audio systems. Unfortunately most subjectivists have little technical knowledge and are quite unable to convert their detection of a problem into a solution. Although their perception of a problem may be genuine, their explanation of the problem or their proposed solution may require the laws of physics to be altered. A particular problem with subjective testing is the avoidance of bias. Technical knowledge is essential to understand the importance of experimental design in avoiding bias. Statistical analysis is essential to determine the *significance* of the results, i.e. the degree of confidence that the results are not due to chance alone. As most subjectivists lack such technical knowledge it is hardly surprising that a lot of deeply held convictions are simply due to unwitting bias where repeatable results are simply not obtained.

Objectivists are those who make a series of measurements and then pronounce a system to have no audible defect. Objectivists frequently have little experience of live performance or critical listening. One of the most frequent mistakes made by objectivists is to assume that because a piece of equipment passes a given test under certain conditions then it is ideal. This is simply untrue because the equipment might fail other tests or the same test under other conditions. In some cases the tests which equipment might fail have yet to be designed.

Not surprisingly the same piece of equipment can be received quite differently by the two camps. The introduction of transistor audio amplifiers caused an unfortunate and audible step backwards in audio quality. The problem was that valve (tube) amplifiers were generally operated in Class A and had low distortion except when delivering maximum power. Consequently valve amplifiers were tested for distortion at maximum power and in a triumph of tradition over reason transistor amplifiers were initially tested in the same way. However, transistor amplifiers generally work in Class B and produce more distortion at very low levels due to crossover between output devices. Naturally this distortion was not detectable on a high-power distortion test. Early transistor amplifiers sounded dreadful at low level and it is astonishing that this was not detected before they reached the market when the subjectivists rightly gave them a hard time.

Whilst the objectivists looked foolish over the crossover distortion issue, subjectivists have no reason to crow with their fetishes for gold-plated mains plugs, special feet for equipment, exotic cables and mysterious substances to be applied to Compact Discs.

More recently the introduction of compression techniques has raised another quality issue where the definition of inaudible has to be carefully qualified. As will be seen in Section 14.3 the artifacts caused by compression are more likely to be audible in stereo and surround sound systems than in mono and even more likely to be audible on high-quality loudspeakers.

14.2 Objective testing

Objective testing consists of making measurements which indicate the accuracy to which audio signals are being reproduced. As the measurements are standardized and repeatable they do form a basis for comparison even if *per se* they do not give a complete picture of what a device-under-test (DUT) or system will sound like.

One important objective parameter is the signal-to-noise ratio (SNR). Figure 14.2(a) shows that this measures the ratio in dB between the largest

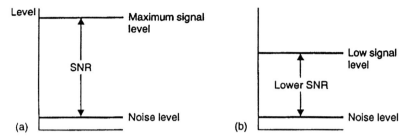

Figure 14.2 (a) SNR is the ratio of the largest undistorted signal to the noise floor. (b) Maximum SNR is not reached if the signal never exercises the highest possible levels.

amplitude undistorted signal the DUT can pass and the amplitude of the output with no input whatsoever, which is presumed to be due to noise. The spectrum of the noise is as important as the level. When audio signals are present, auditory masking occurs which largely prevents the noise floor being heard. Consequently the noise floor is most significant during extremely quiet passages or in pauses. Under these conditions the threshold of hearing is extremely dependent on frequency.

A measurement which more closely resembles the effect of noise on the listener is the use of an A-weighting filter prior to the noise level measurement stage. The result is then measured in dB(A).

Just because a DUT measures a given number of dB of SNR does not guarantee that SNR will be obtained in use. The measured SNR is only obtained when the DUT is used with signals of the correct level. Figure 14.2(b) shows that if the DUT is installed in a system where the input level is too low, the SNR of the output will be impaired. Consequently in any system where this is likely to happen, the SNR of the equipment must exceed the required output SNR by the amount by which the input level is too low. This is the reason why quality mixing consoles offer apparently phenomenal SNRs.

Often the operating level is deliberately set low to provide *headroom* so that occasional transients are undistorted. The art of quality audio production lies in setting the level to the best compromise between elevating the noise floor and increasing the occurrences of clipping.

The frequency response of audio equipment is measured by the system shown in Figure 14.3(a). The same level is input at a range of frequencies and the output level is measured. The end of the frequency range is considered to have been reached where the output level has fallen by 3 dB with respect to the maximum level. The correct way of expressing this measurement is:

Frequency response: −3 dB, 20 Hz–20 kHz

or similar. If the level limit is omitted, as it often is, the figures are meaningless.

There is a seemingly endless debate about how much bandwidth is necessary in analog audio and what sampling rate is needed in digital audio. There is no one right answer as will be seen. In analog systems, there is generation loss. Figure 14.3(b) shows that two identical DUTs in series will cause a loss of 6 dB at the frequency limits. Depending on the shape of the roll-off, the −3 dB limit will be reached over a narrower frequency range. Conversely if

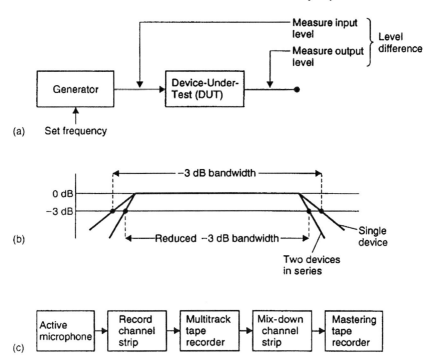

Figure 14.3 (a) Frequency response measuring system. (b) Two devices in series have narrower bandwidth. (c) Typical record production signal chain, showing potential severity of generation loss.

the original bandwidth is to be maintained, then the −3 dB range of each DUT must be wider.

In analog production systems, the number of different devices an audio signal must pass through is quite large. Figure 14.3(c) shows the signal chain of a multitrack produced vinyl disk. The number of stages involved mean that if each stage has a seemingly respectable −3 dB, 20 Hz−20 kHz response, the overall result will be dreadful with a phenomenal rate of roll-off at the band edge. The only solution is that each item in the chain has to have wider bandwidth making it considerably overspecified in a single generation application.

Another factor is phase response. At the −3 dB point of a DUT, if the response is limited by a first-order filtering effect, the phase will have shifted by 45°. Clearly in a multistage system these phase shifts will add. An eight-stage system, not at all unlikely, will give a complete phase rotation as the band edge is approached. The phase error begins a long way before the −3 dB point, preventing the system from displaying even a minimum phase characteristic.

Consequently in complex analog audio systems each stage must be enormously overspecified in order to give good results after generation loss. It is not unknown for mixing consoles to respond down to 6 Hz in order to prevent loss of minimum phase in the audible band. Obviously such an extended frequency response on its own is quite inaudible, but when cascaded with other stages, the overall result *will* be audible.

Some valiant attempts have been made to avoid analog generation loss. Vinyl disks have been made where the disk cutter is fed live from microphones via a minimal mixing process, omitting the generation loss due to tape recording. Alternatively the tape master has been produced normally, but the disk-cutting process is carried out at half speed, shifting all of the cutter response limits up by an octave.

In the digital domain there is no generation loss if the numerical values of the samples are not altered. Consequently digital data can be copied from one tape to another, or to a Compact Disc without any quality loss whatsoever. Simple digital manipulations, such as level control, do not impair the frequency or phase response and if well engineered the only loss will be a slight increase in the noise floor. Consequently digital systems do not need overspecified bandwidth. The bandwidth needs only to be sufficient for the application because there is nothing to impair it.

In a digital system the bandwidth and phase response is defined at the anti-aliasing filter in the ADC. Early anti-aliasing filters had such dreadful phase response that the aliasing might have been preferable, but this has been overcome in modern oversampled converters which can be highly phase linear.

One simple but good way of checking frequency response and phase linearity is square-wave testing. A square wave contains indefinite harmonics of a known amplitude and phase relationship. If a square wave is input to an audio DUT, the characteristics can be assessed almost instantly. Figure 14.4 shows some of the defects which can be isolated. Figure 14.4(a) shows inadequate low-frequency response causing the horizontal sections of the waveform to droop. Figure 14.4(b) shows poor high-frequency response in conjunction with poor phase linearity which turns the edges into exponential curves. Figure 14.4(c) shows a phase-linear system of finite bandwidth, e.g. a good anti-aliasing filter. Note that the transitions are symmetrical with equal pre- and post-ringing. This is one definition of phase linearity. Figure 14.4(d) shows a system with wide bandwidth but poor high-frequency phase response. Note the asymmetrical ringing.

One still hears from time to time that square-wave testing is illogical because square waves never occur in real life. The explanation is simple. Few would argue that any sine wave should come out of a DUT with the same amplitude and no phase shift. A linear audio system ought to be able to pass any number of superimposed signals simultaneously. A square wave is simply one combination of such superimposed sine waves. Consequently if an audio system cannot pass a square wave as shown in Figure 14.4(c) then it will cause a problem with real audio.

As linearity is extremely important in audio, relevant objective linearity testing is vital. Real sound consists of many different contributions from different sources which all superimpose in the sound waveform reaching the ear. If an audio system is not capable of carrying an indefinite number of superimposed sounds without interaction then it will cause an audible impairment. Interaction between a single waveform and a non-linear system causes distortion. Interaction between waveforms in a non-linear system is called *intermodulation distortion (IMD)* whose origin is shown in Figure 14.5(a). As the transfer function is not straight, the low-frequency signal has the effect of moving the high-frequency signal to parts of the transfer function where the slope differs. This results in the high frequency being amplitude

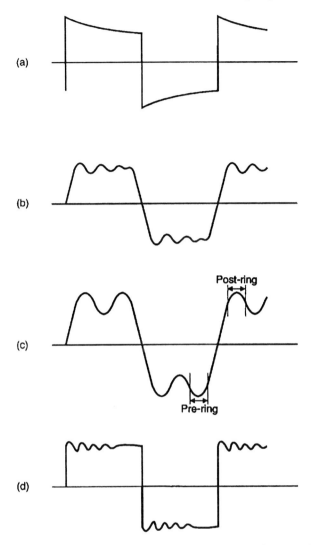

(a)

(b)

(c)

Post-ring

Pre-ring

(d)

Figure 14.4 Square-wave testing gives a good insight into audio performance. (a) Poor low-frequency response causing droop. (b) Poor high-frequency response with poor phase linearity. (c) Phase-linear bandwidth limited system. (d) Asymmetrical ringing shows lack of phase linearity.

modulated by the low frequency. The amplitude modulation will also produce sidebands. Clearly a system which is perfectly linear will be free of both types of distortion.

Figure 14.5(b) shows a simple harmonic distortion test. A low-distortion oscillator is used to inject a clean sine wave into the DUT. The output passes through a switchable sharp 'notch' filter which rejects the fundamental frequency. With the filter bypassed, an AC voltmeter is calibrated to 100%. With the filter in circuit, any remaining output must be harmonic

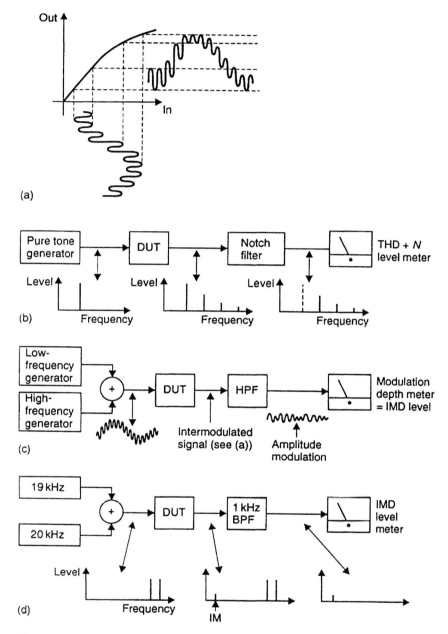

Figure 14.5 (a) Non-linear transfer function allows large low-frequency signal to modulate small high-frequency signal. (b) Harmonic distortion test using notch to remove fundamental measures THD + n. (c) Intermodulation test measuring depth of modulation of high frequency by low-frequency signal. (d) Intermodulation test using a pair of tones and measuring the level of the difference frequency.

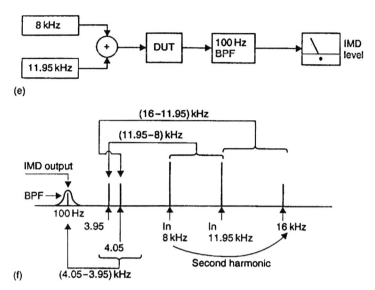

Figure 14.5 *(continued)* (e) Complex test using intermodulation between harmonics. (f) Spectrum analyser output with two-tone input.

distortion or noise and the measured voltage is expressed as a percentage of the calibration voltage. The correct way of expressing the result is as follows:

$$\text{THD} + n \text{ at } 1\,\text{kHz} = 0.1\%$$

or similar. A stringent test would repeat the test at a range of frequencies. There is not much point in conducting THD + n tests at high frequencies as the harmonics will be beyond the audible range.

It is frequently useful to examine the output of the notch filter using an oscilloscope. This *distortion residue* signal can reveal a great deal about the source of the non-linearity, especially in the case of crossover distortion in power amplifiers.

There are a number of ways of conducting intermodulation tests, and of course anyone who understands the process can design a test from first principles. An early method standardized by the SMPTE was to exploit the amplitude modulation effect in two widely spaced frequencies, typically 70 Hz with 7 kHz added at one-tenth the amplitude. 70 Hz is chosen because it is above the hum due to 50 or 60 Hz power. Figure 14.5(c) shows that the measurement is made by passing the 7 kHz region through a bandpass filter and recovering the depth of amplitude modulation with a demodulator.

Alternatively it is possible to test for the creation of sidebands. Two high frequencies with a small frequency difference are linearly added and used as the input to the DUT. For example, 19 and 20 kHz will produce a 1 kHz difference or beat frequency if there is non-linearity as shown in (d). With such a large difference between the input and the beat frequency it is easy to produce a 1 kHz bandpass filter which rejects the inputs. The filter output is a measure of IMD. With a suitable generator, the input frequencies can be swept or stepped with a constant 1 kHz spacing.

More advanced tests exploit not just the beats between the fundamentals, but also those involving the harmonics. In one proposal, shown in (e), input tones of 8 and 11.95 kHz are used. The fundamentals produce a beat of 3.95 kHz, but the second harmonic of 8 kHz is 16 kHz which produces a beat of 4.05 kHz. This will intermodulate with 3.95 kHz to produce a 100 Hz component which can be measured. Clearly this test will only work in 60 Hz power regions, and the exact frequencies would need modifying in 50 Hz regions.

If a spectrum analyser is available, all of the above tests can be performed simultaneously. Figure 14.5(f) shows the results of a spectrum analysis of a DUT supplied with two test tones. Clearly it is possible to test with three simultaneous tones or more. Some audio devices, particularly power amplifiers, are relatively benign under steady state testing with simple signals, but reveal their true colours with a more complex input.

14.3 Subjective testing

Subjective testing can only be carried out by placing the DUT in series with an existing sound reproduction system. Unless the DUT is itself a loudspeaker, the testing will only be as stringent as the loudspeakers in the system allow. Unfortunately the great majority of loudspeakers do not reach the standard required for meaningful subjective testing of units placed in series and consequently the majority of such tests are of questionable value.

If useful subjective testing is to be carried out, it is necessary to use the most accurate loudspeakers available and to test the loudspeakers themselves before using them as any kind of reference. Whilst simple tests such as on-axis frequency response give an idea of the performance of a loudspeaker, the majority produce so much distortion that the figures are not even published.

Consequently it is important to find listening tests which will meaningfully assess loudspeakers, especially for linearity, whilst eliminating other variables as much as possible. Linearity is essential to allow superimposition of an indefinite number of sounds in a stereo image.

As was seen in Chapter 3, the hearing mechanism has an ability to concentrate on one of many simultaneous sound sources based on direction using the *cocktail-party effect*. Sounds arriving from other directions are incoherent and are heard less well. Human hearing can also locate a number of different sound sources simultaneously by constantly comparing excitation patterns from the two ears with different delays. Strong correlation will be found where the delay corresponds to the inter-aural delay for a given source.

Monophonic systems prevent the use of either of these effects completely because the first version of all sounds reaching the listener come from the same loudspeaker. Stereophonic systems allow the cocktail party effect to function in that the listener can concentrate on specific sound sources in a reproduced stereophonic image with the same facility as in the original sound.

Experiments showed long ago that even technically poor stereo was always preferred to pristine mono. This is because we are accustomed to sounds and reverberations coming from all different directions in real life and having them all piled one on top of the other in a mono speaker convinces no one, however accurate the waveform. We live in a reverberant world which is filled with sound reflections. If we could separately distinguish every different reflection in a reverberant room we would hear a confusing cacophony. In practice we hear very well in reverberant surroundings, far better than microphones can, because of the transform nature of the ear and the way in which the brain processes nerve signals.

Clearly the first version of a transient sound to reach the ears must be the one which has travelled by the shortest path and this must be the direct sound rather than a reflection. Consequently the ear has evolved to attribute source direction from the time-of-arrival difference at the two ears of the first version of a transient. Versions which may arrive from elsewhere simply add to the perceived loudness but do not change the perceived location of the source unless they arrive within the inter-aural delay of about 700 microseconds when the precedence effect breaks down and the perceived direction can be pulled away from that of the first arriving source by an increase in level. Figure 7.5 showed that this area is known as the time-intensity trading region. Once the maximum inter-aural delay is exceeded, the hearing mechanism knows that the time difference must be due to reverberation and the trading ceases to change with level.

Unfortunately reflections with delays of the order of 700 microseconds are exactly what are provided by the traditional rectangular loudspeaker with sharp corners. It has been known for some time that these are clearly audible.

Intensity stereo, the format obtained from coincident mikes or panpots, works purely by amplitude differences at the two loudspeakers. The two signals should be exactly in phase. As both ears hear both speakers the result is that the space between the speakers and the ears turns the intensity differences into time-of-arrival differences. These give the illusion of virtual sound sources.

A virtual sound source from a panpot has zero width and on diffraction-free speakers would appear as a virtual point source. Figure 14.6(a) shows how a panpotted dry mix should appear spatially on ideal speakers whereas Figure 14.6(b) shows what happens when stereo reverberation is added. In fact Figure 14.6(b) is also what is obtained with real sources using a coincident

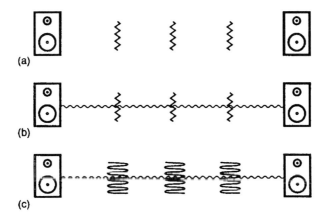

Figure 14.6 Loudspeaker imaging under various conditions. (a) Ideal panpotted dry mix image has point sources. (b) Addition of artificial reverberations to (a). (c) Poor loudspeakers smearing the image due to diffraction.

pair of mikes. In this case the sources are the real sources and the sound between is reverberation/ambience.

Figure 14.6(c) is the result obtained with traditional square box speakers. Note that the point sources have spread so that there are almost no gaps between them, effectively masking the ambience. This represents a lack of spatial fidelity, so we can say that rectangular box loudspeakers cannot accurately reproduce a stereo image.

Non-linearity in stereo has the effect of creating intermodulated sound objects which are in a different place in the image to the genuine sounds. Consequently the requirements for stereo are more stringent than for mono. This can be used for speaker testing. One stringent test is to listen to a high-quality stereo recording in which multitracking has been used to superimpose a number of takes of a musician or vocalist playing/singing the same material. The use of multitracking eliminates intermodulation at the microphone as this only handles one source at a time. The use of a panpot eliminates any effects due to inadequate directivity in a stereo microphone.

It should be possible to hear how many simultaneous sources are present, i.e. whether the recording is double, triple or quadruple tracked, and it should be possible to concentrate on each source to the exclusion of the others.

It should be possible to pan each version of a multitracked recording to a slightly different place in the stereo image and individually identify each source even when the spacing is very small. Poor loudspeakers smear the width of an image because of diffraction and fail this test.

In another intermodulation test it is necessary to find a good quality recording in which a vocalist sings solo at some point, and at another point is accompanied by a powerful low-frequency instrument such as a pipe organ or a bass guitar. There should be no change in the imaging or timbre of the vocal whether or not the low frequency is present.

Another stringent test of linearity is to listen to a recording made on a coincident stereo microphone of a spatially complex source such as a choir. It should be possible to identify the location of each chorister and to concentrate on the voice of each. The music of Tallis is highly suitable. Massed strings are another useful test, with the end of Barber's *Adagio for Strings* being particularly revealing. Coincident stereo recordings should also be able to reproduce depth. It should be possible to resolve two instruments or vocalists one directly behind the other at different distances from the microphone.

Non-ideal loudspeakers act like bit-rate compressors in that they conceal or mask information in the audio signal. If a real compressor is tested with non-ideal loudspeakers, certain deficiencies of the compressor will not be heard and it may be erroneously assumed that the compressor is transparent when in fact it is not. Compression artifacts which are inaudible in mono may be audible in stereo and the spatial compression of non-ideal stereo loudspeakers conceals real spatial compression artifacts.

Compressors raise the noise floor of the audio in such a way that it will be masked. Compressors are tested using the concept of the noise-to-masking ratio (NMR) shown in Chapter 13. If the NMR is always positive, the compression artifacts are inaudible. The same test can be made with a loudspeaker. Figure 14.7 shows that if the spectrum of the input is known, the masking threshold can be calculated. Cumulative decay (waterfall) data can then be used to predict the amount of delayed resonance which would result from that

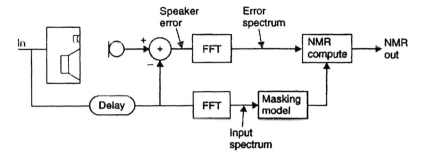

Figure 14.7 Technique needed to determine if a loudspeaker masks its own delayed resonances.

input. The difference between the input spectrum and the decay spectra at various delays will reveal the NMR of the speaker. Clearly this would have to be done with a lot of different program material. Any time the decay spectrum fails to be masked there is a problem.

A further consequence of the NMR concept of loudspeaker performance is what happens when an attempt is made to assess compression systems using loudspeakers? Clearly there is no point listening for unmasked compression artifacts with a loudspeaker in which the delayed resonances are themselves not masked. Consequently using a technique pioneered by the author it is possible to compare loudspeakers by comparing their ability to reveal compression artifacts. Some loudspeakers reveal compression artifacts better than others. The one which reveals the most artifacts is the cleanest speaker.

Two pairs of speakers can be compared using a variable bit-rate compressor which could be bypassed. The inferior speaker only reveals the use of a substantial amount of compression whereas the better speaker reveals the use of a smaller amount. A reference grade loudspeaker reveals the use of the compressor even at its maximum bit rate.

Reference grade speakers should be free of reflections in the sub-700 microsecond trading region so that the imaging actually reveals what is going on spatially. When such speakers are used to assess audio compressors, even at high bit rates corresponding to the smallest amount of compression, it is obvious that there is a difference between the original and the compressed result. Figure 14.8 shows what is found. The dominant sound sources are reproduced fairly accurately, but the ambience and reverberation between them are virtually absent, making the decoded sound much drier than the original.

The effect will be apparent to the same extent with, for example, both MPEG layer 2 and Dolby AC-2 coders even though their internal workings are quite different. This is not surprising because both are probably based on the same psychoacoustic masking model. MPEG-3 fares even worse because the bit rate is lower. Transient material has a peculiar effect whereby the ambience will come and go according to the entropy of the dominant source. A percussive note will narrow the sound stage and appear dry but afterwards the reverb level will come back up. All of these effects largely disappear when the signals to the speakers are added to make mono, removing the ear's ability to discriminate spatially.

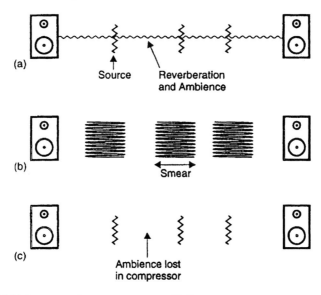

Figure 14.8 (a) Input stereo image to compressor. (b) Compressor output on poor speakers which smear image, masking compression artifacts. (c) Sharp imaging precision speaker reveals ambience and reverb removed by compressor.

These effects are not subtle and do not require golden ears. The author has successfully demonstrated them to various audiences, up to 60 in number, in a variety of untreated rooms. Whilst compression may be adequate to deliver postproduced audio to a consumer with mediocre loudspeakers, these results underline that it has no place in a quality production environment. When assessing codecs, loudspeakers having poor diffraction design will conceal artifacts. When mixing for a compressed delivery system, it will be necessary to include the codec in the monitor feeds so that the results can be compensated. Where high-quality stereo is required, either full bit-rate PCM or lossless (packing) techniques must be used.

In many loudspeakers the amount of distortion rises sharply above a certain SPL. Often this SPL is considerably below the maximum which can safely be produced without damage. Users who are accustomed to this kind of speaker will use operating levels below the onset of increased distortion. When faced with a low-distortion speaker such users often turn it up too high and wonder why they get a headache. The right level is one which gives a natural tonal balance and a speaker should be able to reproduce this level without excessive distortion.

When a pair of reference grade loudspeakers have been found which will demonstrate all of the above effects, it will be possible to make meaningful comparisons between devices such as microphones, consoles, analog recorders, ADCs and DACs. Quality variations between different CD players will be readily apparent. Those which pay the most attention to converter clock jitter are generally found to be preferable. Very expensive high-end CD players are often disappointing because these units concentrate on one aspect of performance and neglect others. In many cases the developers will not have had access to sufficiently good loudspeakers to hear the defects in their creations.

One myth which has taken a long time to be revealed is the belief that a low-grade loudspeaker should be used in the production process so that an indication of how the mix will sound on mediocre consumer equipment will be obtained. If it were possible to obtain an average loudspeaker this would be possible. Unfortunately the main defect of a poor loudspeaker is that it stamps its own characteristic footprint on the audio. These footprints vary so much that there is no such thing as a representative poor loudspeaker and people who make mix decisions on cheap loudspeakers are taking serious risks. It is a simple fact that an audio production can never be better than the monitor loudspeakers used and the author's extensive collection of defective CDs indicates that good monitoring is rare. If it is desired to simulate a low-grade speaker this is best done using filtering.

14.4 Digital audio quality

In theory the quality of a digital audio system comprising an ideal ADC followed by an ideal DAC is determined at the ADC. This will be true if the digital signal path is sufficiently well engineered that no numerical errors occur, which is the case with most reasonably maintained equipment. The ADC parameters such as the sampling rate, the wordlength and any noise shaping used put limits on the quality which can be achieved. Conversely the DAC itself may be transparent, because it only converts data whose quality are already determined back to the analog domain. In other words, the ADC determines the system quality and the DAC does not make things any worse.

In practice both ADCs and DACs can fall short of the ideal, but with modern converter components and attention to detail the theoretical limits can be approached very closely and at reasonable cost. Shortcomings may be the result of an inadequacy in an individual component such as a converter chip, or due to incorporating a high-quality component in a poorly thought-out system. Poor system design can destroy the performance of a converter. Whilst oversampling is a powerful technique for realizing high-quality converters, its use depends on digital interpolators and decimators whose quality affects the overall conversion quality.[1] Interpolators and decimators with erroneous arithmetic have been known.

ADCs and DACs have the same transfer function, since they are only distinguished by the direction of operation, and therefore the same terminology can be used to classify the possible shortcomings of both. Figure 14.9 shows the transfer functions resulting from the main types of converter error.

• *Offset error*   A constant appears to have been added to the digital signal. This has no effect on sound quality, unless the offset is gross, when the symptom would be premature clipping. DAC offset is of little consequence, but ADC offset is undesirable since it can cause an audible thump if an edit is made between two signals having different offsets. Offset error is sometimes cancelled by digitally averaging the converter output and feeding it back to the analog input as a small control voltage. Alternatively, a digital high-pass filter can be used.

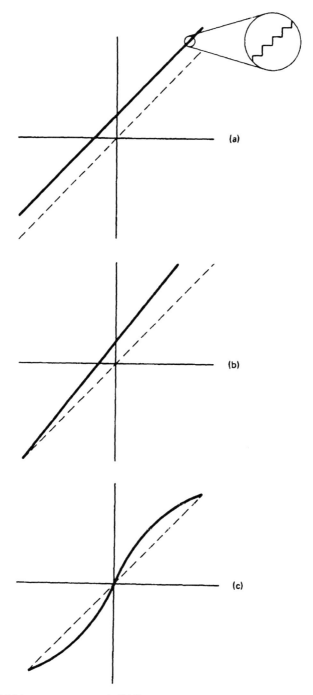

Figure 14.9 Main converter errors (solid line) compared with perfect transfer function (dotted line). These graphs hold for ADCs and DACs, and the axes are interchangeable. If one is chosen to be analog, the other will be digital.

- *Gain error*   The slope of the transfer function is incorrect. Since converters are referred to one end of the range, gain error causes an offset error. The gain stability is probably the least important factor in a digital audio converter, since ears, meters and gain controls are logarithmic.

- *Integral linearity*   This is the deviation of the dithered transfer function from a straight line. It has exactly the same significance and consequences as linearity in analog circuits, since if it is inadequate, harmonic distortion will be caused.

- *Differential non-linearity*   This is the amount by which adjacent quantizing intervals differ in size. It is usually expressed as a fraction of a quantizing interval. In audio applications the differential non-linearity requirement is quite stringent. This is because with properly employed dither, an ideal system can remain linear under low-level signal conditions. When low levels are present, only a few quantizing intervals are in use. If these change in size, clearly waveform distortion will take place despite the dither as can be seen in the figure. Enhancing the subjective quality of converters using noise shaping will only serve to reveal such shortcomings.

- *Monotonicity*   This is a special case of differential non-linearity. Non-monotonicity means that the output does not increase for an increase in input. Figure 14.10 shows how this can happen in a DAC. With a converter input code of 01111111 (127 decimal), the seven low-order current sources of the converter will be on. The next code is 10000000 (128 decimal), where only the eighth current source is operating. If the current it supplies is in error on the low side, the analog output for 128 may be less than that for 127. In an ADC non-monotonicity can result in missing codes. This means that certain binary combinations within the range cannot be generated by any analog voltage. If a device has better than $\frac{1}{2}Q$ linearity it must be monotonic. It is not possible for a one-bit converter to be non-monotonic.

- *Absolute accuracy*   This is the difference between actual and ideal output for a given input. For audio it is rather less important than linearity. For example, if all the current sources in a converter have good thermal tracking, linearity will be maintained, even though the absolute accuracy drifts.

Clocks which are free of jitter are a critical requirement in converters as was shown in Section 8.6. The effects of clock jitter are proportional to the slewing rate of the audio signal rather than depending on the sampling rate, and as a result oversampling converters are no more prone to jitter than conventional converters.[2] Clock jitter is a form of frequency modulation with a small modulation index. Sinusoidal jitter produces sidebands which may be audible. Random jitter raises the noise floor which is more benign but still undesirable. As clock jitter produces artifacts proportional to the audio slew rate, it is quite easy to detect. A spectrum analyser is connected to the converter output and a low audio frequency signal in input. The test is then repeated with a high audio frequency. If the noise floor changes, there is clock jitter. If the noise floor rises but remains substantially flat, the jitter is random. If there are discrete frequencies in the spectrum, the jitter is periodic. The spacing of the discrete frequencies from the input frequency will reveal the frequencies in the jitter.

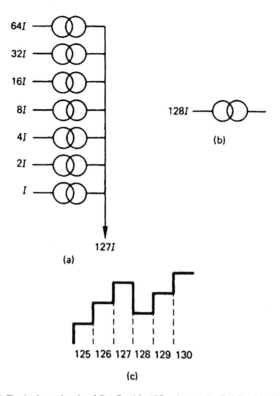

Figure 14.10 (a) Equivalent circuit of DAC with 127_{10} input. (b) DAC with 128_{10} input. On a major overflow, here from 27_{10} to 128_{10}, one current source ($128I$) must be precisely I greater than the sum of all the lower order sources. If $128I$ is too small, the result shown in (c) will occur. This is non-monotonicity.

Aliasing of audio frequencies is not generally a problem, especially if oversampling is used. However, the nature of aliasing is such that it works in the frequency domain only and translates frequencies to new values without changing amplitudes. Aliasing can occur for any frequency above one-half the sampling rate. The frequency to which it aliases will be the difference frequency between the input and the nearest sampling rate multiple. Thus in a non-oversampling converter, *all* frequencies above half the sampling rate alias into the audio band. This includes radio frequencies which have entered via audio or power wiring or directly. Radio frequencies can leapfrog an analog anti-aliasing filter capacitively. Thus good radio frequency screening is necessary around ADCs, and the manner of entry of cables to equipment must be such that radio frequency energy on them is directed to earth. Recent legislation regarding the sensitivity of equipment to electromagnetic interference can only be beneficial in this respect.

Oversampling converters respond to radio frequencies on the input in a different manner. Although all frequencies above half the sampling rate are folded into the baseband, only those which fold into the audio band will be audible. Thus an unscreened oversampling converter will be sensitive to radio frequency energy on the input at frequencies within ±20 kHz of integer multiples of the sampling rate. Fortunately interference from the digital circuitry at exactly the sampling rate will alias to DC and be inaudible.

Converters are also sensitive to unwanted signals superimposed on the references. In fact the multiplicative nature of a converter means that reference noise amplitude modulates the audio to create sidebands. Power supply ripple on the reference due to inadequate regulation or decoupling causes sidebands 50, 60, 100 or 120 Hz away from the audio frequencies, yet does not raise the noise floor when the input is quiescent. The multiplicative effect reveals how to test for it. Once more a spectrum analyser is connected to the converter output. An audio frequency tone is input, and the level is changed. If the noise floor changes with the input signal level, there is reference noise. Radio frequency interference on a converter reference is more insidious, particularly in the case of noise-shaped devices. Noise-shaped converters operate with signals which must contain a great deal of high-frequency noise just beyond the audio band. Radio frequency on the reference amplitude modulates this noise and the sidebands can enter the audio band, raising the noise floor or causing discrete tones depending on the nature of the pickup.

Noise-shaped converters are particularly sensitive to a signal of half the sampling rate on the reference. When a small DC offset is present on the input, the bit density at the quantizer must change slightly from 50%. This results in idle patterns whose spectrum may contain discrete frequencies. Ordinarily these are designed to occur near half the sampling rate so that they are beyond the audio band. In the presence of half-sampling-rate interference on the reference, these tones may be demodulated into the audio band.

Although the faithful reproduction of the audio band is the goal, the nature of sampling is such that converter design must respect EMC and RF engineering principles if quality is not to be lost. Clean references, analog inputs, outputs and clocks are all required, despite the potential radiation from digital circuitry within the equipment and uncontrolled electromagnetic interference outside.

Unwanted signals may be induced directly by ground currents, or indirectly by capacitive or magnetic coupling. It is essential practice to separate grounds for analog and digital circuitry, connecting them in one place only. Capacitive coupling uses stray capacitance between the signal source and point where the interference is picked up. Increasing the distance or conductive screening helps. Coupling is proportional to frequency and the impedance of the receiving point. Lowering the impedance at the interfering frequency will reduce the pickup. If this is done with capacitors to ground, it need not reduce the impedance at the frequency of wanted signals.

Magnetic or inductive coupling relies upon a magnetic field due to the source current flow inducing voltages in a loop. Reduction in inductive coupling requires the size of any loops to be minimized. Digital circuitry should always have ground planes in which return currents for the logic signals can flow. At high frequency, return currents flow in the ground plane directly below the signal tracks and this minimizes the area of the transmitting loop. Similarly ground planes in the analog circuitry minimize the receiving loop whilst having no effect on baseband audio. A further weapon against inductive coupling is to use ground fill between all traces on the circuit board. Ground fill will act like a shorted turn to alternating magnetic fields. Ferrous screening material will also reduce inductive coupling as well as capacitive coupling.

The reference of a converter should be decoupled to ground as near to the integrated circuit as possible. This does not prevent inductive coupling to the lead frame and the wire to the chip itself. In the future converters with on-chip references may be developed to overcome this problem.

In summary, spectral analysis of converters gives a useful insight into design weaknesses. If the noise floor is affected by the signal level, reference noise is a possibility. If the noise floor is affected by signal frequency, clock jitter is likely. Should the noise floor be unaffected by both, the noise may be inherent in the signal or in analog circuit stages.

One interesting technique which has been developed recently for ADC testing is a statistical analysis of the frequency of occurrence of the various code values in data.

If, for example, a full-scale sine wave is input to an ADC having a frequency which is asynchronous to the sampling rate, the probability of a particular code occurring in the output of an ideal converter is a function only of the slew rate of the signal. At the peaks of the sine wave the slew rate is small and the codes there are more frequent. Near the zero crossing the slew rate is high and the probability is lower; the probability of codes being created is nearly equal. However, if one quantizing interval is slightly larger than its neighbours, the signal will take longer to cross it and the probability of that code appearing will rise. Conversely if the interval is smaller the probability will fall. By collecting a large quantity of data from a test and displaying the statistics it is possible to measure differential non-linearity to phenomenal accuracy.

This technique has been used to show that oversampled noise-shaped converters are virtually free of differential non-linearity because of the averaging in the decimation process.

In practice signals used are not restricted to high-level sine waves. A low-level sine wave will only exercise a small number of codes near the audiologically sensitive centre of the quantizing range. However, it may be better to use a combination of three sine waves which exercises the whole range. As the test method reveals differences in probability of occurrence of individual codes, it can be used with program material.

In an analysis of code probability on a number of commercially available CDs, a disturbing number of those tested had surprising characteristics such as missing codes, particularly in older recordings. Single missing codes can be due to an imperfect ADC, but in some cases there were a large number of missing codes spaced evenly apart. This could only be due to primitive gain controls applied in the digital domain without proper redithering. This may have been required with under-modulated master tapes which would be digitally amplified prior to the cutting process in order to play back at a reasonable level.

Statistical code analysis is quite useful to the professional audio engineer as it can be applied using actual program material at any point in the production and mastering process. Following an ADC it will reveal converter non-linearities, but used later, it will reveal DSP shortcomings. It will be interesting to see whether any correlation emerges between perceived quality and the results of tests of this kind.

14.5 Digital audio interface quality

There are three parameters of interest when conveying audio down a digital interface such as AES/EBU or SPDIF, and these have quite different importance depending on the application. The parameters are:

1. The jitter tolerance of the serial FM data separator.
2. The jitter tolerance of the audio samples at the point of conversion back to analog.
3. The timing accuracy of the serial signal with respect to other signals.

A digital interface is designed to convey discrete numerical values from one place to another. If those samples are correctly received with no numerical change, the interface is perfect. The serial interface carries clocking information, in the form of the transitions of the FM channel code and the sync patterns and this information is designed to enable the data separator to determine the correct data values in the presence of jitter. It was shown in Chapter 8 that the jitter window of the FM code is half a data bit period in the absence of noise. This becomes a quarter of a data bit when the eye opening has reached the minimum allowable in the professional specification as can be seen from Figure 14.11. If jitter is within this limit, which corresponds to about 80 nanoseconds peak to peak the serial digital interface perfectly reproduces the sample data, irrespective of the intended use of the data. The data separator of an AES/EBU receiver requires a phase-locked loop in order to decode the serial message. This phase-locked loop will have jitter of its own, particularly if it is a digital phase-locked loop where the phase steps are of finite size. Digital phase-locked loops are easier to implement along with other logic in integrated circuits. There is no point in making the jitter of the phase-locked loop vanishingly small as the jitter tolerance of the channel code will absorb it. In fact the digital phase-locked loop is simpler to implement and locks up quicker if it has larger phase steps and therefore more jitter.

This has no effect on the ability of the interface to convey discrete values, and if the data transfer is simply an input to a digital recorder no other parameter is of consequence as the data values will be faithfully recorded.

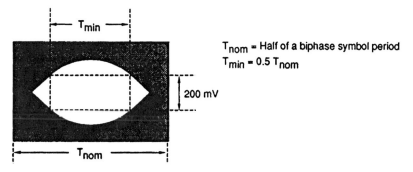

Figure 14.11 The minimum eye pattern acceptable for correct decoding of standard two-channel data.

However, it is a further requirement in some applications that a sampling clock for a converter is derived from a serial interface signal.

It was shown in Chapter 8 that the jitter tolerance of converter clocks is measured in picoseconds. Thus a phase-locked loop in the FM data separator of a serial receiver chip is quite unable to drive a converter directly as the jitter it contains will be as much as a thousand times too great. Nevertheless this is exactly how a great many consumer outboard DACs are built, regardless of price. The consequence of this poor engineering is that the serial interface is no longer truly digital. Analog variations in the interface waveform cause variations in the converter clock jitter and thus variations in the reproduced sound quality. Different types of digital cable 'sound' different and journalists claim that digital optical interfaces are 'sonically superior' to electrical interfaces. The digital outputs of some CD players 'sound' better than others and so on. In fact source and cable substitution is an excellent test of outboard converter quality. A properly engineered outboard converter will sound the same despite changes in CD player, cable type and length and despite changing from electrical to optical input because it accepts only data from the serial signal and regenerates its own clock. Audible differences simply mean the converter is of poor design and should be rejected.

Figure 14.12 shows how a converter should be configured. The serial data separator has its own phase-locked loop which is less jittery than the serial waveform and so recovers the audio data. The serial data are presented to a shift register which is read in parallel to a latch when an entire sample is present by a clock edge from the data separator. The data separator has done its job of correctly returning a sample value to parallel format. A quite separate phase-locked loop with extremely high damping and low jitter is used to regenerate the sampling clock. This may use a crystal oscillator or it may be a number of loops in series to increase the order of the jitter filtering. In the professional channel status, bit 5 of byte 0 indicates whether the source is locked or unlocked. This bit can be used to change the damping factor

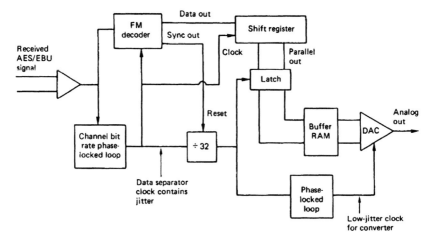

Figure 14.12 In an outboard converter, the clock from the data separator is not sufficiently free of jitter and additional clock regeneration is necessary to drive the DAC.

of the phase-locked loop or to switch from a crystal to a varicap oscillator. When the source is unlocked, perhaps because a recorder is in varispeed, the capture range of the phase-locked loop can be widened and the increased jitter is accepted. When the source is locked, the capture range is reduced and the jitter is rejected.

The third timing criterion is only relevant when more than one signal is involved as it affects the ability of, for example, a mixer to combine two inputs.

In order to decide which criterion is most important, the following may be helpful. A single signal which is involved in a data transfer to a recording medium is concerned only with eye pattern jitter as this affects the data reliability.

A signal which is to be converted to analog is concerned primarily with the jitter at the converter clock. Signals which are to be mixed are concerned with the eye pattern jitter and the relative timing. If the mix is to be monitored, all three parameters become important.

References

1. Lipshitz, S.P. and Vanderkooy, J., Are D/A converters getting worse? Presented at 84th Audio Eng. Soc. Conv. (Paris, 1988), preprint 2586 (D-6)
2. Harris, S., The effects of sampling clock jitter on Nyquist sampling analog to digital converters and on oversampling delta-sigma ADCs. *J. Audio Eng. Soc.*, **38**, 537–542 (1990)

Index

Other audio related titles

Audio for Television

Audio for Television makes the television sound person's life a little easier by outlining all the relevant principles and practices. Newcomers to the field will find it an invaluable, up to date resource and experienced sound people will gain from the explanations of new technology.

Contents: Sound and hearing; Microphones, loudspeakers and stereophony; Analog and digital audio signals; Audio recording; Routing and transmission.

1997, 256pp, 234 x 156 mm, 150 line illustrations, Paperback, 0 240 51464 5

The Art of Digital Audio
Second Edition

The first edition of this book is now regarded as a classic in its field. Now completely re-written to reflect the enormous recent advances in the subject it is even more comprehensive, now covering practical devices such as DCC and MiniDisc.

Partial Contents: Why digital?; Conversion; AES/EBU; Digital audio coding and processing;

Digital Compact Cassette (DCC); Advanced digital audio processing; Digital audio interconnects; Digital recording and channel coding; Error correction; Rotary head recorders; Stationary head recorders; NAGRA and data reduction; Digital Audio Broadcasting (DAB); The compact disc/mini disc.

1993, 490pp, 250 line illustrations, Hardback, 234 x 156 mm, 0 240 51369 X

An Introduction to Digital Audio

Introduction to Digital Audio is the first affordable book to bring all the fundamentals of digital audio to a wide audience. There are reasons instead of facts and practical applications to contrast with the theory.

Contents: Introducing digital audio; Conversion; Some essential principles; Digital coding; Digital audio interfaces; Digital audio tape recorders; Magnetic disk drives; Digital audio editing; Optical discs in audio

1994, 320pp, 150 Line Illustrations, 234 x 156mm, 0 240 51378 9

by John Watkinson

Compression in Video and Audio

This book recognises the wide applications of compression by treating the subject from first principles without assuming any particular background for the reader. An introductory chapter is included which suggests some applications of compression and how it works in a simplified form. In addition a fundamentals chapter contains all of the background necessary to follow the rest of the book.

Contents: Introduction to compression; Fundamentals; Processing for compression; Audio compression; Video compression

1995, 256pp, 150 Line Illustrations, 234 x 156 mm, 0 240 51394 0

Prices available on request. Please direct orders and enquiries to:

Heinemann Customer Services, PO Box 840, Halley Court, Jordan Hill, Oxford OX2 8YW
Tel: +44 (0)1856 314301. Fax: +44 (0)1865 314091
E-Mail: bhuk.orders@repp.co.uk

Focal Press
225 Wildwood Avenue, Woburn, MA 01801-2041
USA. Tel: 781-904-2500. Fax: 781-904-2620

Potential authors

If you are writing a book in which you think we may be interested please contact the Focal Press Publishing Team:

Europe

Margaret Riley, Publisher,
margaret.riley@repp.co.uk
Jennifer Welham, Associate Editor, Photography,
jennifer.welham@repp.co.uk
Tel: +44(0) 1856 314571. Fax:; 44(0) 1865 314572

USA

Marie Lee, Publisher,
marie.lee@bhusa.com
Tel: 781-904-2500. Fax: 781-904-2620

Visit the Focal Press Web Site:

http://www.bh.com/focalpress

New electronic product

SoundStudio
A Simulator for Training Sound Recording for Film & TV

Severiges Television with The University of the West of England

A simulator for training and teaching sound engineers for TV and film, SoundStudio is educational, easy and very realistic. It offers you a unique opportunity to practice the art of listening to sounds. With practical exercises under realistic conditions, the trainee sound engineer can practice what it is like to record both in the studio and on location. The program requires no previous knowledge of either computers or sound engineering. SoundStudio uses modern multi-media technology, providing very high-quality video, sound and graphics. All simulations are in the MPEG video format. The program also provides a theoretical background which describes microphones and other equipment. In a special 'sound-effect studio' the trainee can also practise using sound from the perspective of a narrator. This is a constructive, powerful and easy to understand tool, which is also an easy way to motivate your staff from a cost-effective point of view.

SoundStudio requires: Windows 3.1 or Windows 95, 486/DX2-66 or faster, 16 MB RAM, 300 MB hard disk space, CD-ROM drive (min 2x), 16-bit sound card and MPEG card, that can show video with 256 colours, 640 x 480.

1997, 0 240 51535 8

Digital Audio CD
Student Edition
Markus Erne

Scopein Research, Asrau, Switzerland and Chairman of the Audio Engineering Society, Swiss Section

This audio CD offers an overview of digital audio, demonstrating typical artefacts in digital audio equipment and tests to train the ear. It demonstrates very clearly all areas of digital audio which are covered in the 'Focal Press Music Technology' Series.

Contents: The disk contains examples of the following: Test signals; Sampling processes; Quantization; Jitter; Error concealment; Synchronization; Time varying digital parameters; Data compression; Listening tests.

1998, Compact Disk, 0 240 51501 3,

Visit the Focal Press Web Site........

from Focal Press

Digital Audio CD and Resource Pack
Markus Erne

The disk is a useful teaching tool for lecturers in colleges and professionals in broadcasting and recording studios who will find the teaching resource pack particularly useful. A 50 page manual containing photocopiable hand out sheets and overhead transparency masters, offers the lecturer a complete resource pack, to help support their teaching.

Contents: The disk contains examples of the following: Test signals; Sampling processes; Quantization; Jitter; Error concealment; Sychronization; Time varying digital parameters; Data compression; Listening tests.

1998, 0 240 51502 1, Compact Disk and Resource Pack (ringbound) 50pp, (overheads and handouts).

Please direct orders and enquiries to:
Heinemann Customer Services, PO Box 840, Halley Court, Jordan Hill, Oxford OX2 8YW
Tel: +44 (0)1856 314301. Fax: +44 (0)1865 314091
E-Mail: bhuk.orders@repp.co.uk

Focal Press
113 Wildwood Avenue, Woburn, MA 01801 2041
USA. Tel: 781-904-2500. Fax: 781-904-2620

Potential authors

If you are writing a book in which you think we may be interested please contact the Focal Press Publishing Team:

Europe
Margaret Riley, Publisher,
margaret.riley@repp.co.uk
Jennifer Welham, Associate Editor, Photography,
jennifer.welham@repp.co.uk
Tel: +44(0) 1856 314571. Fax: +44(0) 1865 314572

USA
Marie Lee, Publisher,
marie.lee@bhusa.com
Tel: 781-904-2500. Fax: 781-904-2620

...http://www.bh.com/focalpress

Music Technology Series

The Focal Press Music Technology Series is intended to fill a growing need for authoritative books to support college and university courses in music technology, sound recording, multimedia, and their related fields. The books will also be of value to professionals already working in these areas and who want either to update their knowledge or to familiarise themselves with topics that have not been part of their mainstream occupations.

Series Editor

Dr. Francis Rumsey, Senior Lecturer on the 'Tonmeister' course in Music and Sound Recording at the University of Surrey

Sound and Recording

An Introduction
Third Edition

Francis Rumsey
Tim McCormick
Formerly Deputy Head of Sound at the Manchester Royal Exchange Theatre

Sound and Recording 3/e is designed as an easy-to-read reference for those at an early stage in their careers. It will provide a vital introduction to the principles of sound, perception, audio technology and systems.
Contents: What Is Sound? Auditory perception; A guide to the audio signal chain; Microphones; Loudspeakers; Mixers; Analog tape recording; Noise reduction; Digital recording; Record players; Power amplifiers; Lines and interconnection; Outboard equipment; MIDI timescale and synchronisation.

1997, 384pp, 100 Line Illustrations,
20 halftones, 234X156mm, Paperback,
0 240 51487 4

Sound Synthesis Sampling
Martin Russ
Columnist for Sound on Sound Magazine

This book provides a comprehensive introduction to all the most common forms of analog and digital sound synthesis, as well as covering many of the less commonly encountered techniques used in research and academia.
Contents: Background; Analogue synthesis; Hybrid synthesis; Samplers; Digital synthesis, Using synthesis; controllers; The future of synthesis; Glossary.

1996, 224pp, 100 line Illustrations,
20 halftones, 234x136mm, Paperback,
0 240 51429 7

Sound Synthesis and Sampling
- The CD-ROM
Martin Russ

Based on the book Sound Synthesis and Sampling, the CD-ROM version allows you to actually hear the effects of the techniques discussed in the book - something that the book alone could never achieve This CD-ROM uses audio files and animated diagrams to demonstrate and teach the principles outlined in the book. The user can search the CD-ROM via an interest group category, or simply go straight to the required example to hear the sound, read about the technology and look at the supporting animated waveform and spectrum diagrams. The high quality audio samples can also be played through a CD player.

1997, Compact Disk, 0 240 51497 1

Visit the Focal Press Web Site.......

The Audio Workstation Handbook

Francis Rumsey

The Audio Workstation Handbook has been written for all those needing to understand digital audio and the associated technology used in these workstations.

Contents: Introduction to Computer Systems and terminology; Introduction to Digital Audio; Audio data reduction systems; Data storage media; The digital audio workstation; Recording and editing of audio on disk; networking and interfacing; Video in the audio workstation; MIDI in the audio workstation; Audio in multimedia PC; Audio production; Audio CD mastering.

1996, 286pp, 70 Black & White Line Illustrations, 8 Half Tones, 246 X 189mm, Paperback, 0 240 51450 5

Acoustics and Psychoacoustics

David Howard
Head of Department of Electronics, University of York

James Angus
Senior Lecturer University of York

Acoustics and Psychoacoustics is a comprehensive book which will provide you with a broad-ranging introduction to the subject of acoustics, including the principles of human perception of sound.

Contents: Introduction to sound; Introduction to hearing ; Notes and harmony ; Musical dynamics; Acoustic model for musical instruments; Timbre of musical instruments; Hearing music in different environments; Deceiving the ear; Processing sound electronically.

1990, 224pp, 120 Black & White Line Illustrations, £40 X 100mm, Paperback, 0 240 51428 9

MIDI Systems and Control

Second Edition
Francis Rumsey

The second edition of MIDI Systems and Control has been updated and enlarged to take into consideration the many additions to the MIDI specification.

Contents: Introduction to principles and terminology; Synchronisation and external machine control; Common implementations; Systems control sequences and operating systems; Implementation of MIDI with peripheral devices; Practical systems designs.

1994, 256pp, 150 Line Illustrations, 234 x 156mm, Paperback, 0 240 51370 3

Prices available on request.
Please direct orders and enquiries to:
Heinemann Customer Services, PO Box 840, Halley Court, Jordan Hill, Oxford OX2 8YW
Tel: +44 (0)1856 314301. Fax: +44 (0)1865 314091
E-Mail: bhuk.orders@repp.co.uk

Focal Press
225 Wildwood Avenue, Woburn, MA 01801-2041
USA. Tel: 781-904-2500. Fax: 781-904-2620

....http://www.bh.com/focalpress